Introduction to Human Factors for Organisational Psychologists

Introduction to Human Factors for Organisational Psychologists

Mark W. Wiggins

CRC Press
Taylor & Francis Group
Boca Raton London New York

CRC Press is an imprint of the
Taylor & Francis Group, an **informa** business

First edition published 2022
by CRC Press
6000 Broken Sound Parkway NW, Suite 300, Boca Raton, FL 33487-2742

and by CRC Press
4 Park Square, Milton Park, Abingdon, Oxon, OX14 4RN

CRC Press is an imprint of Taylor & Francis Group, LLC

© 2022 Taylor & Francis Group, LLC

Reasonable efforts have been made to publish reliable data and information, but the author and publisher cannot assume responsibility for the validity of all materials or the consequences of their use. The authors and publishers have attempted to trace the copyright holders of all material reproduced in this publication and apologize to copyright holders if permission to publish in this form has not been obtained. If any copyright material has not been acknowledged please write and let us know so we may rectify in any future reprint.

Except as permitted under U.S. Copyright Law, no part of this book may be reprinted, reproduced, transmitted, or utilized in any form by any electronic, mechanical, or other means, now known or hereafter invented, including photocopying, microfilming, and recording, or in any information storage or retrieval system, without written permission from the publishers.

For permission to photocopy or use material electronically from this work, access www.copyright.com or contact the Copyright Clearance Center, Inc. (CCC), 222 Rosewood Drive, Danvers, MA 01923, 978-750-8400. For works that are not available on CCC please contact mpkbookspermissions@tandf.co.uk

Trademark notice: Product or corporate names may be trademarks or registered trademarks and are used only for identification and explanation without intent to infringe.

ISBN: 978-1-032-13555-7 (hbk)
ISBN: 978-1-032-13559-5 (pbk)
ISBN: 978-1-003-22985-8 (ebk)

DOI: 10.1201/9781003229858

Typeset in Times
by Newgen Publishing UK

Dedication

For Deborah, Matilda, and Jack.

Contents

Preface xvii
About the Author xix

PART I Setting the Scene

Chapter 1 An Introduction to Human Factors 3

    1.1   Introduction to Human Factors 3
    1.2   Human Factors and Organisational Psychology 4
    1.3   A Brief History of Human Factors 6
    1.4   Human Factors and Systems Thinking 10
    1.5   Accident and Incident Analysis 11
    1.6   Management Error 11
    1.7   Cost-Benefit Analysis and Human Factors 12
    1.8   Proactive Management 14
    1.9   Organisational Design and Human Factors 15

Chapter 2 Human Factors in Organisations 17

    2.1   Human Factors in Organisations 17
    2.2   The Nature of Complex Industrial Environments 18
        2.2.1   Interdependence 18
        2.2.2   Rapidly Changing Demands 19
        2.2.3   Uncertainty 20
        2.2.4   Resource Limitations 21
        2.2.5   Competing Demands 21
        2.2.6   Complex Authority Structure 22
    2.3   Error within Stable and Unstable Environments 23
    2.4   Human Factors in Continuous Process Operations 24

Chapter 3 Risk and Uncertainty 27

    3.1   Organisational Complexity 27
    3.2   Characterising Human Behaviour 29
    3.3   Risk and Failure 30
    3.4   The Validity of Risk Estimates 30
    3.5   Risk and Safety Initiatives 31
    3.6   Risk Controls 33
    3.7   Uncertainty and Probability 34
    3.8   Probability Assessment 35
    3.9   The Bowtie Method 36

PART II Human Factors Frameworks

Chapter 4 Human Error-Based Perspectives 41
    4.1 The Nature of Error 41
    4.2 Deconstructing Human Error 43
    4.3 Preventing Human Error 44
    4.4 The Difficulty in Modelling Human Performance 47
    4.5 Safety Management Systems 50

Chapter 5 System Safety Perspectives 53
    5.1 Organisations as Systems 53
    5.2 Organisational Error 54
    5.3 The Sequence of Accident Causation 56
    5.4 The Systemic Model of Accident Causation 57
    5.5 The Origins of System Failure 59
    5.6 Accimap as Representations of System Failure 60
    5.7 The STAMP/STPA Approach 60
    5.8 The EAST Method 61
    5.9 The FRAM Approach 62

Chapter 6 Human Engineering Perspectives 63
    6.1 Systems and Design 63
    6.2 Systems Engineering 65
    6.3 Design Philosophy 66
    6.4 User-Centred Design 66
    6.5 Decision-Centred Design 67
    6.6 Features of Systems Engineering 68
        6.6.1 Operational-Need Determination 68
        6.6.2 Operational Concept 69
        6.6.3 Concept Exploration 69
        6.6.4 Concept Demonstration 69
        6.6.5 Full-Scale Development 70
        6.6.6 Production and Deployment 70
        6.6.7 System Evaluation 71
    6.7 Human Factors in Systems Engineering 72

Chapter 7 Reliability-Based Perspectives 73
    7.1 Reliability Analysis 73
    7.2 Reliability Engineering 73
    7.3 Probabilistic Risk Assessment 74
    7.4 Human Reliability Analysis 75
    7.5 Taxonomies of Human Performance 75
    7.6 Approaches to HRA Modelling 77

Contents   ix

	7.7	Quantitative versus Qualitative 79
	7.8	Contemporary Approaches to Reliability 79
	7.9	System Simulations 80
	7.10	Human Reliability Analysis in the Future 80

PART III Individual Differences and Human Factors

Chapter 8 Information Processing 85

    8.1 Information Processing 85
    8.2 Models of Information Processing 87
    8.3 The Multiple Resources Model of Information Processing 88
    8.4 Limitations of Human Information Processing 89
    8.5 Implications of Information-Processing Research 90
        8.5.1 Checklist Design 91
        8.5.2 Types of Checklists 93
        8.5.3 Difficulties with Checklist Completion 95

Chapter 9 Workload and Attention 97

    9.1 Attention 97
    9.2 The Significance of Attention 97
    9.3 Attention Strategies 98
        9.3.1 Selective Attention 98
        9.3.2 Focused Attention 99
        9.3.3 'Divided' Attention 100
    9.4 Sustained Attention and Vigilance 102
    9.5 Workload 104
        9.5.1 Attention and Workload 105
    9.6 Measuring Workload 106
        9.6.1 Psychophysiological Measures 106
        9.6.2 Subjective Ratings 107
        9.6.3 Secondary Tasks 107

Chapter 10 Situational Awareness 111

    10.1 Situational Awareness 111
    10.2 Level 1 Situational Awareness 112
    10.3 Level 2 Situational Awareness 114
    10.4 Level 3 Situational Awareness 116
    10.5 Team Situational Awareness 117
    10.6 Assessing Situational Awareness 119

Chapter 11 Human Factors and Decision-Making 123

    11.1 Models of Decision-Making 123
    11.2 Descriptive Decision-Making 123
    11.3 Decision Errors 125
        11.3.1 Representativeness Bias 125
        11.3.2 Confirmation Bias 127
        11.3.3 Availability Bias 127
        11.3.4 Anchoring Bias 128
        11.3.5 Misconceptions of Chance 128
    11.4 Prescriptive Decision-Making 129
        11.4.1 The DECIDE Model 129
        11.4.2 Bayes' Theorem 130
        11.4.3 Framing and Risk Assessment 131
        11.4.4 Hazardous Attitudes 132
        11.4.5 Combating the Influence of Biases 133
    11.5 Decision-Making in Complex Environments 133
    11.6 Decision-Making and Reliability Analysis 135
    11.7 Decision-Making under Uncertainty 136

Chapter 12 Fatigue 139

    12.1 Fatigue in Practice 139
    12.2 Circadian Rhythms 139
    12.3 Sleep Disturbance and Stages 141
    12.4 Fatigue and Human Performance 141
    12.5 Shift Work 143
    12.6 Alertness-Maintenance and Sustained Attention 143
    12.7 Individual Differences and Alertness-Maintenance
        (Expertise) 145
    12.8 Fatigue Management Systems (FMS) 145

PART IV *Group Processes and Human Factors*

Chapter 13 Groups and Teams 149

    13.1 The Nature of Groups 149
    13.2 Group Development versus Team Development 150
    13.3 Phases of Team Development 150
    13.4 Facilitating Group Performance 155
    13.5 Team Performance 159
    13.6 Critical Team Behaviours 161

Chapter 14 Human Factors and Leadership 163

    14.1 Leadership in Practice 163
    14.2 Contingency Approaches to Leadership 164

Contents

	14.3	Ineffective Leadership	168
	14.4	Leadership and Safety Culture	169
	14.5	Just Culture and Safety Leadership	170
	14.6	Principles of Effective Human Factors Leadership	170
	14.7	Followership	171
	14.8	Participative Leadership	172

Chapter 15 Communication 175

    15.1 The Nature of Communication 175
    15.2 Communication Errors 176
    15.3 Causes of Communication Errors 179
    15.4 Communication and Team Performance 182
    15.5 Communication across Teams 184
    15.6 Successful Communication 186
    15.7 Semantic and Prosodic Aspects of Communication 188

Chapter 16 Resource Management 191

    16.1 Resource Management 191
    16.2 An Attitude Approach to Resource Management 192
    16.3 Resource Management Redefined 193
    16.4 Resource Management and Human Factors 195
    16.5 The Regulation of Resource Management Initiatives 196
    16.6 Competency-Based Resource Management 198
    16.7 Threat and Error Management 200

PART V Human Factors Tools and Techniques

Chapter 17 Hazard Analysis 205

    17.1 Hazards and Incidents 205
    17.2 The Process of Hazard Analysis 206
    17.3 Minimising the Impact of Hazards 209
    17.4 Hazard Warnings 209

Chapter 18 Cognitive Task Analysis 211

    18.1 Analysing Behaviour 211
    18.2 Task Analysis 212
    18.3 Technique for Error Rate Prediction 212
    18.4 Cognitive Task Analysis 214
    18.5 The Precursor, Action, Result, Interpretation
          Method 216
    18.6 CTA and Symbolic Architectures 216
    18.7 Cognitive Task Analysis and Reliability 218

Chapter 19 Accident and Incident Analysis 221

    19.1 Accident Investigation 221
    19.2 Accident Investigation and Aviation 222
    19.3 The Process of Accident Investigation 223
    19.4 Principles of Investigation 224
    19.5 Information Sources 225
    19.6 The Written Report 227
            19.6.1 Test for Existence 228
            19.6.2 Test for Influence 229
            19.6.3 Test for Validity 229
    19.7 Accident Investigation Protocols 230
    19.8 Probable Cause versus Significant Factors 231
    19.9 Human Factors and Accident Investigation 231
    19.10 Crash 'Recorders' 234
            19.10.1 The Cockpit Voice Recorder 235
            19.10.2 Flight Data Recorder 235

Chapter 20 System Evaluation, Usability, and User Experience 237

    20.1 System Assessment 237
    20.2 Indices 237
    20.3 Productivity 238
    20.4 Organisational Factors and Systems Assessment 240
    20.5 Usability Engineering 242
    20.6 Usability and Generalisation 243
            20.6.1 Usability and User Experience 244
            20.6.2 Documentation and Training 245
    20.7 Archetypes and Personas 246
    20.8 User Testing 246
    20.9 A/B Testing 247
    20.10 Human–Computer Interaction 247
            20.10.1 GOMS and Other Models 249
            20.10.2 Wireframes 251
    20.11 Design Thinking 251

Chapter 21 Human Factors and Ergonomics 253

    21.1 Ergonomics 253
    21.2 Ergonomics and the Normal Distribution 254
    21.3 Workstation Design and the Application
          of Ergonomics 256
    21.4 Training and Ergonomics 259
    21.5 Manual Handling 260
    21.6 Command and Control Systems Evaluation 261
    21.7 A Total Quality Management Approach to Ergonomics 262

Contents   xiii

PART VI   Human Factors in Context

Chapter 22 Human Factors and Automation 265

    22.1   Automated Systems and Human Performance 265
            22.1.1   Automated Systems and Interface Design 265
            22.1.2   The 'Out of the Loop' Syndrome 266
            22.1.3   Cooperative Automated Systems 268
            22.1.4   Adaptable Systems 268
    22.2   Automated Systems and Reliability 269
    22.3   Automation and Cognitive Work Analysis 270
    22.4   Work Domain Analysis 271
            22.4.1   Control Task Analysis 272
            22.4.2   Strategies Analysis 272
            22.4.3   Social Organisation and Cooperation Analysis 272
            22.4.4   Worker Competencies Analysis 273
    22.5   Implications of Cognitive Work Analysis 273

Chapter 23 Human Factors and Aviation Systems 275

    23.1   The Significance of Survivability 275
    23.2   Crash Survivability 276
    23.3   Cabin Safety Hazards 278
            23.3.1   Overhead Lockers 278
            23.3.2   Fire Hazards 279
            23.3.3   Exits and Evacuation 280
    23.4   Process Control 282
            23.4.1   The Origins of Air Traffic Control 285
            23.4.2   The Role of Air Traffic Control 286
            23.4.3   The Flight Progress Strip: A Human
                          Factors Case 286
    23.5   Process Control and Situational Awareness 287
    23.6   Workload Management and Process Control 288
    23.7   The Significance of Displays 290
    23.8   The CDTI: A Case Study in Displays 291
            23.8.1   Pictorial Representation 292
            23.8.2   Display Size and Position 293
            23.8.3   Compatibility of Movement 293
            23.8.4   Luminance and Contrast 294
            23.8.5   Conceptual Compatibility 295
            23.8.6   Colour 295
            23.8.7   Display Mode 295
            23.8.8   Probability Estimation 296
            23.8.9   Decision-Making 296

Chapter 24 Human Factors and Energy 299

- 24.1 Energy Generation and Transmission 299
- 24.2 Accident Causation in Nuclear Systems 301
- 24.3 Human Error in Nuclear Systems 303
- 24.4 Hazard Identification in Nuclear Systems 305
- 24.5 Predictive Strategies 306
  - 24.5.1 Investigation of Causes 307
  - 24.5.2 Immediate Responses 307
- 24.6 Requirements for Hazardous Facilities 307
- 24.7 Management and Nuclear Power 309

Chapter 25 Human Factors and Marine Operations 311

- 25.1 The Marine Environment 311
- 25.2 Causal Network Analysis 313
- 25.3 Occupational Health in the Marine Environment 316
- 25.4 Human Factors Initiatives 316
- 25.5 Human Performance and Off-Shore Platforms 317

Chapter 26 Human Factors and Healthcare 321

- 26.1 Human Factors and Healthcare 321
- 26.2 Resource Management and Teams 322
- 26.3 Organisational Factors and Healthcare 323
- 26.4 Resilience and Healthcare 325
- 26.5 Decisions and Reasoning 326

PART VII   Assessment and Report Writing

Chapter 27 Human Factors Testing Methodology 331

- 27.1 Human Factors Testing 331
  - 27.1.1 Reliability 331
  - 27.1.2 Validity 332
- 27.2 Subjective versus Objective Data 332
- 27.3 Human Factors Methods 333
  - 27.3.1 Operational Analysis 333
  - 27.3.2 Functional Flow Analysis 334
  - 27.3.3 Critical Incident Analysis 334
  - 27.3.4 Task Analysis 336
  - 27.3.5 Fault Tree Analysis 338
  - 27.3.6 Physical Assessment 338
- 27.4 Ethics and Human Factors Testing 339
- 27.5 The Process of Data Acquisition 340

Contents

Chapter 28 Human Factors Assessments 341

    28.1   Human Factors Assessments 341
    28.2   Human Factors Audits..342
    28.3   Standards...343
    28.4   International Standards Organisation (ISO) and Human Factors 344
    28.5   Standardisation and 'Best Practice' 345
    28.6   Designing an Auditing Protocol..346
    28.7   Sampling Procedures 347
    28.8   Human Factors and Occupational Health 349

Chapter 29 Human Reliability Analysis 351

    29.1   Reliability Analysis..351
    29.2   Why a Human Reliability Analysis?...351
    29.3   The Accuracy of HRA..352
    29.4   The Inaccuracy of HRA 353
    29.5   Human Reliability Analysis Techniques 354
    29.6   The Process of Human Reliability Analysis 356
          29.6.1   Define the Problem 358
          29.6.2   Define the Context 358
          29.6.3   Identify Subject-Matter Experts 358
          29.6.4   Knowledge Elicitation Techniques...........................359
          29.6.5   Methods of Representation.......................................360

Chapter 30 Human Factors Report Writing 361

    30.1   Report Writing 361
    30.2   Problem Identification..361
    30.3   Methodology 363
    30.4   Results..363
    30.5   Budgetary Requirements and Anticipated Outcomes 363
    30.6   References..364
    30.7   Approaches to Human Factors Testing 364
          30.7.1   Reactive Human Factors Testing 364
          30.7.2   Proactive Human Factors Testing.............................366
    30.8   Human Factors Reports...367
          30.8.1   Writing Clearly..367
          30.8.2   Research Reports..367

References 369

Index 421

Preface

This text is designed to introduce industrial and organisational psychologists to the discipline of human factors. It also provides a range of tools necessary for the application of human factors strategies and techniques in practice. The text is intended to respond to the growing demand for organisational psychologists to assist in the development and evaluation of initiatives that are intended to optimise the relationship between workers and the operational environments with which they engage.

Introduction to Human Factors for Organisational Psychologists begins with an overview of human factors, its history, and some of the principles on which the discipline is based. It forms Part I of the text that also considers the application of human factors in organisations, together with notions of risk and uncertainty.

In Part II, frameworks for human factors are considered, including error-based and system-safety approaches, presenting human engineering as a construct that offers a structured approach to system or product design. It also articulates the various stages during which human factors assessments and interventions can be introduced.

Part III explores the links between individual differences and human factors, emphasising the role of information processing in enabling successful human performance. Outcomes of information processing, including workload, attention, situational awareness, and decision-making are described in some detail, and are considered alongside dimensions of personality and the impact of fatigue.

Broadening the discussion beyond individual differences, Part IV considers group processes and the impact on team performance, including the role of leadership and followership, especially in dynamic environments. It also explores some of the strategies that might be engaged in assessing teams in different operational contexts.

An important role of organisational psychologists in the context of human factors, lies in the evaluation of systems and organisations with a view towards recommending improvements that will both safeguard the customers and employees, and ensure productivity and efficiencies. Therefore, Part V considers a range of tools and techniques that can be applied by organisational psychologists to acquire human factors-related information, and develop an understanding of the situation or factors that may explain human behaviour. Some of the techniques include hazard analysis, cognitive task analysis, usability and user experience analysis, and accident and incident analysis.

Part VI is intended to illustrate the application of human factors at a more applied level, and particularly as it applies to different organisations or industrial contexts. Therefore, the construct of human factors is considered in the context of aircraft cabin safety, nuclear energy, air traffic control, and the marine environment.

Finally, Part VII describes the processes involved in human factors testing and assessment, drawing on some of the methods described in previous parts of the text, concluding with a description of the components that comprise a human factors report.

About the Author

Mark W. Wiggins, PhD, is a Professor of Organisational Psychology at Macquarie University. He gained his PhD in psychology from the University of Otago, New Zealand in 2001, is a Registered Psychologist in Australia with an endorsed area of practice in Organisational Psychology, and is a Fellow of the Australian Psychological Society. He has a broad range of experience that spans both scholarship and practice. As a research scholar, he is the author or co-author of over 100 publications on topics ranging from diagnostic error amongst physicians and allied health practitioners to cybersecurity in the banking and energy sectors. As a practising organisational psychologist, Dr. Wiggins has acted as consultant to a number of national and international organisations, including the Clinical Health Commission, Energy Queensland, the New South Wales (NSW) Port Authority, Transport for New South Wales, and the United States Federal Aviation Administration.

Part I

Setting the Scene

1 An Introduction to Human Factors

1.1 INTRODUCTION TO HUMAN FACTORS

The term 'human factors' describes a discipline the aim of which is to seek understanding and an optimisation of the relationship between human behaviour and the environment (Salvendy, 1997). It brings together a number of complementary areas of investigation including psychology, engineering, education, and ergonomics. The goal is to provide a holistic approach to problem-solving that deals with a range of potential issues and ultimately yields solutions that optimise human and system performance.

The interest in human factors has tended to parallel developments in technology. More accurately, it is often a failure in relation to the use of new technology that emphasises the importance of an understanding of the role of human behaviour. For example, the introduction of electric motor vehicles has resulted in a number of pedestrian collisions where pedestrians failed to hear oncoming vehicles (Wogalter, Lim, & Nyeste, 2014). Similarly, the introduction of self-drive motor vehicles has highlighted the difficulty that drivers are likely to face in monitoring the behaviour of automated systems over extended periods (Banks & Stanton, 2016).

This tendency towards a 'reactive' approach to the management of deficiencies in the relationship between human performance and technology reflects a broader conflict in the design of human–machine systems where there is often a tension between delivering a product quickly and cost-effectively, and delivering a product that can be assured is safe, efficient, and functional. There are a number of examples where the pressure to production has occurred at the expense of rigorous testing of the potential outcomes, from the use of thalidomide as a sleeping pill for pregnant mothers, to the use of dichlorodiphenyltrichloroethane (DDT) as an insecticide, and the use of asbestos as a building product (Swetonic, 1993). In each case, these products were associated with unacceptable and often fatal outcomes that would likely have been revealed had they been subjected to rigorous testing prior to their introduction to the marketplace.

Despite the potential for unsatisfactory outcomes, the difficulty associated with product development lies in determining the depth and extent of preliminary testing necessary to establish what might be regarded as a reasonable outcome. Erring on the side of caution may result in a product where the costs exceed its value to the market.

Conversely, the failure to test a product sufficiently may result in legal liability for any damages resulting from the product. Therefore, a balance must be struck between the requirement for an extensive and rigorous testing regime, and the necessity for a cost-effective and timely outcome.

In some cases, the potential for imperfect outcomes may be underestimated, thereby resulting in a failure far worse than might have been anticipated during the process of design. For example, the Grand National horse race in 1993 was described as a 'complete debacle' following a failure associated with the starting tape (Wilson, 1995). The 70-metre-long tape failed to rise at a consistent rate, so that some of the jockeys were caught, while others thought that the race had begun. Some of the horses had run at least 4½ miles before officials could stop the race.

Despite the fact that there were no injuries or fatalities associated with the failure, the 1993 Grand National race illustrates how the success or failure of an entire system can often depend on a relatively innocuous component. Consequently, it is important to consider, during preliminary testing, the relationship between various products and the systems with which they will eventually interact.

The perceived importance of preliminary testing differs for different products and systems. For example, failing to establish the usability of a computer game may simply result in the user becoming frustrated. However, failing to establish the usability of a computer system involved in the management of a nuclear power plant might have far more significant consequences. Therefore, the relative impact of the consequences of a failure ought to figure significantly during the preliminary testing phase of a product or system.

1.2 HUMAN FACTORS AND ORGANISATIONAL PSYCHOLOGY

Psychology is most appropriately regarded as a component-discipline within the broader study of human factors. Within psychology, organisational psychology is the specialist field that offers the greatest synergies with human factors, since it is an applied psychology that draws together research outcomes in learning, individual differences, applied cognition, perception, and mental health (Byrne, Hayes, McPhail, Hakel, Cortina, & McHenry, 2014). This intersection of fields of psychology corresponds and contributes to the three primary interventions that are available in optimising the relationship between humans and their environment. These three interventions are:

1. Aspects of the human operator that may be altered to meet the demands associated with a particular task and may include additional training and/or selection strategies;
2. Aspects of the environment that may be altered to meet the particular needs of the human operator and may include redesigning instruments, procedures, checklists, and controls so that errors, and/or the consequences of errors, are minimised; or
3. A combination of both strategies.

These options for interventions constitute the fundamental principles through which the principles of human factors are applied within the operational environment. More

An Introduction to Human Factors

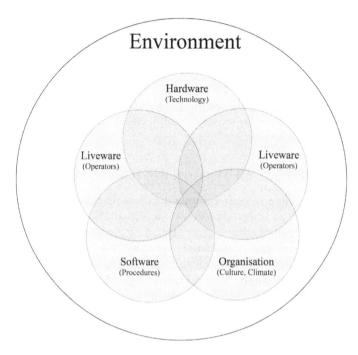

FIGURE 1.1 The SHELLO conceptual model of human factors. (Adapted from Edwards, 1988; Chang and Wang, 2010.)

importantly, they reflect the extent to which human factors, as a discipline, targets the performance of an entire system, rather than any individual component of that system. This relationship is captured in a simple conceptual model developed by Edwards (1988) where the interactions between system components can be identified and targeted for subsequent development. Referred to as the SHELL model, it relates the central liveware, or human operator, to four areas: Software (programs, procedures, checklists, etc.); Hardware (aircraft controls, equipment, etc.); the Environment (the surroundings in which the crew member operates); and Liveware (relations with other people). More recent representations have included 'Organisation' as another area to be considered as part of a revised SHELLO model (Chang & Wang, 2010). Figure 1.1 illustrates the interrelationship between these components.

Consistent with the broader intention of human factors, the SHELL model is based on the assumption that the failure of a system is the product of a mismatch between two or more components of that system. For example, the failure of a driver to read a signed speed limit correctly may be a product of inattention on the part of the driver (Liveware), the poor design of the sign (Software), and/or the location of the signpost (Hardware). Therefore, any 'failure' to respond to the signed speed limit needs to be considered from a number of different perspectives with solutions developed that take into account the interrelationships between these perspectives.

In the case of road traffic speed signs, the impact of inattention could be countered by locating signs in a central position with respect to the direction of travel (e.g.

a head-up display) so that the likelihood of observation is maximised (Hardware), and/or the design of the speed sign could be enhanced by co-locating other features, such as intermittent visual or auditory signals (Software) that would draw a driver's attention. Finally, the importance of speed signs could be highlighted in road safety messages (Liveware), so that drivers are aware of their responsibilities in sustaining their attention during driving. In reality, a combination of strategies, drawing on a range of features, is likely to be most successful, ensuring that the incidence of inadvertent speeding is minimised.

1.3 A BRIEF HISTORY OF HUMAN FACTORS

Historically, the principle of optimising the relationship between humans and their environment has been an important catalyst for technological development. As new technologies have emerged, these were adapted to render difficult tasks easier or enable tasks that were hitherto considered impossible. For example, the introduction of stirrups enabled riders to better retain their balance and control of the movement of horses at speed. This capability freed the riders' hands for other activities, enabling cavalry to adopt the role of mobile archers during battles.

As a formal discipline, human factors first emerged in early stages of World War II (Chapanis, 1999) where the swift development of faster and more powerful aircraft resulted in a need for more extensive training for pilots. Combined with the loss of aircraft and pilots due to the inability to cope with these increased demands, the training required was both time-consuming and costly (Meister, 1999).

In response to the problem, twin strategies were developed soon after WWII, and these formed the basis of the contemporary approach to human factors. The first of these strategies was based on the recognition that operating environments were becoming too complex for the average operator to master. In aviation, the controls and displays were neither systematic in their presentation, nor readable under time-constrained conditions. Fitts and Jones (1947) published a seminal study that described the errors made in the interpretation of aircraft instruments. These included difficulties with legibility, illusions, and the misinterpretation of scales. The outcome was a complete reconsideration of the method through which aircraft cockpit instruments were arranged and designed. Importantly, it was the nature of the environment that was altered to meet the needs of the average operator.

The second strategy that underpins contemporary human factors was based on the need to manage the characteristics of the individual to meet the particular needs of the task. The small percentage of pilots who did graduate from flying training were recognised by casual observers as superior both in their ability to operate the aircraft, and in their ability to process large amounts of information in a stressful environment (Meister, 1999). This led to the development of selection strategies that paralleled developments in organisational psychology. In contemporary industrial environments, strategies involving both personnel selection and system design are engaged to optimise the performance of employees.

In understanding behaviour in complex operational settings, one of the first, systematic assessments of human performance was undertaken in the context of aviation using what was referred to as the 'Cambridge Cockpit' simulator (Davis, 1948).

An Introduction to Human Factors

The equipment was designed to simulate a Spitfire aircraft cockpit and enabled an examination of the effects of fatigue and workload on pilot performance under controlled conditions. Not unexpectedly, severe decrements in skilled performance were reported that were associated with increasing levels of fatigue and workload. Similar results were evident using more advanced simulated environments (Wiener, 1988).

While this preliminary research might have been expected to form a foundation for the future design of human–machine interfaces, many of the principles of design that were developed throughout this early period were not necessarily implemented, even within the aviation industry. The jet-aircraft era that followed World War II, led to a variety of designs, many of which were less than adequate from a human factors perspective. Different manufacturers employed different procedures to operate and interpret aircraft displays and controls, systems were configured differently, and the solution to more complex systems tended to involve the introduction of more complex procedures and checklists, rather than changes to the design of systems. Invariably, this failure to consider poor design in the context of the operating environment only became apparent through aircraft accidents and incidents.

The aviation industry provides a useful case study illustrating the development of human factors as a discipline, as it is a sector that depends on high levels of reliability. Those system failures that do occur are well-reported, and subsequent investigations are particularly thorough, with the outcomes often made available in the public interest. It is also an industry at the forefront of technical innovation, characterised by complex, interdependent components.

In the early years of aviation, technical failures accounted for a significant proportion of aircraft accidents. There was a lack of reliability of engine components, while airframes were constructed of lightweight, fragile materials that were prone to catastrophic failure under the repetitive stresses of flight. However, as aircraft technology became more reliable, particularly in the years following World War II, human error, rather than mechanical failure, became more evident as a key factor associated with aircraft accidents and incidents (Nagel, 1988). In commercial aviation, breakdowns in human performance were highlighted in a number of serious crashes.

In one of the deadliest aircraft accidents in history, a Pan American Boeing 747 and a Royal Dutch Airlines (KLM) Boeing 747 collided at Los Rodeos Airport on March 27, 1977 on the Island of Tenerife in the Canary Islands (Roitsch, Babcock, & Edmunds, 1978). The aircraft were en route to the nearby island of Gran Canaria airport when a terrorist incident caused the diversion of the aircraft to Los Rodeos. Together with a number of other aircraft, they were positioned on the apron of the airport ready to depart once the threat at Gran Canaria had been contained.

Unlike Gran Canaria airport, Los Rodeos is a regional airport and ill-equipped to manage large aircraft such as the Boeing 747. At the time of the accident, the apron was crowded with aircraft to a point where the Pan Am aircraft, now ready to depart, could not negotiate sufficient clearance around the KLM aircraft. The KLM aircraft was instructed to taxi along the length of the runway, reverse direction and then depart towards the east. The crew of the Pan Am aircraft were instructed to follow the KLM aircraft, departing the runway using a taxiway, and take up a position behind the KLM aircraft.

At the time, the airport was covered in a thick sea fog that obscured visibility to a few hundred metres in advance of the aircraft. The air traffic controller, positioned in the control tower, could not see an aircraft once it had departed the apron. Instead, he was relying on the accuracy of the communication exchanges to ensure the separation of aircraft.

At the end of the runway, the captain of the KLM aircraft positioned the aircraft for take-off, now facing the Pan Am aircraft. At this point in the sequence of events, a series of miscommunications occurred between the air traffic controller and the captain of the KLM aircraft that led to its premature departure before the Pan Am aircraft had vacated the runway. The air traffic controller was issuing the KLM aircraft what is referred to as a departure clearance.

The departure clearance comprises a series of instructions that are issued to the crew of an aircraft prior to take-off and that are intended to enable the crew to anticipate actions during a fast-paced and high workload period of the flight. However, the instruction began with the phrase 'after take-off'. The captain of the KLM aircraft, already keen to depart as quickly as possible, heard the keyword 'take-off' and assumed incorrectly that it was a clearance to take-off, rather than simply a departure clearance. He acknowledged the instruction which was not challenged by the air traffic controller.

Unsure that they had actually received a take-off clearance, the first officer on board the KLM aircraft attempted to warn the air traffic controller that the captain had initiated the take-off roll. He used the phrase, 'we are now at take-off', which the air traffic controller assumed referred to the position of the aircraft ready for take-off. Neither the air traffic controller nor the captain of the KLM aircraft could see that the Pan Am aircraft had yet to vacate the runway. However, the pilots of the Pan Am aircraft surmised correctly that the KLM aircraft was taking off and sought to intervene with their own communication exchange. As it coincided almost precisely with the air traffic controller's response to the warning issued by the first officer on the KLM aircraft, only radio static was heard.

As the KLM aircraft advanced down the runway, the Pan Am Boeing 747 came into view. Despite attempting an early rotation, together with the desperate attempts of the crew of the Pan Am aircraft to vacate the runway, the two aircraft collided, leaving 580 passengers and crew fatally injured. Investigators concluded subsequently that the collision occurred because the captain of the KLM aircraft departed in the absence of a take-off clearance. In effect, the captain had misheard the departure clearance as a take-off clearance, and the first officer, although aware of the error, failed to intervene and prevent the departure of the aircraft.

A little over a year later, another aircraft accident occurred involving a Douglas DC-8 that suffered fuel exhaustion and subsequently crashed at Portland, Oregon (National Transportation Safety Board, 1979b). On December 28, 1978, United Airlines Flight 174 was on final approach to Portland International Airport when the flight crew became concerned that the undercarriage may have failed to lock into place. Normally, a small green light would illuminate to indicate that the undercarriage had safely deployed in preparation for landing. However, not only did the light fail to illuminate, but as the undercarriage was deployed, the pilots reported a

An Introduction to Human Factors

sideways movement of the aircraft, consistent with an asymmetrical extension of the undercarriage.

Following established procedures, the flight crew discontinued the approach and were redirected to an area west of the airfield where they could assess the situation and, if necessary, enable the preparation of the cabin for an emergency landing. While the crew, comprising a captain, first officer, and flight engineer initiated a troubleshooting exercise, the flight attendants readied the cabin for what appeared increasingly to be an emergency landing.

As the troubleshooting process continued, fuel was being consumed, in part to prevent the possibility of a fire in the case of an emergency landing. During the process of readying the aircraft for a possible emergency landing, the flight engineer, who was responsible for managing the fuel, noted that the situation was becoming critical. Although he advised the flight crew, the captain failed to appreciate the significance of the fuel state and the aircraft experienced fuel exhaustion and crashed approximately 6 miles from the airport. Two crew members were killed together with eight passengers.

The aircraft accidents at Tenerife and Portland, contributed to what was a watershed in understanding the nature of human performance in advanced technology systems. Rather than unreliable technologies, it became evident, at least in these cases, that there was a need to focus on human behaviour and, in particular, on the quality of the interaction between operational personnel. Edwards referred to these situations as problems relating to the Liveware-Liveware (L-L) interface, where 'Liveware' referred to human actors who are engaged in cooperative activities associated with joint goals.

The practical outcome of this watershed in understanding was the development and introduction of what would become known as Crew Resource Management (CRM) programs. According to Lauber (1984), CRM refers to the use of 'all available resources – information, equipment and people – to achieve safe and efficient flight operations' (p. 20). The application of CRM was most evident in training programs, the aims of which were to facilitate the teamwork skills and knowledge amongst aviation personnel. CRM training initiatives have since been broadened beyond aviation to a range of operational contexts including medicine (Haerkens, Kox, Noe, Van Der Hoeven, & Pickkers, 2018; Moffatt-Bruce, Hefner, Mekhjian, McAlearney, Latimer, Ellison, & McAlearney, 2017), marine operations (Bennett, 2019), and oil and gas operations (Theophilus, Esenowo, Arewa, Ifelebuegu, Nnadi, & Mbanaso, 2017).

Empirical investigations concerning the utility of CRM-based initiatives have been undertaken by a number of agencies, possibly the most notable of which was a team of researchers and practitioners from the University of Texas. Some of the variables considered included cross-cultural perspectives, performance during Line Oriented Flight Training (LOFT), the Advanced Qualification Program, and the use of Quick Access Recorders in debriefing (Helmreich, Merritt, & Wilhelm, 1999).

Consistent with the broadening of CRM beyond the aviation context, assessments of the validity of CRM-based initiatives have been undertaken in other contexts, with improvements reported across a wide range of metrics from employee engagement and turnover to communication and operational performance (Hefner, Hilligoss, Knupp, Bournique, Sullivan, Adkins, & Moffett-Bruce, 2017). In fact, the development of CRM has been recognised as one of the most significant catalysts in improving safety in high consequence, high risk industrial settings, providing the foundation for a range of initiatives, including safety audits and a systems approach to the analysis of accidents and incidents involving human error (Moffat-Bruce et al., 2017).

1.4 HUMAN FACTORS AND SYSTEMS THINKING

Although human factors researchers recognised at the earliest stages, the interaction between humans and the environment in which they functioned, there was a tendency to examine each of the components in isolation. This was probably due, in part, to a pervading view in psychology more generally that 'research' needed to be conducted using highly experimental techniques that required the isolation and manipulation of variables. However, a revision to this perspective that occurred in the mid to late 1980s, and that was broadly coincident with the introduction of CRM, adopted the notion that existing research techniques needed to be adapted to target the interface between humans and the environments within which they operate.

In the context of human factors, the successful performance of a system depends on the relationship between a number of components. The relationship between the components of a system is generally maintained through standard procedures or operations that represent controls on the behaviour of elements of that system (Spiess, 2013). If a system is governed by standard procedures, the successful performance of the system will depend on the accuracy and appropriateness of these procedures. Where a procedure is inappropriate, ambiguous, or non-existent, an error is almost inevitable. For example, in the context of human performance, ambiguous procedures were implicated in an accident involving a Tower Air Boeing 747-136, N605FF, that veered off Runway 4L during a departure at John F. Kennedy International Airport in 1995.

The aircraft was a regularly scheduled passenger/cargo flight departing for Miami, Florida, with 468 passengers and crew aboard. According to the National Transportation Safety Board (1996), the surface weather observations at the time of departure were described as a 700 foot ceiling of broken cloud with 1½ miles visibility with light snow and fog. The flight crew had noted that the taxi surface was particularly slippery and had limited the taxi speed to three knots (about 6 kilometres per hour) prior to departure.

As the aircraft reached 80 knots during the take-off roll, it began veering to the left. The captain applied opposite rudder and nosewheel steering to correct for the error, although the aircraft continued to slide off the runway. Just prior to the aircraft departing the runway surface, the captain selected engines to idle and applied maximum braking. The aircraft came to rest approximately 600 feet to the left of Runway 4L, and approximately 5,000 feet from the threshold of the runway.

A subsequent investigation by the National Transportation Safety Board (1996) concluded that the captain should have rejected the take-off at an earlier stage of the process. This conclusion was supported by references in the B-747 operations manual

An Introduction to Human Factors

and the Tower Air Flight Manual to situations where an aircraft cannot be controlled. However, the NTSB (1996) also recognised that the operating manuals contained ambiguities that may have led to a misinterpretation concerning the definition of situations that 'cannot be controlled'. Consequently, a more concrete definition was recommended so that, if the control of the aircraft could not be corrected with one-half rudder pedal travel, then the take-off should be aborted.

This incident illustrates how a lack of precision in one component of a complex system can significantly impact the overall outcome. It also illustrates the importance of testing interactions between the various components of a system, rather than simply testing each component in isolation. Although the definition for a rejected take-off may have been appropriate in fine weather, and with fewer operational factors, the combination of poor weather, pressure to depart, a strong cross-wind, and a lack of task-appropriate training may have revealed what was otherwise a relatively innocuous oversight. Maurino, Reason, Johnston, and Lee (1995) refer to this type of failure as a *latent condition* and note that they are particularly difficult to identify within a complex system since they often only emerge under specific operational conditions.

1.5 ACCIDENT AND INCIDENT ANALYSIS

Historically, analyses of accidents and incidents represent one of the most common reactive approaches to deficiencies in the relationship between human performance and the operational environment. They tend to be used as a mechanism to both identify the sources of failures, and as a means of assessing whether remedial activities may have been successful (Kletz, 1994; Learmont, 1993).

There is little doubt that accident and incident analyses provide a useful mechanism to identify appropriate countermeasures that are intended to prevent human error. However, the effectiveness of these countermeasures is often limited to situations within which the accident or incident occurred. For example, a countermeasure developed in response to motor vehicle accidents may not be appropriate for the rail context, since the motor vehicle environment embodies a range of sociotechnical issues that are distinct from the issues that impact performance within a rail context. However, the consequence of this perception of the need for domain-specific initiatives is a proliferation of models of accident reporting, each of which may be applicable only within distinct operating contexts.

According to Brown (1995), the term 'accident' describes an event in which an unplanned outcome occurs as a product of inappropriate behaviour. However, Reason (1990) argues that accidents are a function of a more complex interaction in which a number of events and preconditions contribute to an unplanned outcome. Therefore, consistent with a *systems approach*, 'inappropriate behaviour' constitutes only one aspect of a more complex sequence of accident or incident causation.

1.6 MANAGEMENT ERROR

The role of management in maintaining a safe and efficient operating system depends on an understanding of the implications of poor management principles

(Fischer, Frese, Mertens, & Hardt-Gawron, 2018). For example, seeking productivity improvements in the absence of any safeguards is liable to encourage an environment where the significance of other factors, such as safety, may be minimised. Therefore, management embodies a responsibility to ensure that line personnel are sufficiently appraised of the issues, so that no single element of the system is isolated at the expense of another (Dyre, Tabor, Ringsted, & Tolsgaard, 2017).

From a management perspective, the main difficulty lies in determining the extent to which a management decision will have a negative impact on the operational environment. Such is the difficulty in identifying latent conditions that the long-term impact on the system of most management decisions are rarely if ever examined. For example, in the 1980s and early 1990s, the Indian Air Force had a long-standing requirement for an advanced jet trainer that would enable the transition between operations on the 400 knot Kiran and the much faster Migoyan-Gurevich (MiG)-21 fighter aircraft (Mama & Gethin, 1998). However, a lack of funding resulted in a delay in the acquisition of an advanced training aircraft. Subsequently, the MiG-21 conversion unit recorded 11 accidents between 1991 and 1997, of which three were fatal.

The decision to deregulate the airline industry in the United States in the early 1980s had a profound impact on the process through which airlines were managed (Oster, Strong, & Zorn, 1992). Since deregulation, there has been increasing demand for more competitive fares which, in some cases, may have forced airlines to minimise expenditure in other areas, such as safety. While there were no doubt gains in the efficiency of airlines, the implication is that a decision made within a political context may have a significant impact on the management decisions and, unchecked, may constitute the foundation for latent conditions within an organisation.

While Nance (1986) asserts that the decision to deregulate the United States airline industry had a negative impact on safety, the results from statistical comparisons of the frequency of accidents and incidents prior to, and following, deregulation are less conclusive (Oster et al., 1992). Nevertheless, it does raise the notion of 'margins of safety' and the extent to which a system is tolerant of errors.

In systems with a relatively higher margin of safety, failures tend to be less frequent because the incidence of error is identified and corrected, and/or the consequences of errors are minimised prior to a system failure (Kumar, Gupta, Agarwal, & Singh, 2016). Therefore, a reduction in the level of safety within an organisation is not necessarily directly evident in the rate at which aircraft accidents and incidents occur. Other strategies need to be engaged to assess whether there has been a change to the margin of safety, irrespective of whether or not there has been a change in the frequency of incidents or accidents reported.

1.7 COST-BENEFIT ANALYSIS AND HUMAN FACTORS

One of the perennial concerns for the management of organisations is the utility of an investment where there may be no clearly measurable return. For example, it might be argued that a financial organisation with a safety record where there are very few accidents or incidents would be averse to increase funding for safety-related projects, since the probability of an accident or incident within the company is extremely low. However, the nature of accident causation is nonlinear and the absence of accidents or incidents cannot necessarily be used as an indicator of the probability of an accident in the future. Therefore, some variations are required to the typical cost-benefit analysis to take into account less tangible outcomes.

Consistent with this perspective, Sage (1995) suggests that a cost-benefit analysis is only appropriate in cases where there is a clear, observable outcome. Where it is not possible to classify outcomes in purely economic terms, it is more appropriate to use the principle of *cost-effectiveness assessment*. This involves the development of a measure of the utility of an outcome against which non-economic costs and returns can be evaluated. Having identified the utility of the anticipated costs and returns, it becomes possible to assess when costs are likely to yield an appropriate return.

The main difficulty associated with establishing the utility of an investment or outcome is that the 'values' will often be the product of a subjective assessment, rather than an objective analysis as is normally the case for cost-benefit analyses. However, Rouse and Boff (1997) suggest that subjective evaluations are not necessarily any less effective than an objective assessment in establishing an accurate outcome. They recommend the application of an approach consistent with Subjective Expected Utility (SEU) theory. This requires that all non-economic costs (c) and returns (r) be converted to a unitary utility scale from which comparisons can be made. For example, the utility (U) of a particular decision can be calculated using this equation that incorporates a series of costs and associated returns.

$$U_{(c,r)} = U[u(c_1), u(c_2), \ldots u(c_N), u(r_1), u(r_2) \ldots u(r_N)]$$

There are a number of strategies to determine the accuracy of a measure of utility and to identify the parameters that might be examined as part of the decision-making process, including seeking estimates from a range of sources. Similarly, estimates can be drawn from prior experience, such as the costs associated with aircraft accidents or incidents (Bayram, Üngan, & Ardıç, 2017; Laufer, 1987).

Irrespective of the approach taken to increase the accuracy of estimated costs and returns to the utility of a strategy, the process will depend on an understanding of the psychological dimensions of the human decision-making process. Specifically, a number of decision biases may impact the accuracy of the final decision so that either the costs or returns associated with a particular strategy are under- or overestimated (Hilbert, 2012).

One of the most effective mechanisms to avoid the impact of decision biases on estimations involves the application of a systematic approach to cost-benefit analysis. Rouse and Boff (1997) advocate a strategy where:

- Stakeholders are identified for each of the options;
- The benefits and costs are identified for each of the options in attributes;
- The utility functions are identified for each attribute;
- The utility functions are combined across stakeholders;
- The parameters (time) are assessed within utility models;

- The levels of attributes are forecast; and
- The expected utility is calculated for each option.

The accuracy of a cost-effectiveness analysis is ultimately determined by the outcome. Rouse and Boff (1997) describe at least two outcomes where investments in human factors-related initiatives resulted in improvements that significantly outweighed the initial investment. In the first case, a $300,000 initial investment to reduce the human factors errors associated with the use of the pedestal-mounted Stinger missile resulted in an increase in accuracy and a potential saving of $61,000,000 in follow-up launches. In the second case, an investment of $15,000 per year in eye protection equipment at a naval shipyard resulted in savings of $100,000 per year in compensation claims (Rouse & Boff, 1997). These examples illustrate the positive returns that can accrue from an investment that may have appeared less than cost effective during the initial stages of a cost-benefit analysis.

1.8 PROACTIVE MANAGEMENT

The identification of potential benefits that might accrue from an investment in appropriate human factors testing is part of a process that might be referred to as *proactive management*. In this case, rather than waiting and then reacting to system failures as they occur, a proactive approach to the management of safety involves the development of strategies that are intended to pre-empt failures, thereby enabling the development and implementation of remedial strategies.

Effective planning is critical in enabling the anticipation of system failures (Marquardt, 2019). However, the opportunity for planning should not be relegated to management but should involve all members within the organisational structure. The aim is to ensure that the potential problem is clearly understood and engaged from a number of different perspectives, so that the responsibility for the successful implementation of any plans that emerge is shared amongst all employees (Tsai, Jiang, Alhabash, LaRose, Rifon, & Cotton, 2016).

Hollnagel, Cacciabue, and Hoc (1995) suggest that effective planning is only possible when there is a clear understanding of the processes through which a system functions. For example, a highly repetitive task, such as brushing teeth, is generally both efficient and error-free, since the nature of the relationships between components is well understood. However, as systems become more complex, the relationships between the components tend to be subjected to greater variation. Social relationships and interactions between humans and technology can vary significantly, despite the fact that these relationships might be governed by standard procedures and/or expectations. Complex industrial systems also tend to be dynamic, and evolve in the absence of any interaction with an operator. As a result, developing a clear understanding of the nature of a complex industrial system can be extremely difficult.

One of the ways that an understanding of the nature of complex industrial systems can be developed is to create a taxonomy or list designed to capture the relationship between human behaviour and the environment within which humans function. This model may the product of an analysis of previous system failures, where specific deficiencies are identified, but it can also be employed as a predictive strategy to identify potential sources of system failure.

Detailed taxonomies have emerged in aviation (Nitzschner, Nagler, & Stein, 2019), energy (Theophilus et al., 2017), and medicine (Griffey, Schneider, Todorov, Yaeger, Sharp, Vrablik, Aaronson, Sammer, Nelson, Manley, & Dalton, 2019) that serve as checklists to ensure that specific conditions have been considered in the development of new initiatives. 'What-if' scenarios are also considered to identify gaps in procedures or capability that may emerge following a change in the state of a system. For example, in the case of a new air traffic control monitoring system, what would be expected to occur should a software failure render the displays unusable or inaccurate? Similarly, if a major disaster or epidemic was to occur, would local hospitals have the capability to cope with a large number of casualties arriving en masse?

In addition to the identification of potential sources of system failure, a proactive approach to the management of human factors also involves the development of strategies to ensure a positive workplace environment. The potential returns include improvements in productivity and a more positive approach to psychosocial relations (Shikdar & Sawaqed, 2003), leading to a reduced turnover of staff, higher levels of communication and coordination, and an improvement in corporate knowledge and capability (Kerr, Knott, Moss, Clegg, & Horton, 2008)

The relationship between the nature of the workplace environment and human performance is relatively well established. For example, Wilson (1972) observed that, in comparison to a windowless post-operative care unit, half as many patients suffered from post-operative delirium in a care unit within which windows were installed. Similarly, Wise and Wise (1984) noted that employees working in a windowless underground workplace consistently perceived their workplace as markedly inferior to working in a workplace above ground. Part of this perception appears to relate to issues of personal control over the environment, the perceived attributes of sunlight, and feelings of being enclosed. This illustrates the significance in designing workplaces that meet both the psychological and physiological needs of employees.

1.9 ORGANISATIONAL DESIGN AND HUMAN FACTORS

In addition to the physical structure of the workplace, human performance is influenced by the characteristics of the organisational environment. Evidence to support this suggestion can be drawn from a series of studies of coal mines in which technological improvements in the workplace resulted in a shift in the structure of the organisational environment. Operators who once worked in small autonomous teams were subsequently expected to function as large groups where there was a highly centralised structure of authority. According to Hendrick (1997a), the result was an increase in production costs, but also an increase in absenteeism.

The response to the situation on the part of the mining companies was to institute a series of changes to the organisational structure so that teams were less interdependent, and there was a greater opportunity for the selection of co-workers. This strategy resulted in an increase in productivity beyond either of the previous models of organisational design.

Clearly, the appropriateness of a particular organisational structure will depend on a number of factors including the nature of the operational environment and the characteristics of the employees. However, Hendrick (1997a) proposes a model intended to facilitate the development of an organisational design that balances the operational requirements with the needs and perceptions of employees. This process involves:

1. The identification of the goals and anticipated outcomes of the system;
2. The development and weighting of markers to assess the effectiveness of various organisational structures;
3. The design of the organisational structure based on the system complexity, the degree to which the structure should be formalised, and degree to which the decision-making structure should be centralised;
4. A balance between factors associated with the technology, personnel, and other external issues that are likely to impact the organisational design; and
5. A decision concerning the most appropriate organisational structure that is likely to yield a productive outcome over the long term.

This highly structured approach to organisational design provides the basis for the development of an organisational structure that takes into account the needs of workers. From the organisation's perspective, it increases the probability of a positive outcome.

2 Human Factors in Organisations

2.1 HUMAN FACTORS IN ORGANISATIONS

Transportation, energy production, and mining are all industries that can be characterised as high risk, low probability operating environments. The 'high risk' element refers to the extent and severity of the consequences associated with a system failure, while the 'low probability' element refers to the extent to which these failures are likely to occur. On the basis of this observation, these are environments where failures are relatively unlikely. However, when failures do occur, they are generally catastrophic, and tend to have a significant impact.

Within high consequence, low probability operating environments, one of the main goals is the development and implementation of strategies that minimise the likelihood of a failure, or limit the extent of the consequences of failure. However, as systems become increasingly reliable, the development of strategies that further increase system safety becomes increasingly difficult (see Figure 2.1).

The most significant feature associated with Figure 2.1 is the initial reduction in the rate across a relatively short period of time. This is indicative of the impact of relatively significant, systemic changes within the operational environment. For example, within the aviation context, the advent of the jet age and the introduction of sophisticated air traffic control systems were factors that markedly reduced the rate of aircraft accidents. Similarly, the discovery of a vaccine for human papilloma virus has reduced the incidence of cervical cancer amongst women (Franco & Harper, 2005).

An example of the difficulty associated with detecting errors can be drawn from the experience of a maintenance engineer who inserted three master chip detectors into the engines of an Eastern Airlines Lockheed Tristar but failed to confirm that 'O' rings were fitted (Stewart, 1992). The master chip detector consists of a small magnetic probe that protrudes into the oil lines of the engine. It was assumed, by the manufacturer, that any wearing in engine components would result in the release of small metallic fragments progressing through the oil line system. These fragments would eventually be attracted to the metallic probe which could be removed subsequently for inspection.

On the night of May 4, 1983, a maintenance engineer had removed the master chip detectors aboard an Eastern Airlines L1011 Tristar and had been issued three replacement detectors from the stores. Although a tag was attached to each detector

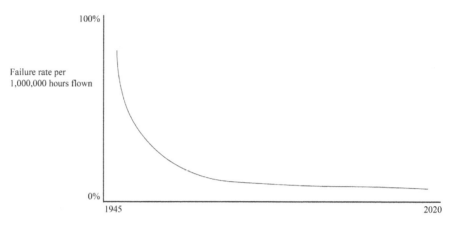

FIGURE 2.1 A conceptual diagram of the failure rate within the aviation industry, since 1960.

indicating that the parts were serviceable, the detectors were not fitted with the 'O' rings necessary to prevent oil leakage through the detector. The result was an in-flight oil leak that resulted in the gradual shutdown of all three engines, and the near ditching of the aircraft near Miami.

Although this incident was significant, and the consequences could have been severe, the development and implementation of remedial effects as a result of this case would probably have resulted in a relatively minimal improvement in overall system safety. In some cases, it is only through incidents or what are referred to as 'near-misses' that opportunities for system improvements become evident.

2.2 THE NATURE OF COMPLEX INDUSTRIAL ENVIRONMENTS

There are a number of features that characterise complex environments, and which contribute indirectly to the incidence of failures. From a systemic approach, these might be termed local factors, since they are a normal part of many operational contexts and do not necessarily result in system failures in the absence of a catalyst. They include a degree of interdependence between components, operational demands that change rapidly, a level of uncertainty as to the reliability and accuracy of system-related information, limits in the availability of resources, competing demands, and a complex organisational structure.

2.2.1 INTERDEPENDENCE

Contemporary systems tend to be highly interdependent, where one component is dependent on the operation of other components. An example of interdependence of systems components can be drawn from the circumstances associated with a relatively minor cabin fire that occurred in an Israel Industries business jet aircraft during the cruise at 35,000ft. Chaffing between wires was the initial seat of the fire and would normally have been dealt with by circuit breakers (Perrow, 1984). However, in this case, the wires led behind a coffee machine that was in use at the time. A short

Human Factors in Organisations

circuit occurred between one of the wires and the coffee machine, and resulted in a larger current, thereby burning the insulating material off a series of wires, including communication wiring and the accessory distribution wiring. The overall result of this interaction was a night emergency landing without landing flaps or reverse thrust, a fire that reignited twice during the descent, and an inoperative antiskid. The aircraft overran the runway as a consequence, although there were no injuries.

This type of situation led Perrow (1984) to suggest that some industries might involve a 'tight coupling' between components. The tighter the coupling, the greater the need to ensure that each component within the system operates at an optimal level.

Tightly coupled systems can generally be characterised as:

- Driven by well-defined choice opportunities;
- Oriented towards well-defined problems;
- Possessing an explicit division of labour;
- Possessing clear procedures for conflict resolution;
- Targeting clear information requirements associated with choices;
- Possessing routines and/or procedures which direct the behaviour of various sub-units within the system; and
- Possessing highly interdependent sub-units.

(Adapted from Crecine, 1986, p. 82)

While tightly coupled systems may lack redundancy, they have an advantage over more loosely coupled systems in providing greater opportunities for errors to be recognised and thereby terminated. In effect, there are more opportunities for a tightly coupled system to 'trip' a system warning or defence, since the failure of a component is immediately evident.

2.2.2 RAPIDLY CHANGING DEMANDS

Successful contemporary organisations must be able to react quickly to changing demands, both from within and outside the industry. These may be commercial demands that emerge from a new entrant to the marketplace, to system failures associated with the manufacture of products. In either case, the structures that characterise the behaviour of an organisation during normal operations can, in some cases, serve organisations well in response to changes in operational demands. For example, when a high reliability organisation shifts from a routine to a high tempo structure, there is an organisational shift from a rank-dependent authority structure to an authority structure that is based on functional skills. This facilitates a horizontal communication structure between groups and enables the provision of impromptu support when and where required.

Anecdotal support for this proposition can be derived from the actions of Sioux City air traffic control, prior to and following the crash of a United Airlines DC-10 in 1989. The aircraft was en route from Denver, Colorado, to Chicago, Illinois, when the first stage disk in the number two engine shattered and severed hydraulic systems one and three. Forces associated with the failure also fractured the number two hydraulic line and resulted in the total loss of the flight control system.

The aircraft was directed towards Sioux City Gateway airport, and came under the control of a relatively inexperienced air traffic controller. Due to the nature of the failure and the rapport that had been established between the air traffic controller and the flight crew, the position was not relegated to a more senior controller as would normally have been the case. Consequently, a relatively inexperienced air traffic controller assumed control of a difficult situation (National Transportation Safety Board, 1990c). This type of adaptation is indicative of an organisational shift based on functional skills, which also facilitated the provision of impromptu support as and where required.

Within high reliability organisations, the shift from a high tempo to an emergency environment is characterised by the application of pre-programmed activities (such as those that might be contained in an emergency plan), and the emergence of functional groups (De Keyser, 1990). In the aviation environment, this strategy is most evident following an accident or incident where individuals will perform according to their specific skills, rather than according to their position. For example, during an evacuation, flight attendants will be required to take on a more authoritarian role in managing passengers than would occur during normal flight. Similarly, in some cases, flight crew may be required to assist with an evacuation, as occurred aboard an Aloha Airlines Boeing 737 in April, 1988 (Barlay, 1990).

A significant portion of the forward fuselage detached shortly after the aircraft levelled out at 24,000ft. The pilots subsequently landed the aircraft safely, although the cabin crew were unable to assist with the evacuation due to the significant injuries sustained as an initial result of the structural failure. Consequently, the captain and first officer undertook the role that would normally have been performed by the flight attendants. This type of behaviour is indicative of the spontaneous emergence of functional roles that tends to occur within high reliability organisations following a critical change in the environmental demands.

2.2.3 UNCERTAINTY

A level of uncertainty is characteristic of most organisations and reflects the fact that risks must be taken to ensure a return (either financial or otherwise) beyond the initial investment. From an organisational perspective, uncertainty refers to factors such as the estimated demand for a particular product or service against the estimated costs in providing the product or service (Saunders, Gale, & Sherry, 2015). Within a railway network, this might equate to the dichotomy that exists between the estimated demand for a route against the costs involved in sustaining the route.

Under most conditions, the management of an organisation requires a level of certainty, and would normally involve some estimate of the risks involved in a particular venture, prior to the authority to proceed with an investment. This process is referred to as risk analysis and is designed to determine the probability of a particular event or series of events. However, the process of risk analysis is also applicable within a human factors context where proposed systems and technology are subject to an analysis of the probability of failure (this is also known as *failure analysis*).

According to Rasmussen (1990a), risk analysis is as much an 'art' as it is a science. This is due to both the number of factors that need to be taken into account, and the nature of the relationship that exists between the various factors and the outcome. In most cases, the nature of the relationship is nonlinear and, therefore, a causal relationship is difficult to establish. This is consistent with the notion that accident causation is more consistent with a chemical reaction than a linear chain of events. The complexity of the relationships is such that it can be exceedingly difficult to predict specific occurrences.

2.2.4 RESOURCE LIMITATIONS

Most contemporary organisations are faced with resource limitations that may constrain either the extent, or the quality of operations. This may result in the substitution of products and services to the extent that safety is jeopardised.

Substitution was a policy clearly evident within Monarch Airlines, a regional airline operating until 1993 in Australia. On one particular flight, the auto-pilot had failed, and the company elected to utilise a two-crew operation, rather than have the equipment repaired (Bureau of Air Safety Investigation, 1994). There was some evidence to suggest that financial considerations prompted this decision, and while this operation itself was not illegal, the pilots were operating within an environment in which they had not formally been trained. Consequently, this type of strategy had the potential to impact adversely the safe operation of the aircraft.

Although resource limitations are generally associated with financial constraints, this is not always the case, and the limitations that occur are not always manifest in tangible resources. Intangible resources within industrial environments include training, safety and regulatory inspections, and regular maintenance schedules. Clearly, the distribution and impact of these resources are more difficult to evaluate than more tangible resources such as staffing or operational costs. Therefore, there is a risk of neglecting the significance of less tangible resources in planning and in-service delivery.

2.2.5 COMPETING DEMANDS

Resource limitations may also be evident as a product of competing demands, whereby an organisation is required to offset one goal against another. For example, in allocating resources to improve safety, funding must be obtained from other sources that may, in turn, reduce the capability of the organisation to compete effectively and thereby recoup the investment. The role of the organisation is to identify those strategies that both facilitate a safe environment and maintain a competitive advantage.

The difficulties associated with competing demands is illustrated by the crash of an Aero Commander, VH-SVQ, operating with Seaview Air while en route from Williamtown to Lord Howe Island off the coast of Australia. Ten weeks prior to the crash, Seaview had been conferred the status of an airline by the then Australian Civil Aviation Authority (Staunton, 1996). Combined with an earlier crash involving a Monarch Airlines' Piper Navaho, the loss of the Seaview Airlines' Aero Commander called into question at the time the competing demands of the CAA as both safety regulator and promoter of the aviation industry.

In a subsequent Commission of Inquiry, Staunton (1996) noted that part of the policy of the CAA was to 'get managers and staff as close as possible to the customers'

(p. 95). While the advantages of this type of approach were recognised, it was also noted that there was the possibility that the role of regulator may be undermined due to the relationship that may develop between the CAA manager (as regulator) and the operator.

This conflict of interest is a clear example of the nature of competing demands within a single organisation. According to Staunton (1996), the policy of the CAA in respect of compliance to regulations involved a concept termed 'graduated response'. This involved a progressive strategy towards compliance, in which education and counselling preceded prosecution. The CAA was not only expected to act as a partner to industry, but CAA officers were encouraged to be 'customer focussed'. Therefore, at one level, the organisation was charged with ensuring compliance, while at the other level, it was charged with meeting the needs of customers.

The same problem belies the Federal Aviation Administration (FAA) in the United States. For example, Nader and Smith (1993) contend that the FAA acted to accommodate airlines, rather than regulate their behaviour. In one example, the Fokker F-100, as operated by American Airlines, included flight attendant stations which violated Federal Aviation Regulations (FAR) that flight attendants should be seated as close to exits as possible. The aircraft had six exits including two forward exits and four over-wing exits. One flight attendant was seated forward, and one at the very rear of the aircraft where there was no exit.

To justify their actions, American Airlines argued, quite accurately, that there was an FAR that required flight attendants to be distributed evenly throughout the cabin. The net result was an extra passenger seat than would otherwise have been the case.

Competing demands also manifest themselves from within an organisation, particularly in the distribution of resources. While this may encourage a level of competition, it might be argued that this may not necessarily result in an improvement in the quality and efficiency of services or products provided. For example, drawing resources from maintenance towards increased marketing may result in a short-term increase in revenue for the organisation but at the expense of reputation in the event of a maintenance-related incident or accident.

2.2.6 COMPLEX AUTHORITY STRUCTURE

As companies increase in size, there is a tendency to create large bureaucracies by simply adding to existing structures. There are a number of issues that may arise from this type of organisational environment including a potential loss of communication, the loss of corporate identity and culture, and the lack of direct accountability.

The loss of communication between various departments within an organisation is a key feature that enables system failures to occur unchecked. An increase in the complexity of an organisation necessitates tighter coupling and the development of rigid, standardised structures for the purposes of communication, as a means of limiting ambiguity between the sender and the receiver. However, the very act of restricting the flexibility of communication structures may lead to errors when flexibility and the 'richness' of a communication strategy is required.

A significant symptom of the loss of communication between sub-units within a system is the loss of corporate identity and culture that may be evident. The role of a corporate culture is to facilitate the development and implementation of organisational norms to direct behaviour. Organisational norms constitute generalised beliefs or understandings held by a group, and which concern the nature of a particular environment (Loh, Idris, Dormann, & Muhamad, 2019). Therefore, norms can enable the process of communication, since an employee 'knows what is expected' without having to be told a particular policy. However, it should be noted that the establishment of group norms depends on significant interaction with members of a group or organisation.

2.3 ERROR WITHIN STABLE AND UNSTABLE ENVIRONMENTS

The nature of human error differs depending on the type of environment within which the human operates. For example, in a tightly coupled system, where the procedures are well defined and consistent, errors are more likely to occur due to the variability that exists in human performance. On some occasions, an operator's performance may appear to be relatively accurate and efficient, while on other occasions, performance may appear to be relatively inaccurate and inefficient. However, in general, an operator's performance is likely to fall somewhere between exceedingly high and low levels of performance. If the level of performance of an operator was recorded over a period of time, the distribution is likely to reflect the pattern depicted in Figure 2.2.

Figure 2.2 depicts a Gaussian (normal) distribution, and while there is some debate as to the explanatory power of this effect, the concept forms the basis of a large number of statistical and biological properties. It also explains the variability that can exist spontaneously within a seemingly stable system. The performance of any human within the system will vary consistent with the notion of the Gaussian distribution. Therefore, errors can occur at any time, and without significant warning. This type of effect is referred to as *stochastic variability*.

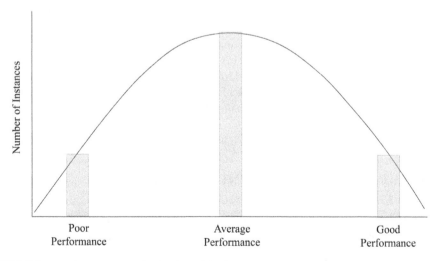

FIGURE 2.2 A conceptual distribution of the frequency of poor, average, and good levels of performance across a 12-month period.

In an unstable operating environment where there is considerable uncertainty, and the demands are constantly changing, different characteristics of errors tend to be evident. The first of these characteristics is related to resource limitations which occur due to rapidly changing demands. For example, resources within an organisation may be allocated according to a perception of the requirements, often months or years in advance. This is typically evident when an organisation is expanding rapidly into new markets with little experience of the demands involved.

The impact of an unstable operating environment was illustrated following the crash of Air Florida Flight 90 into the Potomac River in 1982. The National Transportation Safety Authority (NTSB) noted that the performance of the crew reflected their inexperience in cold weather operations, and that this was directly attributable to the rapid expansion of the airline that had occurred during the preceding years. As a result, pilots within the airline were promoted to senior positions at a more rapid rate than would otherwise have been the case (NTSB, 1982). Clearly, in the case of Air Florida, a conflict existed between devoting resources towards route expansion, and devoting resources to counteract the impact of rapid promotion within the overall pilot population.

The second characteristic of error which is attributable to unstable environments is the impact of human learning and adaptation. Within an unstable environment, there is a strong motivation for certainty and consistency. As a result, it is possible that human adaptation will occur that may increase the probability of a system failure. The main difficulty lies in the degree of flexibility that exists for individual behaviour.

A clear example of this type of adaptation involved a strategy developed by American Airlines and Continental Airlines for the removal of the engine pod of the McDonnell Douglas DC-10. The manufacturer recommended that this procedure be carried out by first removing the engine and subsequently removing the engine pod (NTSB, 1979a). Maintenance engineers at both American Airlines and Continental Airlines developed a procedure that eliminated the two-step process. However, this strategy increased the structural stressors on the engine pylon and eventually resulted in a fracture. The result was the loss of an engine pod following the take-off of a DC-10 from Chicago's O'Hare airport.

The maintenance strategy developed by American Airlines was designed to reduce costs, but it demonstrates how individuals within a system will adapt in the face of uncertainty or ambiguity. This uncertainty is clearly evident in an advisory issued by McDonnell Douglas which stated that it did not 'encourage the practice' (NTSB, 1979a). In other words, the practice was not prohibited: merely discouraged.

2.4 HUMAN FACTORS IN CONTINUOUS PROCESS OPERATIONS

Continuous processes are those operations that require continuous activity on the part of an operator to safeguard and manage the system (Poyet & Leplat, 1993). This activity may vary from psychomotor control, as in the case of a production line, to cognitive control as in the case of an automated vehicle. In each case, the role of the operator is to recognise if and when a change occurs in the system state, diagnose the source of the change, and implement appropriate strategies to rectify the process if and when required. Significant changes have occurred to continuous process operations over recent years, including an increase in the level of automation, the development of larger and more tightly coupled systems, the modification and integration of displays, and/or changes in management style and function.

From the perspective of the continuous process operator, the main role of automation has been to 'distance' the operator from the activity. Therefore, the operator no longer has direct control over the process and must interact with the system through an interface. This reflects a shift from psychomotor control to a more cognitive level of control where the role of the operator occupies the role of 'monitor' rather than 'controller'. This has significant implications for error detection, since operators who have direct control over a system appear to be more accurate in their diagnoses, and are more rapid in their responses to errors than operators who act as monitors.

Within advanced technology settings, automation has been the subject of considerable controversy (e.g. Sarter, 2008). Specifically, the nature and impact of mode errors has been the source of some concern and has been implicated as a causal factor in a number of system failures (Papadimitriou, Schneider, Tello, Damen, Vrouenraets, & Ten Broeke, 2020). As a result, there is a need for systems that are fault-tolerant and that anticipate and advise the operator of failures or potential failures. This is consistent with the principle that automation acts as an expert or advisory system, rather than a system that imposes direct control over a process.

Despite the changes to automation that have occurred over recent years, the difficulty remains that the role of the continuous process operator has shifted from one that involved direct control to one that involves oversight of systems that exercise control. Importantly, the interpretation of the system operation at this level requires access to information displays that enable error detection, diagnosis, and recovery.

While the provision of an array of information displays may enable the integration of information from sub-systems, it may also reduce the speed with which responses can be initiated. Further, the accurate interpretation of information displays requires an accurate mental model and the capability to evaluate and integrate information from disparate sources. However, the complexity of contemporary industrial systems is such that it is often impossible for a single operator to possess a complete knowledge of the system components.

Consistent with the growth of competitive practices within industries, management styles within continuous process operations have changed to reflect a customer focus. This approach is often based on a 'just in time' philosophy which limits 'down time' and the requirement to retain a large stock. Combined, the changes in continuous process operations have impacted the role of the operator and the nature of system errors that are likely to occur.

3 Risk and Uncertainty

3.1 ORGANISATIONAL COMPLEXITY

Prior to the industrial revolution, most societies were organised as self-sufficient agricultural units, between which there was only limited contact. As a result, the consequences of changes to the state of a system tended to be gradual and relatively constrained to a specific geographic population. More importantly, the physical distance between separate communities had the potential to insulate one aspect of society from the adverse consequences of changes that occurred within another.

The level of interdependence between pre-industrial societies constituted a loosely coupled system where changes that occurred within one component had a minimal or delayed impact on the behaviour of another. This was due, in part, to the indirect nature of the relationship between the components, and the variations that could occur due to changes in situational or organisational factors. For example, the requirement for reliable lines of supply often required an invading army to select one route over another, thereby limiting their contact and the resulting external influence to a specific geographic population.

Loosely coupled systems of the type represented in pre-industrial society represent one extreme of a continuum that describes the relationship between system components (see Figure 3.1). At the other extreme are tightly coupled systems in which there is an immediate and direct relationship between the changes that occur in one component and the changes that occur within another (Plant & Stanton, 2016). While contemporary, industrialised society embodies examples of both loosely coupled and tightly coupled systems, it might be argued that there has been a tendency towards the development of more tightly coupled systems than has previously been the case.

Tightly coupled systems are evidence of the interdependence that exists within an industrialised society. For example, a relatively uncomplicated task, such as the provision of hot water to a household on any given day, is dependent on a number of factors including:

(a) The successful design of an insulated hot water system;
(b) The appropriate construction of the hot water system;
(c) The provision of the relevant materials to construct the hot water system;

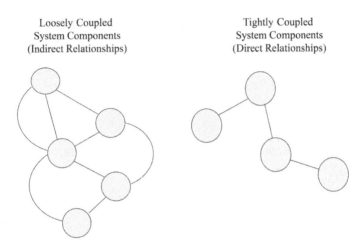

FIGURE 3.1 An illustration of loosely coupled and tightly coupled systems and the nature of the relationship between system components.

(d) The provision of fuel to an electricity generation plant;
(e) The successful design of an electricity plant; and
(f) The capability to conduct electricity over some distance.

If any one of these components was to fail, the resulting change in the system state would have significant implications for both the immediate users of the system and associated users such as designers, operators, and maintenance personnel. Moreover, a system failure within one context may impact systems within other domains, such that a 'domino effect' occurs. An example of this type of situation can be drawn from a series of system failures that were associated with a wheels-retracted landing involving a Continental Airlines Douglas DC-9 at Houston, Texas, in 1996.

Following an investigation of the accident, the National Transportation Safety Board (1997a) identified poor checklist design as one of a number of significant factors involved in the occurrence. The checklist had been designed by the airline, and it was evident that it did not comply with the principles of appropriate checklist design. In combination with an inadequate training regime, the system failure during the design of the checklist culminated in the pilots' failure to adhere to standard operating procedures and the consequent failure to extend the undercarriage prior to landing.

The performance of pilots, like other operators within an industrial environment, is dependent on a series of systems including transport, energy, food distribution, manufacturing, and communication. From an economic perspective, this notion is consistent with the circuitous route of the flow of resources within a closed economy. Each component of an economy depends on the performance of another component. For example, Keynes (1973) suggests that within an economy, the availability of financial resources will determine the inclination to invest and, therefore, the income, output and the levels of employment.

This type of interconnectedness is a characteristic of industrialisation and is frequently associated with an expanding 'region of influence'. For example, the success of multinational organisations has been significantly dependent on the capability to develop markets that transcend regional and national boundaries. The capability for a global region of influence is due, in part, to advances in technology and communication. Products can be shipped rapidly and services can be delivered reliably, even in the most remote locations. However, advances in technology have also provided the basis for an increasingly competitive global economic environment that is not subject to regional boundaries. As a result, the relationship between the producer and consumer tends to be transient, and depends on a combination of factors, one of the most important of which is the reliability of the product and the associated services. Where the quality of a product or delivery is unreliable, consumers will tend to exercise their choice and purchase from competitors.

3.2 CHARACTERISING HUMAN BEHAVIOUR

In the case of isolated tasks examined within a laboratory setting, it may be possible to determine accurately the rate at which human errors are likely to occur. However, it is more difficult to establish the frequency of errors when an operator is immersed within the complexity of an industrial context. In many corporate environments, operators are expected to function within a relatively uncertain environment, where there are competing expectations and demands for consistently high levels of human performance. Similarly, high expectations of human performance exist within the medical and military domains.

High levels of human performance are difficult to maintain consistently, particularly when the relationship between an event and the response of an operator is subject to a wide range of influences. Standard operating procedures, previous experience, personality, and relationships with co-workers each play a role in the types of response that may occur. Even transitory events such as changes in mood or time constraints may influence the behaviour of an operator within a specific situation.

In addition to the immediate or local factors that impact performance, the performance of an operator may also be influenced by organisational factors, such as culture and the availability of resources. Therefore, decisions made by management impose a framework of operating behaviour that either enhances performance or creates an environment where a system failure is inevitable. An example of the former can be drawn from the redesign of a research and development laboratory within the Xerox Corporation. Part of the process of redesign involved the location of a conference table adjacent to a corridor and located within a common area. The table was referred to colloquially as the 'family' table and most of the meetings within the laboratory occurred at the table where the participants could be witnessed and heard by passers-by (Horgen, Joroff, Porter, & Schon, 1999).

The nature of the research and development process at Xerox was such that it often required the input of a number of professionals, many of whom may not have been involved in the original conceptual process. One of the most noticeable features associated with the use of the 'family' table was the cross-fertilisation of ideas that tended to occur and the disintegration of the hierarchy that had previously existed within the laboratory. The main result of the process of redesign was a perceived change in the nature of the organisational culture. The consequence was the capability to develop and introduce a new product into the marketplace within 18 months where five years would have normally been the case.

Where management decisions or inaction impose constraints on human performance, it increases the likelihood that errors or violations will occur. An example can be drawn from the loss of a windscreen from a British Aircraft Corporation (BAC) 1-11 as it climbed through 17,300ft on June 10, 1990. According to the Air Accidents Investigation Branch (1992), the failure was caused by the use of incorrect bolts to fasten the windscreen, a mistake that was due, in part, to the mismanagement of the aircraft general stores. Specifically, it was noted that the carousel from which replacement parts were obtained was mislabelled, and there was no procedure in place to guide the maintenance engineer concerning the acquisition of parts.

3.3 RISK AND FAILURE

Given that human performance is influenced by both individual and organisational factors, it can be difficult to predict accurately the probability that a failure is likely to occur. This type of uncertainty concerning the nature of human performance is not new, nor is the concept of predicting the probability of failure restricted simply to analyses of human performance. Indeed, notions of uncertainty and risk have a history in a range of disciplines including astronomy, meteorology, economics, insurance, finance, politics, and psychology.

Risk is typically conceptualised as the potential for negative consequences such as accidents or incidents (Swuste, van Gulijk, Groeneweg, Zwaard, Lemkowitz, & Guldenmund, 2020). Therefore, it incorporates an element of uncertainty or probability. By reducing the uncertainty associated with a particular event, decision-makers are in a position to better anticipate unexpected or irregular occurrences. Part of this process involves establishing the probability that a failure is likely to occur.

In the aviation and nuclear environments, the probability of a failure has generally been calculated on the basis of a series of anticipated scenarios. The probability estimates are calculated using a combination of historical data and estimates from Subject-Matter Experts (SME). However, the treatment of such data and the mechanisms through which they are combined need to be considered with caution.

According to Rowe (1977), the subjective value of a negative outcome is an important principle associated with risk estimation. When the consequences of a negative outcome are perceived as potentially severe, a decision-maker may take some form of action to reduce exposure to the risk. Conversely, where the outcomes are perceived as relatively minor, action to avoid the risk may not necessarily be taken. This subjective dimension illustrates the difficulty in taking into account individual differences as part of the process of estimating levels of risk.

3.4 THE VALIDITY OF RISK ESTIMATES

Of necessity, estimates of risk in complex operating environments depend on some form of subjective assessment. Therefore, it can be difficult to establish the validity of a model in predicting either the risk of a failure or the potential consequences of a failure. In the case of models that involve complex industrial environments, it is both impractical and cost-prohibitive to engage in empirical analyses and the type of scrutiny that is evident within, perhaps, the scientific community. However, an alternative strategy that might be considered involves the proposition of counter theories.

Essentially, a counter theory is designed to provide an alternative explanation for an event and, thereby, challenge the validity of the original proposal. The debate arising from the proposition of a counter-theory should lead to an outcome in which the most appropriate model is finally accepted (Morgan & Henrion, 1990). The debate can occur at a number of levels within the model, from estimates of the probability of causal relationships, through to the overall interpretation of estimates of risk.

Counter-theories were originally proposed as a means of testing the validity of social policy initiatives where there was a dearth of information concerning the potential outcomes (Feyerabend, 1993). Although there were undoubtedly political overtones associated with the original proposition, at the very least, the process provides an opportunity to consider alternative approaches. Therefore, the development of counter-theories can be useful when the outcomes of a proposal are likely to be somewhat intangible, such as safety or organisational culture.

Despite the attraction of counter theories, there remains a question as to the most appropriate criteria against which the outcomes of a proposition can be examined. In many cases, the decision criteria employed will be influenced by the perception of the policy. For example, the outcome of a decision to commit funding to a safety-related initiative can be considered from two different perspectives. In the first case, it might be perceived in terms of the potential costs associated with an accident, had the funding not been provided. In the second case, it might be perceived as the provision of funds that might otherwise have contributed to the profit margin of the organisation. The differences in the perceptions of the policy lead to the development of more, or less, stringent criteria against which to assess the probability that the policy will be successful.

3.5 RISK AND SAFETY INITIATIVES

Given the complexity inherent in industrial environments, there is usually no guarantee that a particular safety-related initiative will be successful. In some cases, significant amounts of public funds may be invested for little or no reduction in the rate of accidents and incidents. Therefore, safety-related initiatives need to be considered as part of a broader cost-benefit analysis to establish whether or not the likely gains warrant the resources invested.

Where there is only limited information concerning the likely success of a policy initiative, both subject-matter experts and potential users should be interviewed to determine their perceptions of a proposition. Although the results will be subjective, experience in the domain within which the policy is likely to be implemented ought to provide an understanding of some of the potential outcomes that may emerge. Moreover, the use of opinions from subject-matter experts and users is a relatively common investigative strategy in system design and management (Naikar, 2017).

An important part of policy analysis concerns the identification of any assumptions that may be incorporated within a proposal. The extent to which these assumptions can be supported will have significant implications in creating the potential for an initiative to achieve the intended goals. For example, the development of safety culture programmes is often predicated on the assumption that, by changing attitudes, it is also possible to change behaviour within the operational environment (Marquardt, Hoebel, & Lud, 2021). However, the nature of the relationship between attitudes and behaviour is unpredictable, and has been the subject of considerable debate in fields including psychology. Therefore, evidence to support underlying assumptions may be limited.

Assumptions can be made at a number of levels within a policy, from the development of the initiative, through to the methods of evaluation. An example of the former can be drawn from the recommendation by the National Transportation Safety Board that all turbine-powered aircraft with six or more seats should be equipped with an enhanced Ground Proximity Warning System (GPWS) or equivalent warning device (Mortimer & von Thaden, 1999). The recommendation arose following two aircraft accidents involving controlled flight into terrain. Both aircraft were turbine-equipped and there was a clear assumption that, had the aircraft been equipped with an enhanced GPWS, sufficient warning might have been provided to the crew to arrest the descent and prevent the collision.

In the United States, aviation-related recommendations from the National Transportation Safety Board are made to the Federal Aviation Administration (FAA). The FAA is expected to consider the merits of a recommendation and determine whether it should be developed as a regulation and integrated into government policy. In the case of the installation of enhanced GPWS, there are competing demands that must be considered. The installation and maintenance of the system will incur costs to the operator while there is no absolute guarantee that the installation of system will prevent similar accidents in the future.

In many industrial environments, policy initiatives will involve some element of uncertainty. For example, the decision by a rail company to open a new transport route or purchase new rail carriages will have outcomes that are difficult to predict with absolute assurance. A calculated judgement must be made concerning the potential returns, both financially and safety-related.

Given the potential consequences of some policy initiatives, calculated judgements ought to be based on precise estimates of the level of probability of likelihood. In the case of the installation of the enhanced GPWS in turbine-equipped aircraft, an accurate estimate of the proportion of accidents that might be prevented will enable a cost-benefit analysis to be conducted and should facilitate the decision-making process. However, this process requires the identification of those elements of an initiative that are subject to uncertainty, and the development of explicit estimates of probability.

Although it can be difficult to establish precise levels of probability, there are a number of strategies available that can be used to facilitate the calculation of accurate estimates. These include:

1. Referring to related empirical or social policy research;
2. Obtaining advice from experts;
3. Conducting a pilot study; and/or
4. Considering similar initiatives in similar contexts.

There are three types of probability that need to be identified as part of policy analysis in the context of human factors. The first pertains to specific quantities, such as the proportion of failures that are likely to occur due to human error. Differences between the type of assessment or the nature of the domain may result in different outcomes. In addition, extrapolations may need to be made where the data do not directly relate to the environment under examination.

The second type of probability pertains to the relationship between the elements of a model, such as the relationship between attitudes and behaviour. In the case of human error, there are a number of different theoretical perspectives that purport to explain the basis of human performance. Each approach may have some merit, although it is unlikely that a single, universal model of human performance will emerge. Therefore, each model will entertain some level of uncertainty in predicting the nature of the relationship between components.

The level of disagreement between experts is the final type of probability that needs to be considered as part of policy analysis. Experts within a domain will generally provide different perspectives, although it may be possible to reach a general level of agreement. However, where differences between experts do occur, it is important to ensure that these differences are recognised and are incorporated as part of an analysis that enables the assessment of risk.

3.6 RISK CONTROLS

Risk controls are strategies that are implemented to manage risk to the greatest extent possible. They include procedures, training, and technological solutions. They are intended to prevent an error from impacting the system more broadly, identify that an error has occurred (with a view to correction), and/or intervene and correct an error that has occurred. Typically, a combination of risk controls is implemented based on the assumption that risk controls are imperfect.

The application of risk controls is especially evident in the banking industry where assessments of risk differentiate successful from unsuccessful investments (Ellul & Yerramilli, 2013). In this case, the difficulty lies in a highly incentivised environment where there are significant individual and organisational rewards for behaviours that generate returns on investments. In the absence of strong risk controls, there is tendency to extend risk on the assumption that the returns on the investment will be significant (Lim, Woods, Humphrey, & Seow, 2017).

In the banking industry, where exposure to risk often occurs at the level of individual employees, organisational culture has emerged as an important strategy to control risk (Carretta, Fiordelisi, & Schwizer, 2017). In addition to metrics that are intended to calculate the level of risk exposure, organisational culture is considered an overarching framework that guides the level of risk to which an organisation is prepared to be exposed.

A positive organisational culture is one that reinforces an appropriate tolerance for risk given the circumstances. It is both encouraged and reinforced through a combination of training, exemplary leadership, and complementary procedures and tools. It promotes adherence to expectations of performance, despite difficulties in anticipating the types of situations that are likely to be confronted by employees.

Organisational culture has also emerged as an important risk control in the construction industry where it can be difficult to implement technological solutions that might assist in the identification and prevention of errors or inappropriate behaviour (Lingard & Holmes, 2001). This is an organisational context where there appears to be an implicit acceptance within the industry of an inherent level of risk associated with construction activity. Nevertheless, there remains an impetus for proactive approaches to safety through the implementation of risk controls (Hallowell, Hinze, Baud, & Wehle, 2013)

In mining, a significant research effort has been undertaken to identify and consolidate successful risk controls that can be applied across a range of sites and organisations. Referred to as RISKGATE, it comprises a knowledge base that summarises successful interventions to manage and control hazards in mining (Kirsch, Shi, & Sprott, 2014). For example, preventative controls might include a maintenance program, regular inspections, and frequent tests. By sharing knowledge and employing a combination of evidence-based strategies, there is an opportunity for a proactive approach to system safety.

3.7 UNCERTAINTY AND PROBABILITY

Quantifying the level of uncertainty associated with an activity provides a structure within which the probability of an event or set of circumstances can be estimated in the absence of additional information. This is particularly important for decisions regarding human behaviour, since the probability of an event is generally reliant on subjective estimates. Therefore, the probability depends on both the characteristics of the event, and the nature of the information on which the estimates of probability are based (Grünwald, 2018).

Where there is a greater level of certainty concerning the frequency with which events occur, the probability of an event can be determined for a given period as the frequency of the event, divided by the exposure to the conditions under which the event occurs (Viner, 1996). For example, an organisation may record 50 injuries amongst baggage handlers over a given year. If 100 baggage handlers accumulated 35 hours per week for 48 weeks of the years, the cumulative exposure to the environment would be 168,000 hours. Therefore, the probability of an injury per hour over a given year could be calculated as:

$$P = \frac{Frequency}{Exposure} = \frac{50}{168000} = .0003$$

where P denotes the probability of an event

Given that the probability of an injury is .0003, it is possible to calculate the impact of an increase in the number of baggage handlers on the frequency with which injuries are likely to occur in the future. If the number of baggage handlers was to increase to 150 for the following year, the cumulative exposure to the environment would also increase to 252,000 hours. If it is assumed that the probability of an injury will remain constant, the frequency of injuries over the year can be calculated as:

Frequency = P × Exposure = .0003 × 252,000 = 75.6

Irrespective of the amount of historical data available to predict the probability of an occurrence, an element of stochastic variability will always exist in systems that are exposed to extraneous influences. For example, human performance will differ over a period of time, depending on a range of organisational and social factors, many of which are outside the influence and control of an organisation. As a result, an element of error must be incorporated as part of an assessment of probability.

3.8 PROBABILITY ASSESSMENT

The reliance on historical data to predict future occurrences is consistent with a reactive approach to the management of system safety. However, it must be noted that there are many situations where it is impossible to await the accumulation of historical data prior to the development of an assessment of the probability of a system failure. The use of expert testimony is a more proactive approach, although this information is subject to uncertainty that must be taken into account during the process of assessment.

While it is possible to quantify the probability of an occurrence, this process does not necessarily indicate whether the risk is acceptable or unacceptable under the circumstances. This is often a subjective response in which it may be tempting to make a judgement based on relatively arbitrary criteria. To overcome this problem, Wyler and Bohnenblust (1991) developed a model that integrates the probability of a failure with the potential impact of the failure (see Figure 3.2). Although the model

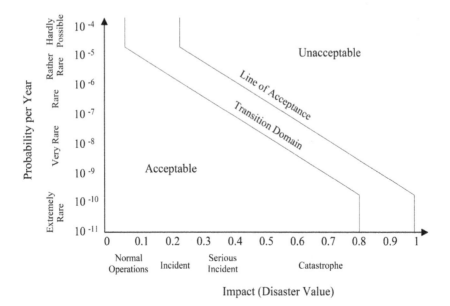

FIGURE 3.2 Probability–impact diagram. (Adapted from Wyler and Bohnenblust, 1991.)

incorporates a number of arbitrary features, it embodies the relationship between probability and the relative impact of an occurrence. Where the probability of a failure is extremely rare, relatively more serious outcomes are 'acceptable' from the perspective of risk management. For example, since the probability of a meteorite colliding with an aircraft at altitude is extremely unlikely, it might be considered unreasonable to expect that preventative measures be developed to ensure that the possibility of this type of occurrence is averted.

Where a failure is relatively more likely to occur, it is expected that preventative measures will be developed to minimise the impact of even the most minor of incidents. Therefore, the range of 'acceptable' consequences is significantly less than if the event was less probable. However, the area of transition denotes the variability that may occur in the relationship between risk and acceptable and unacceptable consequences of a failure.

The validity of the probability-impact model will ultimately depend on the relative acceptability to the wider community of the consequences of a failure. From a rational perspective, the response should be consistent with the model. However, values and attitudes towards system failures depend on a number of factors, including media coverage, the age and nature of the injuries to the victims, and the extent to which negligence may have been involved. Therefore, the community may perceive system failures that might have been the product of extremely rare events as ultimately preventable.

3.9　THE BOWTIE METHOD

Identifying the risks associated with an activity involves an assessment of both the precursors of an event and the consequences. Aiding this process is a method that begins with the event that is the target of an analysis. For example, marine pilots often transfer to vessels at sea using a ladder deployed in the vicinity of a smaller transfer vessel. Typically, the ladder is wet and may not be fixed to the vessel, leading pilots to lose their footing and potentially fall into the sea or onto the deck of the transfer vessel.

Using the Bowtie method of risk identification and assessment, various 'what if' scenarios are employed to target scenarios that might occur under different conditions (de Ruitjer & Gruldenmund, 2016). For example, in the case of the marine pilot, the precursors to a slip and fall might include windy conditions, a poorly maintained ladder, a significant sea swell, and/or poorly maintained boots. In combination, one or more of these conditions may result in a slip and fall from the ladder. However, protective measures can be introduced that are designed to prevent the conditions that might result in a fall.

In the case of the weather conditions, limits might be applied at an institutional or organisational level, based on the direction and strength of the wind and sea swell. In this case, the pilot is not exposed to the risk of a fall. Similarly, if a fall does occur, the transfer vessel might have moved away from the ship so that, should the pilot fall from height, impact with the deck of the transfer vessel will be avoided. Depicting these conditions resembles a bowtie (see Figure 3.3).

Risk and Uncertainty

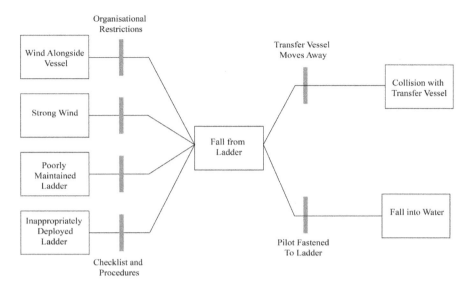

FIGURE 3.3 An illustration of the Bowtie method of risk identification and assessment.

The Bowtie Method is intended as a proactive strategy that enables an assessment of both the event and associated controls (McLeod & Bowie, 2018). In the context of risk management, controls are strategies that are intended to manage risks, ideally by capturing and thereby preventing an activity or event that is likely to lead to an incident or accident. In Figure 3.3, the risk controls are illustrated as vertical bars. In this case, the implementation of checklists and the appropriate application of procedures is intended to reduce the likelihood that a pilot ladder will be poorly maintained or deployed incorrectly, thereby minimising the likelihood of a fall.

Part II

Human Factors Frameworks

4 Human Error-Based Perspectives

4.1 THE NATURE OF ERROR

As the reliability of complex technical systems has increased, the impact of human error on system failures has become increasingly evident. For example, in aviation, human error has typically been regarded as a significant factor in at least 70 per cent of accidents and incidents (Kharoufah, Murray, Baxter, & Wild, 2018). Human error accounts for a similar proportion of industrial accidents, although Bridges, Kirkman, and Lorenzo (1994) suggest that all accidents other than those involving natural disasters involve human error at some level.

At the simplest level, a human error is defined by the outcome. For example, the failure to perform an activity that was required is referred to as an error of omission (Reason, 2002). A failure that involves the application of an error other than that required is referred to as an error of commission. In either case, the outcome is not what the operator intended.

Although the term 'human error' has often been used as a label to identify instances where operators failed to conform to appropriate procedures, the use of the term belies the complexity of system failures within the operational environment. A range of both internal and extraneous factors influence the performance of human operators, each of which may have a deleterious impact on the system. Rasmussen (1982) suggests that the term 'human error' best describes a breakdown in the relationship between the operator and the system. Therefore, any remedial response must involve a consideration of both the nature of the operator and the nature of the system.

The nature of the operator is most commonly understood by undertaking an analysis of the characteristics of the errors that occur within various operational environments. The aim of this process is to identify the similarities that exist between classes of error, thereby enabling conclusions to be drawn concerning the broader nature of human error. Nagel (1988) adopted this type of strategy using a three-stage model that coincides with a simplified model of information processing. Using this approach, conclusions could be drawn concerning the stage of information processing at which most errors are likely to occur, and the severity of the consequences of these errors.

Rasmussen (1982) has also developed a three-stage, Skill-Rule-Knowledge (SRK) model of human performance, the aim of which was to identify generic mechanisms by which human errors occur. The first of these stages involves skill-based behaviour in which nonconscious behaviour and routines are applied to resolve familiar problems. In less familiar tasks, conscious rules may be applied, and this type of behaviour constitutes the second, rule-based stage of the three-stage model.

In cases where the task is unfamiliar, operators are forced to rely on knowledge-based behaviour and the ability to resolve novel situations using generic problem-solving strategies. At this stage, human performance is governed by a detailed analysis of the situation, followed by a deliberate strategy of information acquisition and management. Therefore, the process of problem resolution will depend on both the skills of the operator and the characteristics of the problem.

According to the SRK model of human performance, human error is conceived as part of a broader notion that involves the progression towards expertise and the development of efficient and effective performance. At the initial stage of skill acquisition, behaviour is likely to be characteristic of the knowledge-based stage. However, expertise is characterised by the automatised behaviour reminiscent of the skill-based stage. Therefore, errors at the knowledge-based and rule-based stages might be considered the product of unsuccessful 'experiments' on the part of the operator to improve performance and progress towards skill-based behaviour. An error will only occur if the environment does not allow the operator to observe that an error has occurred, and/or it is not possible to reverse the error in advance of any adverse consequences.

This perception of human error is based on the assumption that operators do not intentionally seek adverse outcomes. Rather, errors are interpreted as products of a genuine attempt to manage the system as effectively and appropriately as possible under the circumstances. This is consistent with Reason's (1990) conceptualisation of error as either an unintentional deviation from a planned strategy, or the application of an inappropriate plan.

The unintentional nature of human error suggests that there are a number of underlying psychological and system-related factors that influence human performance and error within operational environments. The implication is that human error is based, in large part, on a combination of the demands imposed by the operational environment, and the limitations of the operator to process and manage task-related information accurately, and in a timely manner. Therefore, the successful management of a system, which depends on an operator's capacity for perception, decision-making, and responses, is influenced by the level and quality of training and operator selection, and the design and operation of the systems with which the operator interacts (see Figure 4.1).

Consistent with the perspective that human error is an outcome of the interaction between a combination of factors, Bridges et al. (1994) proposed the Human and Organisation Error (HOE) classification scheme where system elements, human factors, organisational factors, and regulatory issues each contribute to the process of accident causation (see Table 4.1). From a systems perspective, successful operator/human performance is understood to depend on a suitable level of communication and the provision of accurate and reliable information through an appropriate human–system interface. Organisational factors and regulatory issues include the provision of sufficient resources, the development and enforcement of operating procedures,

Human Error-Based Perspectives

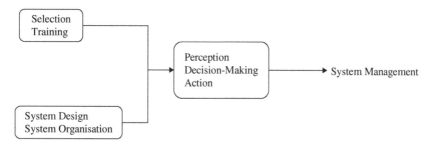

FIGURE 4.1 An illustration of the relationship between operator and organisational factors on elements of information processing and the subsequent management of the system.

TABLE 4.1
An Adapted Version of the Human and Organisational Error Classification Scheme Proposed by Bridges et al. (1994)

Regulatory	Organisation	Human	System
Staffing	Staffing	Violations	Communication
Communication	Communication	Communication	Human–System Interface
Policy	Operating Policy	Job Design	
Surveillance	Surveillance	Information Processing	
Violations	Job Design	Knowledge	
	Morale	Experience	
	Violations	Human–System Interface	
	Maintenance		
	Knowledge		
	Experience		

and appropriate levels of knowledge and experience amongst staff. Finally, the human factors associated with this model are consistent with previous models where training, knowledge, and information processing are targeted as significant elements. Combined, the elements of this model suggest that human error is a complex phenomenon where the characteristics of both the system and the operator contribute to human error and accident causation.

4.2 DECONSTRUCTING HUMAN ERROR

Where the model proposed by Bridges et al. (1994) provides a broad overview of accident causation, it does not necessarily yield a generic model of the types of errors that may occur together with the antecedents of these errors. A more generic approach was advocated by Reason (1990) where human error or 'unsafe acts' are separated according to *intended* or *unintended* actions. In the case of the former, the outcome is typically a mistake or a violation. In the case of the latter, the outcome is either a slip or lapse, the causes of which are generally associated with a memory failure.

According to Reason (1990), mistakes are due to a number of factors that can broadly be defined as either rule-based or knowledge-based errors. This type of classification appears to be consistent with the delineation between rule-based and knowledge-based behaviour proposed by Rasmussen (1982). However, Rasmussen (1982) contends that the type of behaviour that occurs will differ according to the relative familiarity with the environment in which the operator is functioning. Where the operator is more familiar with the environment, rule-based errors will tend to occur. In less familiar environments, knowledge-based errors tend to be evident. This contrasts with Reason's (1990) approach where behaviour is not necessarily dependent on the relatively familiarity of the task.

Despite the differences that appear between generic models of human performance and error, invariably there is a reliance on memory as a mediator between successful and unsuccessful performance within the operational environment. For example, Shiffrin and Schneider (1977) suggest that successful human performance is dependent on the retention of task-related information in long-term memory and the capability to recall and manipulate this information successfully in working memory.

Given the role of memory in the management of human performance, it might be asserted that the basis of human failure falls broadly into two categories:

(a) That appropriate task-related information is available in long-term memory, but is either misapplied or is not activated due to the impact of situational factors; or
(b) That appropriate task-related information is not available in long-term memory due to a lack of experience or training.

Support for this delineation can be derived from Woods' (1989) suggestion that the sources of information-processing errors in the helicopter environment comprise:

(a) Inert knowledge, in which situation-relevant knowledge is not accessed as or when required from long-term memory; and
(b) Missing or incomplete knowledge.

In the case of both sources of error, individual and situational factors have the potential to impact the nature of human performance. Therefore, given the same exposure of training, and in a similar task-related situation, it remains possible that the behaviour of two operators will differ. Where there is likely to be less control exercised over individual and situational factors, as in complex industrial organisations, anticipating human behaviour becomes even more problematic.

4.3 PREVENTING HUMAN ERROR

As early as 1964, the US Navy recognised the significance of human error in the safety of its operations. As a mechanism to understand of the nature of human error amongst Navy personnel, ten principles were developed, specifically oriented towards the investigation of human error. These are listed in Table 4.2, and illustrate

TABLE 4.2
A List of Ten Principles Involved in Human Error Investigations Developed by the US Navy in 1964

1. The most obvious human error-related factors are not always the most significant in precipitating human error.
2. There is usually no one simple solution to human error problems.
3. Beware of general solutions to 'classes' of problems.
4. Humans do not always act as they are expected to, particularly during critical situations.
5. Intelligent operators can (and often will) attempt to 'beat' controls which they perceive as unrealistic and unworkable.
6. Reported incidents of human error usually trigger an influx of incidents, which were previously unreported.
7. Avoid using generalised human error statistics as a gauge of productivity and/or morale.
8. Avoid superficial corrective action such as the 'notification of responsible supervision'.
9. Do not ignore what are perceived as minor, unimportant problems.
10. Encourage direct observation of the actual working environment.

Source: 'Human Error Principles' (1964).

the ongoing difficulty that large, dynamic organisations experience in managing the incidence of human error.

The principles listed in Table 4.2 were originally used as a basis for investigations. However, they also apply as indicators to operators concerning the recognition and prevention of human error in complex operations. Therefore, the principles could be applied proactively to prevent the incidence of human error and create a level of psychological ownership of safety-related initiatives among operators.

In contemporary organisations, a level of safety is expected while maintaining productivity. Therefore, there is often a perceived conflict between the resources necessary for production and the resources necessary for safety. It is only relatively recently that complex industrial organisations have begun to perceive the two goals of safety and production as inextricably linked, such that an investment in safety is associated with improvements in production. This might be evident through fewer injuries, lower insurance premiums, a reduction in staff turnover, and/or greater levels of creativity.

While it might be argued that a complementary approach to safety and production is only possible in those organisations that possess sufficient resources, there are a number of strategies available that are cost-effective, and have the potential to lead to improvements in safety. For example, the development of confidential reporting systems is a relatively cost-effective mechanism whereby decision-makers can receive information from line-managers and operators (Donaghy, Doherty, & Irwin, 2018). In the absence of this type of communication, decision-makers may be tempted to draw assumptions concerning the adequacy of safety.

In addition to resourcing proactive safety initiatives such as confidential reporting systems, redundancy or defence mechanisms are also required within organisations. These defences include audits, systematic analyses of performance, training, and advocacy. However, these need to be implemented appropriately, since poorly executed strategies have the potential to be perceived as superficial and may not necessarily identify the antecedents of human error.

Since the precipitating factors leading to errors may not be immediately obvious, the impact of human error on the performance of a system can be underestimated. Where some relationships may be evident, such as the failure to stop at a red light, other dependencies may be less obvious. The design of the system, the extent to which preliminary analyses were undertaken, and/or organisational culture may each play a role in precipitating error. From the perspective of the accident investigator, the difficulty involves identifying which of a number of diverse and seemingly unrelated factors may have contributed to the failure.

In high-consequence environments, the process of accident investigation is designed to identify the antecedents of a failure and thereby enable the development of strategies to prevent a similar occurrence in the future (Benner, 2019). Therefore, it is generally regarded as a *reactive* approach to the management of safety, since a failure is necessary as the trigger for the subsequent analysis. This type of retrospective approach is also bound within a relatively narrow frame of reference where relationships between significant factors are identified as part of the investigative process.

Despite the relatively widespread application of accident and incident investigation in domains including transport, construction, and energy generation, there are challenges associated with this type of strategy, especially in managing system safety. For example, an operating system may embody a number of hazards, none of which is sufficient, in isolation, to trigger a system failure. Nevertheless, the safety of the system needs to be investigated, since there remains a risk of a failure. Therefore, a safety management strategy that simply employs accident and incident investigation in the absence of other strategies is unlikely to prevent system failures over the longer term.

In contrast to a reactive approach to the management of system safety, a *proactive* strategy involves a regular examination of the system to identify and test the relative impact of potential hazards and develop appropriate mechanisms to manage their impact. Hazard and Operability Analysis (HAZOP) is a useful example of this type of approach in which hypothetical scenarios are constructed to identify any potential sources of error (Dunjó, Fthenakis, Vilchez & Arnaldos, 2010). The aim is to develop an ongoing process of review and analysis to prevent errors and/or mitigate the consequences should they occur.

Understanding the nature of hazards is an important part of the process of error prevention and mitigation. However, the other issue that needs to be understood is the complex relationship that exists between hazards and the operation of a system. Reason (1997) suggests that this relationship is nonlinear, to the extent that relatively minor changes in one part of a system may have a significantly disruptive and unpredictable influence on system behaviour. Therefore, the relatively innocuous decision by an operator to light a cigarette, or take a nap, or make a cup of coffee may trigger a complex set of events that eventually culminates in a system failure.

Human Error-Based Perspectives

The non-linearity of system failures presents the most significant difficulty in identifying and managing hazards. Where a product or system may be innocuous in one context, it may present a significant hazard within another. Therefore, there is a need to establish when a particular aspect of a system or product is likely to become hazardous, and when and how an intervention may be applied most effectively.

A hazard is said to exist when there is a potential for a property or situation to cause harm to the normal operation of a system (National Occupational Health and Safety Commission, 1996). Where some hazards may be relatively obvious, others may be more difficult to anticipate, so that even the most diligent of organisations is likely to experience a failure at some point. However, this should not necessarily dissuade organisations from pursuing a strategy of hazard identification, since the process has the allied benefit of creating an organisational culture where safety is perceived as a significant organisational objective.

Where a proactive approach to system safety may appear to be an attractive alternative, the success of strategies such as hazard analysis is dependent on an understanding of the nature of human behaviour and the extent to which human behaviour changes and responds to shifts within the system state. An example can be drawn from everyday life where a commuter drives the same route, taking approximately the same time each morning to arrive at the office. However, on one occasion, overnight road works may have been completed later than expected so that traffic is forced to reduce speed. Given that the commuter has travelled the same route, and with relatively the same experience each day, the road works are likely to be unexpected. Where sufficient warning is provided to indicate that traffic is slowing, and the commuter reacts appropriately, a collision may be avoided. Therefore, the prevention of a system failure (collision) is dependent on: (a) the provision of sufficient and timely information of a change in the system state; and (b) an appropriate response from the commuter.

From the perspective of a system designer, the provision of sufficient and timely information is likely to depend on the nature of the operator. If the operator fails to appreciate the significance of a change in the system state and/or the information is provided in a form that is unusable by the operator, then the consequences of a system failure are likely to remain unresolved. Consequently, the management of a system failure in a complex socio-technological environment is fundamentally dependent on an understanding of the particular capabilities and behaviour of human operators in a range of diverse environments.

4.4 THE DIFFICULTY IN MODELLING HUMAN PERFORMANCE

The interest in modelling human performance comes from the desire to anticipate and, thereby, improve the process through which humans interact with the operational environment. Where inappropriate or inadequate performance is anticipated, *defences* can be employed to prevent the occurrence, and/or prevent the adverse consequences of occurrence. These defences can include training, advisory systems, warnings, alarms, and guards.

Most researchers agree that human error tends to occur where there is a mismatch between the demands of the situation and the capabilities of the operator. Therefore, predicting human error depends on an understanding of the process of change in situational demands and how this impacts information-processing. Where a simulation can be developed, it may be possible to identify those system demands that are likely to exceed the capabilities of operators, thereby providing the information necessary to develop remedial or mitigating strategies that can be applied to support the operator.

Despite the apparent simplicity of the antecedents of human error, the relationship is always subject to individual and situational variations. More importantly, it may be difficult to predict when these changes will occur and the extent to which they will impact a system. This problem is further compounded when organisational factors such as a lack of leadership or poor communication undermine individual capabilities.

Although there can be difficulties in modelling human performance, Woods (1989) has suggested that there are at least three strategies that could be applied to give some indication of the potential sources of error. The first of these involves a relatively simple approach in which an expert in the field of human error examines a system and identifies the potential sources of error. These sources of error might be situational or organisational factors that may impose unreasonable demands on the operator. For example, in considering a new design for a train cab, an expert might identify those systems or devices that are most likely to impact workload and reduce the attentional resources available.

Although this approach has the advantage of simplicity, the effectiveness depends on the skills and knowledge of the expert, and the information arising from the process is primarily qualitative. Therefore, it can be difficult to establish conclusively the risk or probability that a failure will occur. However, this type of information may be pertinent when cost-benefit analyses are undertaken to determine whether or not the development of a defence mechanism is cost-effective.

The second approach to human error modelling involves the development of an algorithm that takes account of the complexity and variability associated with task performance. For example, Kieras and Polson (1985) considered the complexity of a task as a product of the demands on working memory and the number of procedures required to complete the task. On the basis of this algorithm, it was possible to model the demands imposed by a human–computer interface under error-free conditions. Where the demands exceeded the perceived capabilities of the user, the system could be redesigned, or alternative procedures devised

This cumulative approach to human error modelling is obviously dependent on the accuracy of the algorithm employed and is likely to be most applicable in those environments where the operator's performance is relatively constrained. However, it also depends on a subjective assessment of the load imposed on working memory. In some cases, the demands on working memory may be underestimated to the extent that the potential for error is also underestimated.

Where some approaches to error modelling can only be applied in relatively constrained operating environments, it is possible to develop more complex simulations and still maintain the accuracy of the outcomes. One example was developed by Harrald, Mazzuchi, Spahn, Van Dorp, Merrick, Shrestha, & Grabowski (1998) that simulates the complex relationship between situational and individual factors within the maritime environment.

Human Error-Based Perspectives

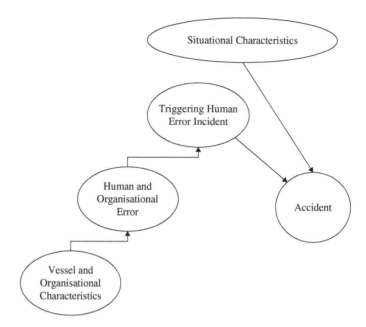

FIGURE 4.2 An illustration of the relationship between antecedents and accident causation in the maritime environment. (Adapted from Harrald et al., 1998.)

The model developed by Harrald et al. (1998) was based on the assumption that human error in the maritime domain can be predicted by range of key indicators including the characteristics of the vessel, organisational factors, and the nature of the operator. Situational factors, like changes in the weather conditions, are assumed to increase or decrease the likelihood that an error will progress to an accident (see Figure 4.2).

The probability of a human error is calculated by seeking expert comparisons of the relative probability that specific types of error would occur for each vessel in the fleet. The five types of errors used were:

1. Diminished ability;
2. Hazardous shipboard equipment;
3. Lack of knowledge, skills, or experience;
4. Poor management practices; and
5. Faulty perceptions or understanding.

The responses from experts are subsequently quantified and incorporated as part of a system simulation. This provides the basis against which to test the effectiveness of a series of risk reduction mechanisms. In the case of each simulated trial, the risk of an accident can be quantified, thereby identifying both the most effective risk reduction mechanisms and the types of conditions under which they were effective.

Despite the complexity of the model proposed by Harrald et al. (1998), the effectiveness of the system remains dependent on both the accuracy of expert judgements and a number of unfounded assumptions concerning the relative distribution of human errors. Therefore, in the absence of long-term evaluations, there are challenges in determining whether this type of error modelling is capable of providing accurate and reliable information within the operational environment.

4.5 SAFETY MANAGEMENT SYSTEMS

In complex industrial environments, identifying and anticipating specific sources and trajectories of human failure can be difficult due to the range and strength of the interdependencies between components. Therefore, the management of human error in these contexts requires that broad areas of risk are identified, in addition to specific instances or cases. The goal is to identify potential opportunities for error and devise strategies that can be implemented that safeguard the system. Typically, this involves collecting, analysing, and interpreting data on a continuous basis to identify and consider sources of risk (Stolzer, Friend, Truong, Tuccio, & Aguiar, 2018).

Safety Management Systems (SMS) comprise processes and strategies that are intended to codify and standardise approaches to the management of safety within an organisation. Sources of risk are identified and assessed using a range of tools, the intention of which is to identify potential opportunities for unsafe practices. Organisational factors that impact system safety include training, leadership, organisational policies and practices, access to resources, and organisational culture (Ismail, Doostdar, & Harun, 2012).

In many jurisdictions, SMS have become legal requirements, against which the performance of an organisation can be assessed. However, the implementation of an effective SMS involves more than simply focusing on processes and procedures. It requires a focus on both the technical and non-technical aspects of organisational behaviour. For example, operators are more likely to report hazards if they trust that management will quickly recognise, investigate, and implement appropriate remedial strategies (Álvarez-Santos, Miguel-Dávila, Herrera, & Nieto, 2018).

The contribution of organisational factors to perceptions of system safety differs, depending on the context. For example, amongst construction workers working on a large apartment building in Malaysia, the critical issues that were identified included a lack of safety-related knowledge amongst workers and the availability of well-designed Personal Protective Equipment (PPE) (Ismail et al., 2012). In the construction industry in Singapore, the emphasis tends to be directed towards the role of safety policies, commitment, and accountability, the responsibility for which lies with organisational leadership (see Figure 4.3) (Kim, Rahim, Iranmanesh, & Foroughi, 2019).

A positive, productive SMS operates alongside Total Quality Management (TQM) principles and standards that comprise ISO9000 (Álvarez-Santos et al., 2018). However, despite the fact that the critical elements of an SMS are defined within the context, the effectiveness of SMS can difficult to establish objectively. This is due, in part, to the challenges in attributing causality, since organisational safety performance is a product of a range of factors, each of which contributes differently to the overall outcome. However, in high reliability organisations, establishing the

Human Error-Based Perspectives

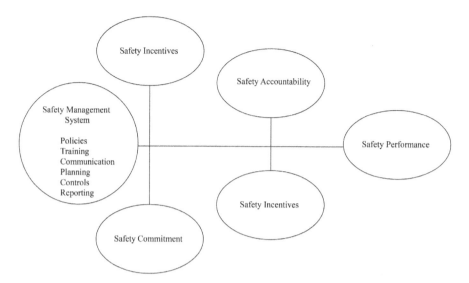

FIGURE 4.3 An example of the components, in combination with a Safety Management Systems (SMS) that contribute to safety performance. (Adapted from Kim et al., 2019.)

effectiveness of an SMS is further complicated by the challenges in deriving safety-related data such as incidents or accidents where few occurrences might occur (Li & Guldenmund, 2018).

Where occurrence data are insufficient or unreliable, subjective data tend to be acquired to establish the perceived appropriateness or effectiveness of an SMS. Typically, data are acquired from a range of employees and other sources within an organisation, using questionnaires that are intended to acquire information concerning the various aspects of the SMS, the implementation of strategies, and the perceived effectiveness of the outcomes in the workplace (Nikulin & Nikulina, 2017). A range of tools is employed for data analysis, including multiple regression, sequential equation modelling, and data envelope analysis.

At its simplest, the goal associated with safety management systems is to enable the identification of changes that may have occurred and that have the potential to impact the likelihood of an occurrence (Hale, Heming, Carthey, & Kirwan, 1997). These changes might include the introduction of new technologies or systems, the introduction of new procedures, and/or changes in personnel. The identification of these changes enables the implementation of appropriate risk controls to contain or capture failures that might occur as a result of the vulnerabilities that emerge.

Safety management systems also enable a more proactive strategy whereby changes that are contemplated within an organisation can be modelled to establish whether vulnerabilities are likely to emerge. The outcomes of this process are intended to improve organisational decision-making, enabling opportunities to consider a range of options and ensuring that the precursors of system failure are managed to the greatest extent possible (Jazayeri & Dadi, 2017).

Like all data-driven systems, the value of a safety management system lies in both the quality and the type of the data derived. While objective data such as on-time arrivals, the frequency and length of delays, the availability of spare parts, the frequency of repairs, and/or the availability of suitability qualified staff are important indicators of the health of a system, subjective perceptions can also be useful, particularly from a within-person perspective.

Pulse surveys that are taken at regular intervals can provide a useful barometer of the impact of changes in the workplace. For example, psychological characteristics such as perceived job demands, perceived resources available, psychological safety, and perceived morale, all contribute to performance in the workplace (Kim et al., 2019). Changes in these characteristics at the within-person or within-team level can indicate deteriorations that may otherwise be imperceptible.

5 System Safety Perspectives

5.1 ORGANISATIONS AS SYSTEMS

Like many contemporary systems, commercial organisations are becoming increasingly complex and are operating within a highly competitive economic environment. Their success depends on a wide range of factors including technology, human operators, and procedures, each of which interacts on a periodic basis to effect positive outcomes. These outcomes can be assessed using both economic and safety indicators. Fundamentally, it is this balance between economic gains and the maintenance of safety that characterises the nature of the system.

The complexity involved in commercial organisations has become such that a greater number of specialist operators is required to ensure that the system functions efficiently and safely. However, technology also plays an increasingly specialist role in the management of system, to the extent that operators are divesting themselves of active participation in the system. The result is a lack of direct control and an increasingly interdependent operating environment. Nevertheless, the operator generally retains the ultimate responsibility for the safety and efficiency of various aspects of the system.

The dilemma associated with the responsibility of the operator is highlighted by the crash of an Air New Zealand DC-10, Flight TE 901, into Mount Erebus, Antarctica in 1979. According to Vette & McDonald (1983), the crash was precipitated by a programming error in the Inertial Navigation System (INS) that was thought to have been corrected, although the new flightpath tracked directly over Mount Erebus. Programming the INS was a procedure that was normally carried out by the navigation section within Air New Zealand, and the errors and corrections that were made by the navigation section within Air New Zealand were unknown to the pilots involved (Vette & McDonald, 1983).

An allied issue associated with this accident involved the descent of the aircraft to 6,000ft, well below the 14,000ft required by Standard Operating Procedures (SOP). This apparent violation was a key element in recommendations advanced by the Transport Accident Investigation Commission (1980), that the accident was due to 'pilot error'. However, Vette & McDonald (1983) argue that the situation is more complex, since the company had developed television advertisements in which aircraft were witnessed operating below 14,000ft. Therefore, it might have been surmised that this was a normal operating practice to afford passengers the clearest view of the Antarctic terrain.

Irrespective of the ultimate responsibility for the crash, the fact remains that human error is often a product of a systemic failure. In the case of Flight TE 901, a fault lay within the system, and manifested itself in an observable error. Therefore, there appears to be a relatively strong relationship between operator performance and the characteristics, values, and expectations of the organisation.

5.2 ORGANISATIONAL ERROR

The impetus for the systemic approach towards the explanation of human error is based on the principle that human error is inextricably linked to the operational environment within which the error occurs. Those features of the operational environment that may impact error include tangible elements, such as procedures and guidelines, and less tangible elements, such as organisational culture. In combination, these features direct human behaviour within a range of diverse environments.

Errors can occur at any number of stages within a system and create an environment within which operational errors become almost inevitable. For example, the failure of an organisation to provide sufficient equipment or training is likely to encourage the development of system 'work-arounds' and may thereby lead to errors. The errors that occur at an operational level are generally referred to as active failures, since the consequences of these actions tend to be immediate. These types of *errors* usually involve operators, and include events such as:

- A failure to follow standard operating procedures;
- Overlooking task-related information; and
- Load-shedding due to high workloads.

It should be noted that these examples of active failures are descriptors, rather than explanations *per se*. They simply describe an active error on the part of the operator, and fail to indicate the circumstances under which the failure occurred. An understanding of the circumstances requires a more detailed analysis of the organisational environment.

Associated with active failures, latent conditions lie dormant within a system until they are activated by a series of local factors. Reason (1993) adopts the metaphor of a resident pathogen to explain the behaviour of latent conditions, since they represent a disposition towards system failure. Latent conditions may include:

- A lack of communication between departments;
- Changes in long-standing procedures; and/or
- Changes in responsibility.

It should be noted that latent conditions do not necessarily result in active failures. However, the impact of a latent condition remains resident within a system over a period of time. Reason (1990) refers to the relationship between active failures and latent conditions as akin to a chemical reaction, mediated by a particular combination of natural elements or local factors. In isolation, the elements remain inert. However, when these elements are combined, the consequences are often severe.

The vast majority of accidents that occur are not the product of an active failure that occurs in isolation. Rather, accidents tend to be the product of a combination of factors, and might have been prevented, had any of these factors not been present at the time.

An example of the combination of latent conditions, active failures, and local factors can be drawn from the loss of the *Herald of Free Enterprise*, a roll-on/roll-off ferry that capsized in 1987 with the loss of 193 lives. Contrary to normal practice, the ferry had departed Zeebrugge, Belgium with its inner and outer bow doors open (Lee & Chung, 2018). This resulted in a massive inflow of water soon after the ferry had departed port and the ferry quickly capsized. This accident, in particular, illustrates the significance of latent and local factors in precipitating failures within complex systems.

The responsibility for closing the doors rested with the assistant bosun. He had opened the doors on arrival at Zeebrugge, and having been relieved by the bosun, the assistant bosun fell asleep in his cabin (Johnston, 1995). He was subsequently awoken when the ship began to capsize. While this might appear to have been a relatively clear case of negligence, it should be noted that there was no procedure in place to cross-check whether the bow doors had been closed. Moreover, other crew members had taken the responsibility to close the doors on previous occasions.

While the ultimate responsibility for closing the bow doors rested with the chief officer, regulations required that he both check that the bow doors were closed, and simultaneously take his position on the bridge in preparation for 'Harbour Stations'. However, it was not possible to see the bow doors from the bridge, nor was there any electronic indicator.

More importantly for the captain of the vessel, the standing orders issued by the company stated that:

01.09 Ready for Sea

Heads of Departments are to report to the Master immediately they are aware of any deficiency which is likely to cause their departments to be unready for sea in any respect at the due sailing time. In the absence of any such report, the master will assume, at the due sailing time, that the vessel is ready for sea in all respects.

(Reproduced from Allinson, 1993)

Clearly, this situation involved a complex series of latent conditions, including:

- The standard operating procedures;
- The lack of indications on the bridge; and
- The lack of redundancy.

However, these conditions only became influential once a series of local factors emerged, including that:

- The assistant bosun was asleep; and
- Other crew members failed to check the doors.

If either of these local factors had not been present, the accident would probably not have occurred. However, in combination with the latent conditions that already remained resident within the system, the result was an active failure on the part of the assistant bosun.

Latent conditions are the most troublesome of problems, since they are difficult to identify in the absence of a system failure. Consequently, the approach taken within many complex industrial systems is to identify and rectify systems once an active failure occurs.

5.3 THE SEQUENCE OF ACCIDENT CAUSATION

Ultimately, the identification and removal of latent conditions requires an understanding of the sequence of accident causation, and the capacity to identify the relationships that exist between performance at the operational level and the decisions taken at the organisational level within a system. Reason (1990, 1997) provides a theoretical basis to explain the nature of this relationship, and suggests that a system can be divided into a number of discrete sections, each of which contributes to performance at the operational level.

At the pinnacle of any organisation are the high-level managers and decision-makers. The responsibility of high-level decision-makers is to set system goals and develop strategies that are likely to achieve these goals. However, decision-makers, like other humans, are fallible, and strategies may be developed that have a negative impact on the performance of operators at lower levels within an organisation. For example, prior to the overrun of Qantas Flight QF1, a Boeing B747-400, in driving rain at Bangkok, Thailand, the company had adopted a standard operating procedure where thrust reversers were normally set at idle for landing.

Under normal conditions, following the landing of a commercial aircraft, the direction of engine thrust is reversed to increase deceleration. Qantas' recommendation effectively required the minimal use of reverse thrust, minimising costs associated with normal operations, including the maintenance costs associated with reversers (Australian Transport Safety Bureau, 2001). However, the decision to employ reverse thrust rested with the flight crew who had to assess the situation and determine the appropriate configuration for landing.

On approach to Bangkok International Airport on 23 September 1999, the flight crew of QF1 decided to adopt a normal configuration for landing, despite extensive water contamination on the runway. Immediately prior to touchdown, the captain advised the first officer, as the pilot-flying, to execute a go-around. However, as the throttles were advanced, the aircraft touched down on the runway, and the captain retarded the throttles and cancelled the command for a go-around. In the absence of reverse thrust, the brakes were applied, although they were ineffective due to the water contamination and the aircraft subsequently overran the runway.

The responsibility to implement the decisions formulated by high-level managers falls to line management (also known as middle management). In many cases, line managers are required to balance the successful implementation of strategies with limitations associated with training, resources, workload, and/or time pressure. Failures at this level of an organisation might include a lack of sufficient or appropriate training, the lack of reasonable facilities or equipment, and/or a lack of sufficient communication between organisational departments (Cohen, Francis, Wiegmann, Shappell, & Gewertz, 2018).

Any failures occurring at the line management level of an organisation produce preconditions that remain resident within the operating environment. For example, a lack of adequate resources may result in higher workload, culminating in poor morale amongst staff, and a lack of motivation to complete tasks productively and safely. In some cases, an increase in incidents involving occupational health and safety may be indicative of these preconditions (Maurino et al., 1995).

The failures that occur within a system will generally be most evident during productive activities where the operator interacts with the environment to produce some form of valued outcome. An active failure occurs when the outcome is inappropriate or the outcome does not occur according to the organisational goals.

To safeguard against system failures, a series of defences is usually developed, either spontaneously or intentionally, throughout a system. System defences serve to either mitigate the effects of an error or reduce the probability of an occurrence. In minimising the probability of an occurrence, physical defences might include guard rails, fences, or some degree of automation (Hollywell, 1996). Less tangible defences might include cross-checks during checklists or co-signatures on maintenance work that has been completed.

Defences that mitigate the outcome of an error include seat belts, smoke hoods, hard hats, and gloves. These defences do not stop an error from occurring. Rather, they ensure that the consequences of the error are less severe than would otherwise have been the case.

5.4 THE SYSTEMIC MODEL OF ACCIDENT CAUSATION

Reason's (1990) model of organisation failure is illustrated in Figure 5.1. Each of the failures that occurs at either the high-level decision-maker or the line management phase is regarded as a latent condition or resident pathogen, since they do not necessarily result in a catastrophic failure in isolation. Typically, a failure occurs when a series of preconditions occurs simultaneously, thereby creating a window of opportunity for a failure.

Resident pathogens can present differently, depending on the nature of the operational environment and the context within which the organisation functions (Swuste, Groeneweg, Van Gulijk, Zwaard, & Lemkowitz, 2018). For example, an investigation

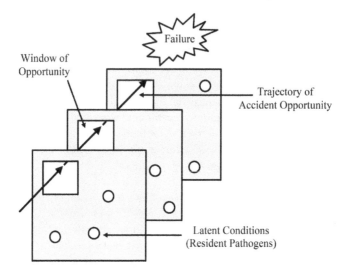

FIGURE 5.1 A graphical depiction of the sequence of organisational failure.

into the causes of a gas leak at the Union Carbide factory in Bhopal, India in 1984, revealed a range of preconditions and resident pathogens, including:

- Poor maintenance;
- The proximity of the plant to housing;
- Failed safety systems;
- Incompetent management; and
- Poor government decisions (Baram, 1993).

While the active failure was the contamination of a tank of methyl isocyanate and the subsequent release of lethal gas, it was an outcome of a series of failures that occurred both within the organisation and amongst government regulators. Throughout the surrounding area, more than 2,500 people died from chemical inhalation.

In 1986, the *Challenger* space shuttle disintegrated soon after launch leading to the deaths of seven members of the United States space program (Kletz, 1994). The primary cause was identified as a failure in the O-ring on a solid rocket fuel booster, manufactured by a sub-contractor to National Aeronautics and Space Administration (NASA): Morton Thiokol.

During the night prior to launch, concerns were raised about the O-ring and the low temperatures that would exist at the launch site (Jenkins, 1988). The management team were aware of the defect associated with the O-ring, but argued that it was extremely unlikely that the O-ring would fail, since it had never failed previously.

According to the subsequent Presidential Review Committee concerning the *Challenger* accident, it appeared that the issue uppermost in the minds of the management team was the launch of the shuttle, rather than the safety of the crew on board (Jenkins, 1988). However, it was also noted that:

- There was a lack of standardised criteria for decision-making;
- There was a lack of understanding concerning the appropriate input for such a decision;
- There was an absence of a chain of command, particularly during the final decision process; and
- There was a lack of a clear mechanism to assess the outcome of the discussion to launch.

Like the gas leak in Bhopal, the decision to launch the *Challenger* space shuttle was a product of a combination of factors, many of which were 'resident' within the system for many months prior to the launch. For complex industrial organisations such as NASA, the difficulty involves identifying the potential for latent conditions before they manifest in the form of active failures.

5.5 THE ORIGINS OF SYSTEM FAILURE

System failures tend to be most evident in organisations in which there is a lack of individual accountability, pressures due to management and/or production, and/or social structures such as decentralisation. Recognising the potential impact of these elements, a number of organisations have developed strategies that are intended to encourage staff members to undertake psychological ownership of the safety within the organisation (Curcuruto & Griffin, 2018). The aim in establishing a level of ownership is to facilitate Safety Citizenship Behaviour (SCB) without the imposition of external demands, such as additional paperwork or supervision.

Improving SCB is designed to offer greater autonomy to individual workers, recognising that managers may hold different priorities. This is illustrated in an analysis of the crash of Downeast Airlines Flight 46 in the United States in 1979 that resulted in the deaths of 17 passengers and crew (Nance, 1986). On initial investigation, the accident appeared little more than a pilot continuing an approach below minimum visual requirements and subsequently colliding with terrain. However, further investigation revealed a company that was operated by an extremely demanding manager who consistently derided his pilots to ensure that the passengers arrived, despite delays and bad weather. It was surmised that an organisational culture had developed where pilots would feel a level of pressure to continue a flight, despite their own concerns for the safety of the flight.

The type of overt management pressure evident at Downeast Airlines is not necessarily the only situation where management pressure can impact the performance of operators. Operators can also be subjected to covert pressure that might range from subtle suggestions on the part of senior staff, to a risk-tolerant culture that is embedded and normalised within an organisation. Given the choice, operators will often accede to the expectations of management, engaging in behaviours that they recognise personally as unnecessarily hazardous.

Given the obvious costs associated with system failure, a number of approaches have been developed to model the interactions between the components of systems. The goal is to enable the identification of potential threats that might otherwise be overlooked or obscured. The timely and accurate identification of threats creates opportunities for remediation or mitigation, safeguarding the system.

5.6 ACCIMAP AS REPRESENTATIONS OF SYSTEM FAILURE

Accimap is a means of representing the relationships between different components of a system that may have contributed to an accident or incident. It is based on the assumption that the events that immediately preceded the occurrence were themselves preceded by the decisions, actions, and activities of other actors. The intention is to identify the sequence of events that afforded the opportunity for the failure, thereby enabling interventions that might prevent a similar occurrence in the future (Waterson, Jenkins, Salmon, & Underwood, 2017).

Typically, there are at least six levels that contribute to an accimap, beginning with the role of government and regulators who develop and enact policy and legal obligations (Svedung & Rasmussen, 2002). The organisation plays a role in adopting and implementing regulations within the legal framework. Within an organisation, management provides the guidelines, training, and resources necessary to undertake activities safely and successfully. Staff engage directly with the system, drawing on the capabilities provided by management to undertake the work necessary to ensure the operation of the system.

Failures at any one of the stages can permeate through to the nature of the work undertaken by staff, the outcome of which can be an error. However, each of the stages is, in turn, influenced by external factors, including changes in economic conditions, the demands of customers, and political change. In the context of accimaps, the challenge lies in identifying laws, regulations, and policies that might inadvertently have created the conditions associated with an error at the work stage of the process.

Accimaps have provided an important foundation for subsequent developments in models that are designed to characterise and thereby enable the anticipation of system failures (Salmon, Hulme, Walker, Waterson, Berber, & Stanton, 2020). Importantly, they clarify the interdependence between decisions at different levels of a system, extending responsibility beyond operators to include management and even policy-makers.

5.7 THE STAMP/STPA APPROACH

Capitalising on the systemic nature of accident and incident causation, the Systems-Theoretic Accident Modelling and Processes (STAMP) is based on the proposition that a system failure reflects the ineffective management of fluctuations in the state of the system. It is based on the assumption that adverse events will occur from time to time that impact safe and efficient operations (Leveson, 2016). These adverse events might include the failure of system components, human error, and/or environmental threats. Exercising control over these events involves the development and implementation of robust constraints, including redundancy, appropriate procedures, physical guards, and training (Leveson, 2015).

In the context of STAMP, constraints are exercised through a hierarchy that imposes controls on activities with the intent to reduce exposure to risk. For example, in most jurisdictions, legislation requires that drivers and passengers secure their seatbelts prior the movement of a motor vehicle. This is a *risk control* that is intended to reduce injuries should the vehicle be involved in a collision and is a behaviour constrained through enforcement. Together with education, this has proven a reasonably effective strategy to ensure compliance, although the level of control can be augmented further with the introduction of a technical solution that prevents a vehicle from being driven without the seatbelts secured.

An important target for STAMP is the interface between different parts of the system, including the transition from development to operational activity (Carayon, Hancock, Leveson, Noy, Sznelwar, & van Hootegem, 2015). The challenge occurs when one part of the system invokes changes that are either not communicated or are miscommunicated to different parts of the system. For example, an operator may identify a weakness in the design of a product but develops a work-around to ensure that operational performance is sustained. While the risks associated with the work-around have not been tested, the demands for operational performance, together with the difficulties in reporting design-related issues, may lead to the continued use of the work-around to a point where it becomes normalised. This is referred to as asynchronous evolution and creates unintended hazards that, unchecked, may lead to a system failure (Leveson, 2015).

Ensuring that systems operate as intended requires that components of the system each retain an accurate understanding of the processes through which a system functions (Leveson, 2004). Among operators, this constitutes their mental model and ensures a degree of consistency between operators, designers, and technical systems, enabling the coordination of activities.

5.8 THE EAST METHOD

Like the STAMP approach, the Event Analysis of Systemic Teamwork (EAST) method is designed to capture the complex relationships that comprise a system. However, where STAMP targets structures of system control, the EAST approach targets distributed cognition, or the extent to which information held by different components of the system is shared (Walker, Gibson, Stanton, Baber, Salmon, & Green, 2006). Using Social Network Analysis as a framework, components are referred to as agents and they can be human, organisational, or technical. The successful operation of the system is presumed to depend on a combination of the information held by agents, the activities that they undertake, and the communication that occurs between agents.

Modelling the interactions using EAST enables a visual and computational representation of the dependencies that exist within a system. Under different conditions, the nature of these relationships changes, potentially revealing vulnerabilities, particularly where links fail between components (Stanton, 2014). This approach can also be employed proactively by purposely 'breaking' links to establish the impact on system performance (Stanton & Harvey, 2017).

Underpinning EAST is an assumption that human performance varies, just as the performance of technological systems can vary. Therefore, the critical aspect of system behaviour is not that a single element may fail, but to what extent the system is vulnerable should a link between system components fail. In this case, it offers an opportunity to identify opportunities for redundancy, where other components of a system supplement the loss of information.

EAST has also been employed to characterise the complexity of systems, potentially leading to simplification. For example, impediments to the timely and efficient transmission of information can be identified and removed or redesigned so that information from one part of a network can progress seamlessly to other parts of the network. This is particularly important in time-critical contexts such as automated driving (Banks & Stanton, 2019).

5.9 THE FRAM APPROACH

Like EAST, the Functional Resonance Analysis Method (FRAM) represents interconnections between the components of systems. These components can be human, organisational, or technical, and each embodies a function in the context of the system (Hollnagel, 2012). In developing a representation of the interrelationships between these functional components, it is important to capture the processes as they occur in reality (Clay-Williams, Hounsgaard, & Hollnagel, 2015).

Each function is represented by a hexagon that incorporates six features, including time, control, resources, preconditions, and outputs. These features embody a degree of variability, particularly in the context of outputs (Hollnagel, 2012). This variability leads to changes in the performance of a system, and can explain both desirable and undesirable outcomes.

In response to variability at different stages of a system, there is an assumption that operators will adjust their behaviour to attend to the perceived demands that are imposed by the system (França, Hollnagel, dos Santos, & Haddad, 2019). For example, in the absence of the requisite tools necessary to replace a failed component, an engineer may seek an alternative solution to meet the imperative to return a vessel to service as quickly as possible, even though the solution is not necessarily ideal.

In the context of safety, FRAM has been employed to explain the precursors to system failures, although there is some acknowledgement that these precursors can be transient. However, there is an increasing emphasis on the application of FRAM to characterise successful performance, particularly in the context of threats. For example, modelling the successful response of a hospital in managing the demands of an epidemic enables comparisons against less successful hospitals, offering opportunities for improvement.

6 Human Engineering Perspectives

6.1 SYSTEMS AND DESIGN

As a concept, system safety is based on the assumption that an interrelationship exists between the various components that comprise an operational environment. The nature of this interrelationship is such that changes in one part of a system are likely to impact other parts of the system. The challenge involves predicting the nature of these relationships to identify opportunities to improve performance and prevent failures.

In the context of human behaviour, the difficulty lies in predicting reliably the responses under various operational conditions. For example, under normal conditions, an aircraft pilot would expect to read back accurately a clearance issued by an air traffic controller. However, under some circumstances, errors can occur. In the following exchange, an aircraft is maintaining 23,000ft and the co-pilot requests to climb to 31,000ft.

- The controller replies: 310 (31,000 ft) is the wrong altitude for your direction of flight: I can give you 290 (29,000 ft), but you will have to negotiate for higher.
- The co-pilot responds: Roger, cleared to 290 (29,000 ft), leaving 230 (23,000 ft).
- No challenge from the controller.
- At 24,000 ft, the controller queries the altitude of the aircraft and responds: I did not clear you to climb; descend immediately to 23,000 ft. You have traffic at eleven o'clock, fifteen or twenty miles.

(Cushing, 1994, p. 28)

This exchange illustrates how a simple misinterpretation can lead to behaviour with potentially significant consequences. It also illustrates the sporadic nature of errors, since the controller would normally have been listening to the pilot's reply to confirm a particular instruction. This is a type of error that is very difficult to predict and, therefore, to prevent.

System safety is synonymous with appropriate design. Therefore, both the likelihood and the consequences of errors should be considered during the design of a system. This requires:

- The inclusion of failsafe systems;
- The capability to cope with rough handling by an operator;
- A consideration of the average capabilities of the operator;
- Protection against inadvertent or uncontrollable human error;
- The protection of occupants; and
- A reliable system.

In accounting for differences in the way in which a system or device will be used, designers consider the user as integral to the design process. However, despite the opportunities afforded by a user-centred approach to design, devices continue to be developed that fail to meet the needs of users. This probably reflects the challenge that designers face when seeking to integrate human factors principles into the design process.

Designing a novel system is a difficult activity that is often impacted by the resources available, the time available for development and evaluation, and the range of preferences of prospective users. For example, Vicente, Burns, & Pawlak (1997) describe a case where a customer requested taller control panels in a control room because the original control panels 'looked too short'. Since modifications and testing take time, there is often a tendency to rely on informal human judgement, rather than empirical evidence, as the basis for the final product.

Although there may be short-term opportunities afforded by a rapid design process, there may be costs over the longer term. Hendrick (1997b) recounts a situation in which a particular style of de-boning knife used in abattoirs was resulting in a relatively high incidence of tendinitis and tenosynovitis amongst users. The original design of the grip was horizontal, as is the case with a typical kitchen knife. This was subsequently altered to a pistol grip as part of a redesign of the knife. In addition to a reduction in work-related injuries, the redesigned knife increased production by up to 6 per cent, resulting in an increase in profit.

In many cases, the challenge in identifying the potential for human error lies in the capacity for operators to adapt to poor designs, with failures occurring sporadically. For example, in the majority of cases, cargo handlers were perfectly capable of closing the rear cargo door of the McDonnell DC-10 aircraft, despite its unusual design as a door that opened outwards, rather than the typical 'plug' type door that was kept in place by cabin pressure.

The door was operated electrically and was secured using a series of four latches, with the final locking pin inserted into place by an external lever. In the case of the Turkish Airlines Flight 981, a scheduled flight from Paris to London on March 3, 1974, the latches were not in place when the door was closed and the baggage handler forced the locking pin into place using his knee (Bradley, 1995). This gave the impression that the cargo door was secured and also prevented the warning light from illuminating in the cockpit.

While climbing through 11,300ft, the cargo door of the aircraft detached from the fuselage and caused a rapid decompression and the collapse of the cabin floor with the loss of 346 passengers and crew. Soon after the accident, the Federal Aviation Administration responded with an Airworthiness Directive (AD) which required the redesign of the system to a failsafe mechanism.

In complex systems, waiting for failures to occur and then responding is referred as the 'science of muddling through' (Lindblom, 1959). The analogy is of cracks in a dam wall that are constantly being repaired. As one crack is repaired, another is likely to emerge in another location. In managing the cracks, a more effective approach might involve developing an understanding of the nature of the dam wall and the factors that might precede cracking. This would involve:

- The identification of the factors that might contribute to a failure (such as water pressure); and
- An estimation of the probability that these factors would occur (such as during a flood).

This is an approach consistent with the notion of *Danger* as the product of a combination of the *Hazards* (or factors) involved in a particular activity and the *Risk* (or probability) that the hazards will occur (Salvendy, 1997).

$$\text{Danger} = \text{Hazards} \times \text{Risk}$$

This relationship suggests that reducing the danger of a failure involves either reducing the frequency or severity of hazards, and/or a reduction in the likelihood of a negative outcome.

6.2 SYSTEMS ENGINEERING

Systems engineering is a methodical process that enables the identification of hazards and risks associated with a system or device. It begins with the development of an understanding of the needs of the potential users of a system. This process of consultation yields a series of specifications that can be employed in developing a design. It also provides the basis for the construction of the system, a process that requires a multidisciplinary approach and that is revised continuously (Stoop, de Kroes, & Hale, 2017).

The significance of systems engineering in the design and development of a safe and efficient operational environment is highlighted by incidents in which systems have failed. For example, in June 1997, two Union Pacific freight trains collided in Devine, Texas, as a result of an inadequate dispatching system (NTSB, 1998).

The error occurred when a dispatcher issued a clearance to the 9186 South train to proceed from Gessner (north of Devine), but failed to communicate the caveat 'not in effect until after the arrival of 5981 North at Gessner'. The failure to communicate this information occurred despite the fact that the information was displayed on a computer screen at the dispatcher's console. Under Union Pacific regulations, a dispatcher is required to read a clearance as it is displayed on the computer screen. In addition, the train crew are required to read back a clearance so that it can be compared against the information on the dispatcher's computer screen.

The NTSB (1998) noted that the Union Pacific regulations were insufficient to ensure a reasonable level of redundancy in the case of an error on the part of the dispatcher. Furthermore, the workload experienced by the dispatcher at the time of the accident was likely to have contributed to the error. Therefore, it might be concluded that the system of dispatching developed by Union Pacific failed to take into account the range of conditions that may have been experienced by users and their resultant behaviour.

6.3 DESIGN PHILOSOPHY

Systems engineering is a largely conceptual process, since a hypothesis (inference) can only be tested through the development and implementation of the actual system. Until this stage, the relationships between components are simply estimated on the basis of previous experience (Sun, Houssin, Renaud, & Gardoni, 2019). Consequently, it might be argued that an effective design process involves a recognition of those aspects of a system that are unlikely to succeed, rather than establishing, with certainty, that the system is likely to achieve the system specifications.

Historically, the progress of system design was often a response to a system failure. For example, the development of vehicles for the space industry was characterised by a series of initial failures from which engineering guidelines were developed (Zimmerman, 1998). This process of trial and error is a feature of many domains, particularly during the initial stages of design and development. However, as experience in the construction of systems increases, there is an opportunity for the development of standards from which subsequent designs can emerge. This process is intended to reduce the frequency with which system failures occur and provide guidelines for designers concerning the minimum criteria for the successful operation of the system.

6.4 USER-CENTRED DESIGN

The principle of user-centred design reflects the requirement to consider the satisfaction of the user's needs as one of the key design goals associated with systems engineering (Stanney, Maxey, & Salvendy, 1997). It contrasts with the traditional, task-centred approach to systems design where the user was simply considered as one of a number of resources to implement an appropriate outcome.

Arguing the benefits associated with a user-centred design can be difficult, since humans are capable of achieving favourable outcomes, irrespective of the adequacy of the design of a system (Elie-Dit-Cosaque, & Straub, 2011). The result is a level of performance that, over time, may be generally at a satisfactory level. However, when a change occurs to the system state, an inadequate design can increase the probably of an error. The relationship is illustrated in Figure 6.1 in which a change in the system state results in a rapid loss in system performance for traditional approaches to design, but asymptotes for user-centred approaches to design.

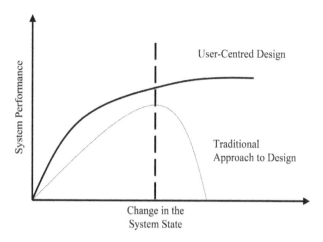

FIGURE 6.1 A conceptual diagram of the relationship between system performance and changes in the system state for traditional and user-centred approaches to system design.

Although there are certainly greater operating efficiencies afforded by user-centred designs, the most significant advantage associated with this type of approach is the increased capability of the system to cope with unexpected changes in the system state. Specifically, the provision of an adequate amount of information, together with an appropriate level of control over the system, increases the probability of a positive outcome in response to a system failure (Endsley & Kiris, 1995; Wei, Macwan, & Wieinga, 1998).

6.5 DECISION-CENTRED DESIGN

Decision-centred design constitutes one of the most important aspects of a user-centred approach to systems engineering. It involves an analysis of the decision requirements associated with the task, and the development of systems that respond to the user's needs (Klein, Kaempf, Wolf, Thordsen, & Miller, 1997; Schnittker, Marshall, Horberry, & Young, 2019). The aim is to ensure that the user is provided with sufficient information in a form that facilitates the decision-making process and increases the probability of a favourable outcome (O'Hare, Wiggins, Williams, & Wong, 1998).

In many operational environments, users are faced with formulating decisions based on uncertain and/or unreliable data. For example, military commanders functioning in an early-warning capacity are charged with determining the intent of various operational targets. Information, such as the trajectory and altitude of the target, will determine the extent to which it constitutes a threat. In this case, the role of a system designer is to provide task-related information in a form that is appropriate and timely, given the situation.

To determine the type of information required to formulate a decision, a Cognitive Task Analysis (CTA) can be applied to understand the nature of a task. The goal is to identify various decision points, and then determine the cognitive basis for a decision. The process involves the acquisition of information pertaining to cues, the relationships between different variables, and the use of analogy and situational awareness (Klein et al., 1997; Papautsky, Strouse, & Dominguez, 2020). From a design perspective, the aim of CTA is to provide sufficient information to enable the development of a system that meets the decision requirements of the user in a variety of situations. This is particularly important in situations where technology is intended to augment the capabilities of the user in resolving decisions.

Despite the significance of the human–technology relationship in systems design, it is one of a number of factors that ought to be considered as part of systems engineering. Other factors that need to be considered include the cost-effectiveness of the system, the availability of the appropriate technology, and the required accuracy of the system. This information, together with the needs of users, constitutes the most comprehensive approach to process of system design.

6.6 FEATURES OF SYSTEMS ENGINEERING

Although the significance of the user should not be understated, the process of systems development also incorporates a consideration of other factors, including operating costs and the use of technology. The accurate identification of these factors requires a sequential approach that begins with the determination of the need for the system and culminates with a systems evaluation. This process is designed to maximise the opportunity to identify and respond to any deficiencies that might exist within the system.

6.6.1 Operational-Need Determination

The recognition of the need for a system can occur via two main routes. It can either be developed as part of a strategic plan on the part of high-level decision-makers, or it can be generated at the operational level, in recognition of the need for a system to improve or extend existing practices (Czaja, 1997; Mauborgne, Deniaud, Levrat, Bonjour, Micaëlli, & Loise, 2016). In the case of high-level decision-makers, the operational need will be determined by organisational factors such as the requirement to improve services or products, or the desire to advance new products. However, at the operational level, the need is typically determined in response to an existing system or procedure. For example, the investigation of an accident or incident may reveal flaws within an existing system that ought to be improved.

The need for a system will be determined by a combination of factors including:

- The extent to which the proposal improves on an existing system;
- The advantages that could accrue from such a change;
- The extent to which technology might be employed; and
- The extent to which the impact of any constraints might be minimised or eliminated (Chapanis, 1996).

Human Engineering Perspectives

6.6.2 Operational Concept

The development of an operational concept is an extension of the initial process of identifying a need. It is designed to provide a conceptual understanding of the intended role of the system, and the conditions under which the system is expected to function. The outcome of this process is an operational profile that incorporates:

- A description of the concept;
- The conditions under which the system is expected to function;
- The operational limitations of the system;
- The nature of the user–system interface;
- A description of any preliminary user training that may be required; and
- An analysis of the strategies necessary to maintain the system.

Ideally, the operational profile should be depicted both graphically and in the form of a written response. The overriding objective is to ensure that the information is communicated in sufficient detail to enable the exploration of the concept in further detail.

6.6.3 Concept Exploration

The exploration of a concept involves the development of detailed solutions to meet the design criteria specified at the operational concept level. It is a design process that demands a level of creativity and a willingness to explore novel alternatives. It also requires that suggestions are subjected to detailed analysis and testing prior to their inclusion within the system.

In many cases, the operational concept will require initiatives at a number of different levels within a system (Ball, Evans, & Dennis, 1994). For example, the operational concept may require a level of efficiency that demands a fundamental shift in the existing user-training process. This reflects the interdependence of system components and it may ultimately yield modifications to the operational concept to incorporate a balance between the various system demands.

6.6.4 Concept Demonstration

The aim of the concept demonstration is to ensure that the system meets the specifications detailed in the design concept. However, the process of preliminary testing also provides an opportunity to consider the relationship between system components and identify any limitations that may become apparent, particularly in operational performance.

Simulation plays a significant role in the demonstration of a construct as it enables a controlled analysis of the relationship between system components. However, in addition to the more structural components, the function of a system should also be simulated at a cognitive level. This is expected to reflect a more accurate assessment of the nature of the system and the factors that are likely to impact system performance (Cacciabue and Hollnagel, 1995).

According to a linear approach to system function, the accuracy of a simulation is determined by the extent to which the various parameters that comprise a system can be estimated accurately. Where there is a level of control over these parameters, a simulation can be used to examine the relative impact on the outcome. However, the complexity of human performance means that it is often very difficult to create parameters of sufficient accuracy and reliability to yield useful data that reflect human performance in practice (Carayon, Wooldridge, Hose, Salwei, & Benneyan, 2018).

Despite the difficulties associated with modelling all aspects of human performance, a number of general principles have been developed that can be used to assist the process of systems design. For example, Kieras and Bovair (1984) have established that a successful response to changes in a system state is dependent on the accuracy of the user's mental model of the system. Therefore, the user's mental model of a system could comprise the foundation for the design and development of an interface between the user and the system. This contrasts with a more task-centred approach where the design of an interface is generally based on the characteristics of the system.

6.6.5 Full-Scale Development

Where the simulation of a system provides a number of useful guides for the development of concept specifications, it is only when the system has been fully developed that the dynamic relationships between system components become fully evident. Therefore, there is a requirement for continuous operational testing to ensure that the system design continues to meet the original specifications.

Operational testing is an essential aspect of the process of systems design. It provides a mechanism to evaluate the relationship between system components in a setting that is relatively consistent with the operational environment. This enables the identification of deficiencies that might not have become apparent during the previous stages of design.

Despite the advantages afforded by testing a system within an operational setting, it is unlikely to reveal all of the interrelationships that exist between components. For example, users are less likely to develop 'work-arounds', or short-cuts, while under observation or in response to novel systems. However, 'work-arounds' tend to be a relatively common occurrence within the operational environment (Blijleven, Koelemeijer, & Jaspers, 2019).

6.6.6 Production and Deployment

Having established that a system functions successfully at an operational level, the production version is developed to the final stage and is eventually implemented within the operational environment. A major part of this process involves training potential users to ensure that they have skills and knowledge necessary to operate the system as intended.

The effectiveness of the training process is an element of systems engineering that is often neglected, although it has the potential to significantly impact both the accuracy and the efficiency with which the system is used. Inadequate training has been implicated in a number of accidents and incidents where users have either failed to recognise the significance of the information that they were receiving, and/or they failed to access the appropriate information to solve a problem (Strauch, 2017).

An example of the impact of training on system performance can be drawn from the crash of an Airborne Express Douglas DC-8 during a post-modification functional evaluation flight on December 22, 1996. The flight crew were conducting a test of the stall warning indication when the pilot initiated an inappropriate stall recovery technique that resulted in the loss of control of the aircraft. The aircraft subsequently collided with terrain, killing the crew.

According to the National Transportation Safety Board (1997b), neither pilot had been exposed to the stall manoeuvre in the DC-8 in the configuration that was used in the accident flight. Moreover, the flight simulators used by Airborne Express were incapable of accurately simulating the behaviour of the aircraft in this type of stall condition. Since guidelines were limited concerning the types of manoeuvres to be conducted during evaluation flights, the pilots were ill-prepared to manage the aircraft effectively.

6.6.7 System Evaluation

The evaluation of a system following its implementation enables an assessment of the system outcomes against the concept specifications. The process is based on a recognition that, in most cases, it is not possible to test all aspects of a dynamic system prior to its implementation within the operational context. However, system evaluation also provides a means of acquiring additional data to improve subsequent designs.

There are a number of strategies available to acquire and manage data pertaining to the effectiveness of a system. Each strategy has advantages and disadvantages depending on the areas of interest within the system. For example, a traffic management system might be evaluated from two perspectives:

- The number of movements per hour; or
- The number of accidents within a given period.

The data arising from each of the two strategies yield different types of information pertaining to the operation of the system. In the case of the number of movements per hour, there is an emphasis on the efficiency of the system. However, in the case of the number of accidents within a given period, there is an emphasis on the frequency with which the system fails. Taken together, the data arising from the evaluation provide information concerning the balance between efficiency and safety.

An alternative strategy for system evaluation involves the development of a baseline of performance against which system performance can be compared following implementation. This provides a robust method of assessing the effectiveness of the system provided that the features that are subject to investigation represent an accurate reflection of the performance of the system.

6.7 HUMAN FACTORS IN SYSTEMS ENGINEERING

As a process, systems engineering itself is subject to human capabilities and limitations. Consequently, errors can occur during the design process amongst both users and designers where there is a failure to perceive relevant information, a failure to process the relevant information, and/or a failure to respond to information. These failures can be attributed to cognitive biases, where responses are influenced by non-rational modes of thinking.

One of the most pervasive biases that is likely to impact performance within a team-oriented environment is referred to as 'group-think'. Identified by Janis (1972), the term refers to the tendency for a small group to reach consensus, rather than infringe on the harmony of the group. This bias can occur, irrespective of the nature of the decision, and may lead to errors that impact system design.

Where a design team has been subdivided into separate sections, errors may occur due to the requirement for a greater level of coordination than might otherwise be the case. For example, large projects require that the different design sections are aware of the guidelines being employed by other sections within the design team. Where assumptions are made between different sections, and/or differences occur between the group goals, biases are liable to emerge which may influence the decision-making process (Kaba, Wishart, Fraser, Coderre, & McLaughlin, 2016).

Organisation factors also have an impact on the design process, particularly in the constraints imposed. For example, design specifications generally include a time limit, following which a system must be at the stage necessary for implementation within the operational environment. However, the uncertain nature of the design process means that it can be very difficult to estimate precisely the duration of systems design to ensure that the outcome is both appropriate and effective. This potentially leads to the planning fallacy where estimates concerning the time required to complete a task are underestimated (Love, Sing, Ika, & Newton, 2019).

In addition to temporal constraints imposed on the design process, there are fiscal constraints that will tend to either limit the scope of the design process or limit the extent to which the design is evaluated prior to implementation. The danger is that a system may reach operational status in a form that is less than satisfactory for the user.

7 Reliability-Based Perspectives

7.1 RELIABILITY ANALYSIS

The origins of reliability analysis can generally be linked to engineering and the development of complex industrial systems. The lack of reliability of the technology at the time necessitated the development of a methodology that established the probability that a system might fail, thereby providing an opportunity to establish alternative strategies to guarantee the supply of products or services. The principle of system redundancy was a significant outcome of this process, although the overall confidence in technology and engineering was tempered by a number of systems failures including the sinking of the *Titanic* in 1912 and the disintegration of the Hindenburg Airship in New Jersey in 1937 (McClellan & Dorn, 1999).

Due to the emphasis historically on the reliability of engineered solutions, it was only in the late 1960s that human performance began to be examined systematically as part of an assessment of system reliability. The shift in emphasis coincided with an increase in the reliability of technical solutions that tended to reveal human performance as a potentially less reliable component of the system. This transition was especially evident in the aviation industry following the development of more reliable power plants and structural components. Where prior to 1960, mechanical or structural failure often contributed to aircraft accidents, following 1960, human error tended to be identified more frequently as a significant factor (Erjavac, Iammartino, & Fossaceca, 2018).

7.2 RELIABILITY ENGINEERING

As a discipline, reliability engineering was designed to integrate the process of systems engineering and probability analysis (Dougherty & Fragola, 1988; Zio, 2009). The intention was to use established models of the behaviour of systems and identify the probability of a system failure under various conditions. By identifying the probability of a failure, it becomes possible to design or redesign a system, such that it takes into account the impact of the forces to which the structure will be subjected (Saleh & Marais, 2006). Therefore, the structure has a level of redundancy and safety.

Reliability engineering is predicated on a relatively stable relationship between events. For example, increasing the tension on a wire in one domain ought, given the same environmental circumstances, to have the same impact as increasing the tension on the same type of wire in another domain. Consequently, the inherent structure of the material and its relationship with external events will remain relatively consistent, taking into account any differences that may arise as a result of the manufacturing process.

Combining materials to form a structure will increase the complexity of the engineering model and the interaction between events. However, it should still be possible to develop a reliable model of any structure, given sufficient time and resources. Although this set of circumstances is unlikely to occur in practice, the principle enables estimates to be made on the basis of extrapolations and known relationships (Ortmeier, Schellhorn, Thums, Reif, Hering, & Trappschuh, 2003).

Extrapolations can be made about a product by understanding the constituents from which the product is constructed. In the engineering domain, mixing two raw elements will result in a consistent reaction over a number of trials, given a consistent testing environment. However, consistency in human behaviour is less predictable. In the memoirs of A. V. Roe, he recalls experiments where an aircraft was towed by a motor vehicle to become airborne (Taylor & Munson, 1975). Part of this process required the 'tower' to release the aircraft so that it could land safely. Despite repeated warnings concerning the consequences of failing to release the towline, inexperienced 'towers' habitually held on to the towline, the result of which was inevitably a crash.

Given that reliability engineering incorporates the deconstruction of a system and the development of an understanding of the relationship between constituents, it is probably unsurprising that initial efforts to understand human behaviour as part of this process also involved a strategy of deconstruction. The aim of this process was to develop a universal model that could be incorporated into a broader analysis of the risks associated with a system.

7.3 PROBABILISTIC RISK ASSESSMENT

Probabilistic Risk Assessment (PRA) is a broad term that refers to the process of establishing the probability of undesirable consequences that may arise from large-scale system failures. It incorporates a range of strategies that examine the features of a system including the design of the plant, practices and procedures, and human behaviour. The aim of PRA is to integrate the information arising from the various analyses to formulate an overall assessment of the risk, and thereby enable the appropriate distribution of resources.

PRA is most closely associated with the nuclear industry and developed, in part, due to public and scientific concerns regarding the potential consequences of systems failures involving nuclear power plants (Modarres, Zhou, & Massoud, 2017). Unlike other industries, a system failure at a nuclear facility has the potential to impact a significant proportion of the surrounding population. In the worst-case scenario, a system failure would result in the release of radioactive gas and the contamination of both the population and the surrounding area for a considerable period of time.

The impact of nuclear contamination was illustrated on March 11, 2011 when the Fukushima Daiichi nuclear power plant north of Tokyo in Japan experienced a series of three nuclear meltdowns and the release of radioactive material up to 20 kilometres from the site, affecting approximately 154,000 residents (Kato, Onda, Saidin, Sakashita, Hisadome, & Loffredo, 2019). As a result, there has been a concerted effort amongst researchers and practitioners to develop strategies to anticipate the probability of a system failure.

Initially, the development and implementation of PRA focused on Individual Plant Examinations (IPE) and the need to quantify the probability of specific types of system failures. However, many of the issues and principles that have been identified as a result of an IPE can be applied to other nuclear facilities and, indeed, to other high-reliability industrial environments (Dougherty, 1993). Consistent with this perspective, the PRA strategies that have been developed for the nuclear industry are generally applicable in other high-reliability environments.

Irrespective of the industrial environment, a characteristic common to most high-reliability organisations has been a recognition that human performance has a significant impact on the probability of a system failure (Le Bot, 2004). However, the assessment of human performance has generally been regarded as the least developed and most problematic element of PRA (Dougherty & Fragola, 1988; Cacciabue & Vivalda, 1991). The difficulty lies in developing a model that effectively captures the transient and uncertain nature of human performance that occurs within the operational environment.

7.4 HUMAN RELIABILITY ANALYSIS

Human Reliability Analysis (HRA) refers to a collection of methodologies for data acquisition and management, the broad aim of which is to describe the relationship between an operator(s) and a system. The outcome of this analysis is an understanding of the characteristics of human error, the likelihood that errors will occur, and the extent to which safety-related initiatives will impact the frequency of error and/or the severity of the consequences. This information is used in conjunction with the data arising from a broader PRA to establish the overall risks associated with a system.

Amongst some researchers, HRA represents the weakest point in the process of PRA (Mosneron-Dupin, Reer, Heslinga, Strater, Gerdes, Saliou, & Ullwer, 1997). The implication is that, in the absence of improvements in the accuracy of HRA, engineers are justified in developing automated systems that limit the input of the operator. A similar argument has influenced the development of automated systems in the transport environment and remains the subject of considerable debate (Banks & Stanton, 2016; Banks, Plant, & Stanton, 2019; Stanton & Young, 1998).

7.5 TAXONOMIES OF HUMAN PERFORMANCE

One of the difficulties associated with contemporary approaches to HRA involves the lack of consensus concerning a universal model of human performance. The development of human performance models in reliability engineering has tended to mirror the historical evolution of human performance models that occurred within psychology. For example, Berliner, Angell, and Shearer (1964) proposed a behavioural model which was broadly consistent with contemporary theoretical perspectives that existed within psychology (Skinner, 1974). The model focused on observable behaviour such as scanning, reading, calculating, and activating. The underlying assumption was that each behaviour would incorporate a level of reliability that could be used either to design novel systems, or to assess the overall reliability of an existing system.

Although a model based on observable behaviour can be useful as a classification scheme, it does not necessarily take into account the cognitive aspects of human performance such as mental representation, diagnosis, and situational awareness. It might be argued that these and other cognitive features associated with a task will have an equally significant impact on subsequent behaviour. A similar observation has been made in the differences between behavioural task analysis and cognitive task analysis, where the latter incorporates both the behavioural and the cognitive features associated with a task (Papautsky et al., 2020).

In response to the limitations to Berliner et al.'s (1964) model of human performance, Rasmussen, Pedersen, Carnino, Griffon, Mancini, and Gagnolet (1981) proposed a model that incorporates both cognitive and behavioural elements. From a psychological perspective, the model is consistent with an information-processing approach where there is a process of stimulus detection, information analysis, and the execution of a response (see Figure 7.1).

One of main implications associated with the human performance model proposed by Rasmussen et al. (1981) is the distinction between skill-based, rule-based, and knowledge-based behaviour (see Chapter 4). Skill-based behaviour is characterised by the detection of a change in the system state and the execution of an immediate response. This type of behaviour is consistent with principle of automatic

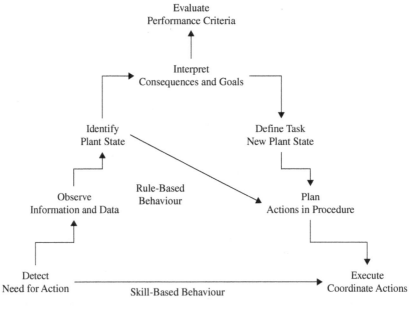

FIGURE 7.1 A model of human performance for reliability analysis. (Adapted from Rasmussen, 1982.)

Reliability-Based Perspectives

information processing in which responses appear to be nonconscious (Shiffrin & Schneider, 1977).

In comparison to skill-based behaviour, rule-based behaviour involves a greater level of information processing, and is mediated by the activation of conscious rules (Rasmussen, 1983). Having isolated the change in the system state and identified the nature of the change, a rule is recalled and a plan is developed to execute an appropriate response (see Figure 7.1). These rules may be prescribed by an organisation in the form of standard operating procedures or may be consciously developed by the operator through interaction with the system.

Knowledge-based behaviour invokes the highest level of information processing and is normally activated in the absence of prior experience. It requires the acquisition of information deliberately, and the analysis of this information to solve a problem (Rasmussen, 1983). In comparison to rule-based and skill-based behaviour, knowledge-based behaviour invokes significant cognitive resources, such that human performance may be impeded.

The Skill-Rule-Knowledge (SRK) model of human performance has had a major impact on the development of analyses of human reliability. Nevertheless, alternative models of human performance have also been developed and this reflects the difficulty in progressing from a conceptual framework to quantifiable estimates of reliability. For example, performance within a high-reliability, high-consequence environment might require the application of rule-based or skill-based behaviour. However, in the case of a systems failure, knowledge-based behaviour may need to be applied, particularly in the case of unusual or unforeseen circumstances. The difficulty lies in quantifying the differences between the human performance requirements, and the extent to which such differences need to be taken into account from a reliability perspective.

7.6 APPROACHES TO HRA MODELLING

One of the main aims associated with HRA is the capability to predict when, where, and how human performance will impact the performance of a system. This capability depends on an appropriate model of the relationship between the operator and the system. Dougherty (1993) suggests that there are four distinct approaches to modelling human performance in complex systems: procedural, temporal, influential, and contextual.

Procedural models such as the Technique for Human Error Rate Prediction (THERP) assume a linear relationship between events such that behaviour 'one' is followed by behaviour 'two' that is followed by behaviour 'three' (Swain & Guttman, 1983). This is the least complicated of the approaches and reflects a linear, behavioural perspective of human performance (Dougherty, 1993). A task is identified and is deconstructed into its behavioural constituents, each of which can be considered in isolation. The outcome of this process is an estimate of the cumulative reliability of an activity or task.

Although the deconstruction of a procedure or task has the advantage of simplicity, the extent to which it reflects human performance is unclear. Human behaviour is not necessarily linear in sequence, and is influenced by cognitive features such as perception, situational awareness, and mental models. Therefore, estimates of reliability that are obtained following the behavioural deconstruction of a task may not reflect the overall level of reliability that is subsequently observed within the operational environment.

An alternative approach to procedural models involves the deconstruction of a task on the basis of a *temporal* dimension. This approach is designed to identify the number and the types of tasks that are expected to be performed within a given time period. Models such as the Time Reliability Correlation (TRC) are based on the assumption that reliability is a function of the probability that a task is performed successfully by a specific time. Therefore, reducing the time period, or increasing the number of tasks to be performed within a given period, will likely reduce the reliability of the system.

Like the procedural approach to HRA, the temporal approach does not necessarily account for cognitive factors involved in the performance of a task. In addition, there is a risk of assuming that the performance of one task is equivalent to another, despite the fact that different cognitive resources may need to be employed. Finally, a temporal approach to HRA does not take into account the possibility that events can be anticipated by some operators and will not necessarily depend on the onset of a specific cue to initiate a response.

Despite the difficulties associated with procedural and temporal approaches to HRA, both have been used extensively as part of risk assessments in the nuclear power plant environment (Sakurahara, Mohaghegh, Reihani, Kee, Brandyberry, & Rodgers, 2018). In contrast, the *influential* approach to HRA has been applied less frequently due probably to its relative complexity. Unlike the procedural and temporal approaches, influential approaches such as the Success Likelihood Index Methodology (SLIM) are designed to take into account Performance Shaping Factors (PSF) that may have an impact on human performance.

Performance Shaping Factors include the characteristics associated with the situation, the nature of the task, and issues related to human performance. Typically, a relationship is presumed to exist between PSFs and the probability of human error associated with an activity (Yin, Liu, & Li, 2021). The difficulty involves identifying the relevant PSFs associated with a task and establishing their relative impact.

The extent to which individual PSFs are taken into account as part of influential approaches tends to be limited, despite the assumption on which these types of models are based. This possibly reflects a holistic approach where the impact of PSFs is taken into account in the form of a few 'global' terms, rather than seeking to identify the relative impact of each individual PSF. The result is a trade-off between parsimony and the accuracy of the outcomes of the HRA.

An alternative to the influence approach is provided by the *contextual* approach to HRA in which the analytical process takes into account the impact of individual PSFs (Onofrio & Trucco, 2018). However, where influence models depend on the quantification of the impact of PSFs, the nature of contextual approaches is such that the need for the quantification of PSFs is minimised to some extent. Rather, a qualitative approach may be used to describe the relative impact of PSFs on human performance (Mosneron-Dupin et al., 1997).

7.7 QUANTITATIVE VERSUS QUALITATIVE

Traditionally, HRA has involved the quantification of probabilities that culminates in an overall value that can be used for purposes such as organisational decision-making. However, the difficulties associated with accounting for PSFs have resulted in a shift in emphasis, so that some models of HRA include qualitative data as part of the broader process of risk assessment. Nevertheless, the issue remains the subject of debate.

The main difficulty associated with the application of qualitative data involves the integration of these data with quantitative information (Strand & Lundteigen, 2016). For skill-based or rule-based behaviour, it may be possible to employ quantitative techniques in isolation, and determine the cumulative reliability of human performance. However, in the case of knowledge-based behaviour, there are a number of factors that may impact an outcome, and the relative influence of each factor can depend on features that are difficult to quantify objectively, such as organisational culture.

As part of the movement towards a mixed-methods approach to HRA, Mosneron-Dupin et al. (1997) have suggested that, in some cases, the term 'probability' should be discarded in favour of the term 'likelihood' that better conveys the nature of qualitative data. This is particularly the case where subjective estimates are made concerning the nature of human performance. In many cases, the impact of PSFs is to be determined by subjective assessments on the part of experts, rather than by objective empirical techniques.

7.8 CONTEMPORARY APPROACHES TO RELIABILITY

One of the most significant changes associated with contemporary approaches to HRA involves the extent to which organisational factors are being incorporated as PSFs. Although the impact of organisational factors on human performance is considered significant amongst theorists, the nature of the relationship has been difficult to assess proactively. Typically, the relationship between an organisation and the performance of an operator has only been identified once an error has occurred.

Among aircraft accident investigation authorities, the then Bureau of Air Safety Investigation[1] (BASI) in Australia was one of the first to recognise the impact of organisational factors and incorporate organisational failures into event analyses (Maurino et al., 1995). Although the results of these analyses are qualitative and largely subjective, successive investigations revealed a systemic link between decisions made at the organisational level and unsafe acts that occurred within the workplace (Bureau of Air Safety Investigation, 1994, 1996).

In the nuclear environment, Strater and Bubb (1999) used a contextual approach to HRA to identify the impact of organisational PSFs on human performance within a particular power plant. Although the relative impact of each organisational PSF is difficult to compare, the factors identified as part of this analysis included:

- The inappropriate construction of the system;
- A tightly coupled system that was intended to be redundant;

- An inappropriate technical layout; and/or
- The unavailability of redundant systems.

According to Bagnara, Di Martino, Mancini, and Rizzo (1991), the extent to which such socio-organisational factors impact on human performance is often under-represented in HRA. The difficulty appears to lie in the use of the term 'performance shaping factors', which implies that the operator and the factors that shape performance are independent. However, operators rarely function in isolation from an organisation. An organisational structure establishes goals, guides and constrains behaviour, and provides feedback concerning performance. Therefore, the intrinsic relationship that exists between operators and an organisation needs to be incorporated into the theoretical models of human performance that form the basis of HRA.

7.9 SYSTEM SIMULATIONS

As technology has developed, a greater opportunity has emerged for the development of computer-based systems that have the capacity to simulate the interaction between humans and complex industrial environments. Complex computer-based simulations present a significant opportunity to develop cost-effective, accurate and reliable assessments of human performance in a diverse range of operational settings (Harrald et al., 1998; Li & Mosleh, 2019).

Although system simulation has been used with some success in the nuclear and maritime industries, it should be noted that the models of human performance on which these systems are based will have a significant impact on the validity of the subsequent outcomes. For example, Furuta and Kondo (1993) developed a system that was designed to simulate knowledge-based performance following changes in a system state. The system was based on a cognitive model of human performance derived from Rasmussen (1986). However, an analysis of human behaviour in response to an abnormal condition in a nuclear power plant suggested that the simulation did not take into account cognitive features such as 'forgetting' and changes that may occur in assumptions that underpin operators' mental models.

7.10 HUMAN RELIABILITY ANALYSIS IN THE FUTURE

Despite the potential application, HRA has yet to be applied extensively outside the energy and transportation industries. Where the nuclear industry has an extensive repertoire of proactive approaches to system safety, other industries, such as construction and health, have relied on more reactive approaches such as event analyses (accident investigation). It is only recently that more proactive measures have been developed.

The failure to adopt HRA as a proactive form of safety management more broadly probably relates to the complexity of many work settings compared to the relatively controlled environments that characterise the nuclear industry. Invariably, they tend to be more dynamic, uncertain, time-constrained, and subject to high levels of risk. However, they are often also high-consequence environments that are significantly dependent on the safety of the system. Domains such as health and mining are tightly coupled and there is a significant reliance on high levels of human performance to ensure that systems function appropriately. As a result, HRA would appear useful in these environments provided that accurate models of cognitive human performance can be identified.

NOTE

1   Now referred to as the Australian Transport Safety Bureau (ATSB).

Part III

Individual Differences and Human Factors

8 Information Processing

This chapter examines information processing as a framework for understanding human performance in complex organisational settings. Different models of information processing are explained in the context of human factors, and limitations to human information processing are considered as an explanation for variations in human performance that are evident in the workplace. Finally, the chapter outlines some of the implications of human factors issues in the context of information-processing research initiatives.

8.1 INFORMATION PROCESSING

Investigations of human information processing are typically characterised as 'reductionist', since psychological functions are 'reduced' to a series of fundamental components that can be subjected to experimental analysis. The value of this approach lies in the deconstruction of a complex structure to a series of individual components, isolating the relationships between elements and enabling conclusions to be drawn concerning causal relationships.

The value of the reductionist approach to investigations of information processing is illustrated in the study of memory and the recall of digits. From a contemporary human performance perspective, the recall of items from long-term memory has become increasingly important due to the need to retain and recall passwords (Morrow, Leirer, Carver, & Decker Tanke, 1998; Yildirim & Mackie, 2019). While there are a number of issues that might be considered as part of this research, possibly the most significant issues concern the need to be able to recall different passwords that are sufficiently memorable, but are nevertheless 'strong', over a period of time and with infrequent use.

Approaching the problem from an empirical perspective, the methodology might involve testing the accuracy of participants in response to passwords with differing numbers of items. However, since the recall of these items is in response to a particular stimulus, such as a website or similar application, it might be useful to determine the extent to which the average person can recall items accurately across a range of different contexts. These contexts might be defined according to the number of

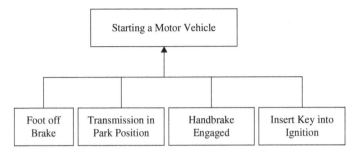

FIGURE 8.1 A deconstruction of a simple task involving the process of starting a motor vehicle with automatic transmission.

different passwords that need to be retained, their similarity, and/or the frequency with which they are engaged. By systematically increasing the demands and observing the outcomes, it becomes possible to establish experimentally the ideal number of passwords to be retained, their optimal size, and frequency with which they need to be applied to ensure that they are recalled accurately.

There are three main techniques that are used to reduce otherwise complex behaviours for the purposes of research. The first is based on the behavioural features of a task, and is similar to a task analytic technique whereby a task is deconstructed into its constituent elements. Each phase of the process can be analysed systematically and in isolation. The process is illustrated using the starting procedure for a motor vehicle with automatic transmission (see Figure 8.1).

The second approach to task decomposition involves the division of performance from a functional perspective. For example, most information-processing models involve a memory component and a processing component (Dorner & Schaub, 1994; MacKenzie, Ibbotson, Cao, & Lomax, 2001; Naikar, 2017). In the case of the memory component, the emphasis is placed on information retention, while the processing component involves the transfer of information from one form of memory to another.

Memory is an important component of information processing and the two most common forms are 'working' or short-term memory and long-term memory (Logan, 1988). However, there are a number of other forms of memory that impact human performance including:

- Episodic Memory, in which memory is based on episodes such as the 9/11 attack in the United States or the loss of the *Challenger* space shuttle;
- Semantic Memory, in which memory is based on an underlying meaning, such as the functional operation of an engine;
- Declarative Memory, which includes memory for facts such as the date on which industrial and organisational psychology was first established as a discipline; and
- Procedural Memory, which includes memory for relationships or rules such as using an indicator prior to initiating a turn in a motor vehicle (Baddeley, 1998).

Information Processing 87

The final approach to the deconstruction of tasks occurs on the basis of a temporal (time) dimension. In isolating the time required to respond to a stimulus, it is possible to isolate and thereby infer the existence of cognitive structures (Neri, Shappell, & DeJohn, 1992).

The three methods of task deconstruction are often used in combination and a useful example can be drawn from Sternberg (1977). Sternberg (1977) designed an experiment to determine the process through which information is retrieved from long-term memory. The problem was approached by initially teaching a group of students a series of digits, that, through rehearsal, were presumed to have been transferred into long-term memory. Within each memory set, there was a different number of digits ranging from one to eight. Sternberg (1977) sought to examine the time it took to recall the series of numbers that were in the original memory sets in comparison to the reaction times associated with a series of dummy numbers.

Sternberg (1977) assumed that there were two alternative explanations as to what might occur when an individual was asked to recall whether a number was in the original memory set. First, the individual might examine all of the numbers in memory, irrespective of whether or not the number was found: a process referred to as an exhaustive search. Secondly, the individual might examine the numbers in the original memory set, and if a match is found, would respond prior to searching the remaining numbers in the set: a process referred to as a self-terminating search.

The results indicated that the average response latencies (time to respond) increased linearly, irrespective of whether or not the original or the dummy numbers were used. If the effect was self-terminating, it might be expected that the response latencies to 'yes' would be relatively lower than the response latencies to 'no'. The fact that they were not suggests that memory retrieval is based on an exhaustive search of a memory set, irrespective of whether or not the desired item is located.

In the context of human performance, the outcomes of Sternberg (1977) have implications for situations where there is a requirement for responses in time-critical environments. In effect, the outcomes suggest that the larger the original memory set, the longer the period with in which a desired memory item will be located. Therefore, an optimal response is more likely to be achieved if the memory set is relatively small. From an applied perspective, this requires the design of systems where the range of possible responses is constrained, and/or operators are provided with memory aids such as checklists.

8.2 MODELS OF INFORMATION PROCESSING

Irrespective of their origins, the majority of models of information processing comprise three fundamental elements: an encoding process, a comparative process, and a selection process. Although this is a relatively simplistic approach, this type of three-stage model captures the main features of information processing. More detailed analysis will further subdivide these elements, consistent with the reductionist approach to the analysis of information processing.

According to a simplified model of information processing, the encoding stage involves the transfer of physical stimuli such as sound or light, into a neural stimulus, largely based on electrochemical signals (Proctor & Vu, 2016). Once the information has been successfully encoded, a comparison is made between existing information and the novel information. Should there be some type of match, and a suitable response is required, the appropriate response is selected from long-term memory.

A more complex example of an information-processing model is provided by Wickens and Flach (1988) in which the process of information acquisition and management is assumed to be moderated by attentional resources. This addition to the simplified model is designed to explain some of the effects associated with increasing levels of workload. Similarly, the addition of a short-term memory store is designed to explain the apparent filtering of information that can precede the encoding stage. While models of information processing are useful from a conceptual perspective, the validity of these models is difficult to test empirically. In general, different aspects of the model are tested in isolation, consistent with the reductionist approach.

8.3 THE MULTIPLE RESOURCES MODEL OF INFORMATION PROCESSING

The multiple resources model is based on the assumption that there are different types of short-term memory stores each of which is associated with different types of information. For example, visual information is presumed to be processed by a memory store that differs from the memory store that processes aural or tactile information (Logan, 1988; Tindall-Ford, Chandler, & Sweller, 1997). This distinction is designed to explain the outcomes of empirical research where there is an indication that different types of information can be processed simultaneously, while information of the same type can create conflicts that limit the amount of information that reaches the encoding stage (Wickens, 1991, 2008).

The multiple resources model also represents an attempt to mediate between two perspectives concerning the mechanism of information processing. The first of these perspectives is based on the assumption that information is processed serially, or in sequence (Fitts & Posner, 1967). In this case, it is assumed that there is a limited amount of information that can be processed at any point in time. Any additional information is unable to be processed.

In contrast to serial processing, the 'parallel' perspective of information processing is based on the assumption that different pieces of information can be processed simultaneously. The multiple resources model incorporates elements of both the serial processing and parallel processing perspectives, depending on the type of information that is being processed. Where information is derived from different resources, there is some evidence to suggest that the information can be interpreted in short succession.

From an applied perspective, the implications of the multiple resources model lie in the presentation of information in multiple modes to ensure that critical information is recognised and interpreted. For example, visual warnings might complement auditory warnings, based on the assumption that they comprise distinct processing modalities.

8.4 LIMITATIONS OF HUMAN INFORMATION PROCESSING

Irrespective of the model of information processing, there are a number of structural elements associated with human information processing that limit the accuracy and efficiency with which humans are capable of responding to particular stimuli. The first of these limitations involves the initial encoding of information from the environment. For example, the capacity to detect changes in the environment is often linked to the extent to which the change is anticipated (Nijstad, Stroebe, & Lodewijkx, 1999). This level of expectancy allows the allocation of cognitive resources to both identify when the stimulus emerges and prepare an appropriate response. However, it should be noted that expectancy can also degrade human performance in situations where a critical situation is unlikely to occur or occurs gradually. There are a number of examples where this situation has occurred, including the crash of the Air New Zealand DC-10 in the Antarctic in 1979. The crash occurred during a sightseeing flight from Auckland to Antarctica.

The Inertial Navigation System (INS) in the aircraft had been programmed so that the aircraft flew directly over Lewis Bay and Mount Erebus. However, the flight crew were under the assumption that they would track via McMurdo Sound. The track via McMurdo Sound would enable the aircraft to descend to afford the passengers a better view of the Antarctic environment. Therefore, on reaching what they thought was McMurdo Sound, the flight crew executed a standard descent to 6,000ft and proceeded on course (Vette & McDonald, 1983).

As they were in unfamiliar territory, the crew confirmed their position using visual references. As is typically the case when referring to visual references, the flight crew tried to identify features and cross-reference these features with a map. However, the geographical nature of Lewis Bay was similar, albeit on a smaller scale, to the geographical nature of McMurdo Sound. Therefore, the flight crew saw, in effect, what they expected to see as they cross-referenced their position. They were unaware that they were tracking directly towards Mount Erebus with an elevation of 12,000ft (O'Hare & Roscoe, 1990).

In addition to the positive and negative effects of expectancy, human information-processing capacity is also limited by short-term memory. Since short-term memory is both capacity-limited and subject to transience, the amount of information that can be processed at any given time is limited. Where short-term memory has reached capacity, the individual is said to be 'overloaded', and information or task shedding tends to occur (Orlandi & Brooks, 2018). This involves the loss of information that might otherwise be required for the effective management of a task.

The propensity towards information overload is normally associated with inexperience or a lack of familiarity within a particular domain. For example, de Groot (1966) noted that, in comparison to novice chess players, experts could recall more accurately the positions of chess pieces placed on a chessboard after they were removed. This suggests that, as individuals become more experienced, they develop a more efficient method of retaining and recalling information. By processing information more efficiently, experts have a greater residual capacity in short-term memory to manage additional tasks (Norman, Brooks & Allen, 1989; Sturman, Wiggins, Auton, Loft, Helton, Westbrook, & Braithwaite, 2019). For example, an experienced driver

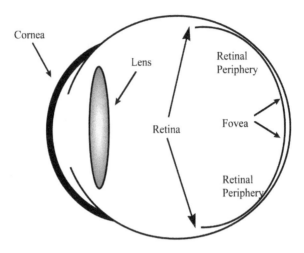

FIGURE 8.2 A simplified diagram of the cross-section of the eye.

might be capable of discussing a complex issue while maintaining effective control of a motor vehicle. This capability is less evident among inexperienced drivers, presumably due to the increased demands on working memory (Yuris, Sturman, Auton, Giacon, & Wiggins, 2019).

The encoding of information during the initial stages of processing is subject to the stimuli reaching a particular threshold for detection (Carr, McCauley, Sperber, & Parmelee, 1982). For example, a very faint light in the distance may remain undetected until there is luminance sufficient to trigger the rods in the periphery of the retina (see Figure 8.2). Similarly, other types of stimuli must reach a threshold before they can be encoded. Where an operator is fatigued, the threshold at which activation occurs increases, such that a stimulus will require a greater level of intensity than normal to trigger a response. At an applied level, this requires some consideration of the level of luminance necessary for a visual stimulus, the level of loudness necessary for an auditory stimulus, the level of concentration necessary for an olfactory stimulus, and the level of pressure for a tactile stimulus, taking into account the context within which the stimuli need to be perceived. With higher levels of ambient noise, there is a risk that critical information will be masked by other stimuli.

8.5 IMPLICATIONS OF INFORMATION-PROCESSING RESEARCH

The principles that underpin information processing have been applied across a range of contemporary systems, including smart phones, computer interfaces, and computer websites.

In general, the goal is to reduce the demands on information processing by simplifying tasks and obviating the need to retain information in short-term memory. These systems also capitalise on humans' capacity for learning by adopting standardised processes for interaction across systems. However, they can also act as a defence against errors that might be attributed to the limitations of human information processing.

8.5.1 Checklist Design

Checklists were initially advocated in response to errors where operators failed to undertake activities that were necessary to ensure the efficient and safe operation of a device or system. For example, in the aviation environment, checklists initially emerged in the 1930s in response to a crash involving a Boeing 299, the forerunner to the famous Boeing B-17 Flying Fortress. To prevent damage caused by wind, the aircraft was equipped with locks that prevented the excessive movement of the rudder and elevator while the aircraft was parked (Ayers, 2021). Prior to departure, the pilot had forgotten to remove the lock as was normal practice and the aircraft crashed on departure. The accident highlighted the need for checklists or aide-memoire to assist operators in recalling increasingly complex sequences of activities, particularly in time-constrained conditions or under periods of high workload.

As happens in many situations, incidents often trigger the introduction of new procedures that are intended to prevent a reoccurrence. However, rather than rethinking and revising existing practices, there is an inclination to introduce additional procedures that are simply added to existing procedures. The outcome is a proliferation of activities that may, in some cases, conflict with other activities, leading to confusion and error.

In the United States Airforce, the proliferation of regulations was recognised as an impediment to efficient performance, and effort was invested to reconceptualise and reduce the number and size of regulations, which included operational checklists. The revised checklists improved efficiency without any deleterious impact on safety. This was possibly due the reduction in the demands on short-term memory that were associated with introduction of the revised checklists, together with a process that was better aligned with the limitations of human information processing.

The application of information-processing principles is especially evident in the development of checklist design in high-consequence contexts such as medicine and transportation. Checklists are used to provide a level of standardisation between operators, and prevent errors of omission or commission. However, while checklists represent one of the most common methods of preventing human error, their initial development failed to account for human information processing.

Sponsored by the National Aeronautics and Space Administration (NASA), Degani and Wiener (1990) examined the application of checklists, both in their use and design. The focus on checklist design and implementation arose as a result of a number of aviation-related incidents and accidents where checklists had been misused. For example, omissions of checklist items preceded the crash of both a McDonnel Douglas MD-80 at Detroit Metro Airport in 1987 (NTSB, 1988), and a Boeing B-727 at Dallas Forth-Worth in 1988 (NTSB, 1989).

As early as 1969, the National Transportation Safety Board (NTSB) in the United States had noted the significance of checklists following the crash of a Pan American Boeing 707 at Anchorage, Alaska. The flight crew had failed to set the flaps prior to take-off, and the aircraft overran the runway. As a result of this accident, the NTSB recommended the development of a research agenda that would determine the optimal mechanisms for checklist presentation.

Among the outcomes of this research was the observation that contemporary aircraft checklists were designed differently, depending on the organisation and the domain in which they are applied (Degani & Wiener, 1990). It was also noted that the majority of checklists examined were intended to achieve one or more of the following objectives:

- To aid in recalling the process of configuring the aircraft;
- To provide a standard foundation for verifying aircraft configuration, irrespective of the psychological and physical condition of the flight crew;
- To provide convenient sequences for motor movements and visual fixations across the cockpit panels;
- To provide a sequential framework to meet internal and external cockpit operational requirements;
- To allow mutual supervision (cross-checking) amongst crew members;
- To enhance a team (crew) concept for configuring the plane by keeping all crew members 'in the loop';
- To dictate the duties of each crew member to enable optimal crew coordination and the logical distribution of cockpit workload; and
- To serve as a quality control tool by flight management and government regulators over the pilots in the process of configuring the plane for the flight.

(Adapted from Degani & Wiener, 1990)

While the application of checklists is clearly associated with improvements in performance, their effectiveness lies in both the appropriateness of the design of the checklist, and the organisational culture within which the checklist is expected to be applied. For example, checklists are unlikely to be applied regularly and effectively in organisations where the application of checklists is regarded as burdensome or distracting. Similarly, checklists that are perceived as cumbersome and that impede performance are less likely to be implemented in practice. Therefore, like other systems, the design of checklists should engage users to ensure that the benefits of checklists are appreciated and that their implementation complements, rather than constrains, operational performance.

Although much of the early research on checklists targeted the aviation context, the principles established are equally applicable within other operational environments, including healthcare and other complex, high-consequence settings (Clay-Williams & Colligan, 2015). Checklists have been introduced and their effectiveness established in anaesthesiology, surgery, mining, marine, and energy distribution and transmission. In medicine, the World Health Organisation has established a surgical safety checklist that is advocated for use in hospitals. However, like other checklists, the universal application and implementation of the checklist is variable, even within hospitals (Rydenfält, Johansson, Odenrick, Åkerman, & Larsson, 2013). This is despite evidence that the application of a checklist is associated with a reduction in perioperative mortality (Ramsay, Haynes, Lipsitz, Solsky, Leitch, Gawande, & Kumar, 2019).

8.5.2 Types of Checklists

While there are a number of different types of checklists available, paper-based checklists remain relatively common in a range of industrial contexts. A typical example of a checklist issued in the 1980s by Northwest Airlines for the McDonnell Douglas MD-80 aircraft is provided in Figure 8.3. The MD-80 is a twin-engine, jet aircraft that seated approximately 155 passengers in an all-economy configuration. The aircraft and its successors are normally employed on short/medium-haul routes.

Other types of checklists in common use are the scroll checklist (see Figure 8.4), the mechanical checklist, the electro-mechanical checklist (see Figure 8.5), and the vocal checklist. The integration of computer-based or application-based systems has also enabled the development of augmented checklists. These are commonly divided into display and pointer checklists without feedback, and display and pointer checklists with feedback. The latter offers a level of redundancy whereby the system checks the status of the particular function and refers this information back to the console.

Evaluations of augmented checklists suggest that they are superior to paper checklists, during both normal and abnormal operations, and result in fewer errors of omission. However, there may be an increase in response latency as a result of computer-aided checklists, and the extent to which this occurs may be related to training and/or the design of the interface.

A combination of checklists may also be used within the same operational setting. For example, both computer-aided and paper checklists are used in aviation and healthcare. In the aviation context, Electronic Centralised Aircraft Monitoring

---

MD-80

EXTERNAL ELECTRIC & PNEUMATIC SOURCE – START

PNEUMATIC X-FEEDS................................BOTH CLOSED
PNEUMATIC AIR SOURCE.....................CONNECTED & ON
PNEUMATIC X –FEEDS..OPEN
PNEUMATIC PRESSURE (25 PSI MIN)........................CKD

AFTER ENGINES STABILISED

PNEUMATIC X-FEEDS................................BOTH CLOSED
ELECTRIC POWER..CKD
EXTERNAL ELECTRIC & PNEUMATIC........DISCONNECTED
COMPLETE – AFTER START CHECKLIST

(from NTSB, 1988, p. 138)

FIGURE 8.3 An example of part of the checklist issued by Northwest Airlines for the McDonnell Douglas MD-80. (Adapted from NTSB (1988, p. 138)).

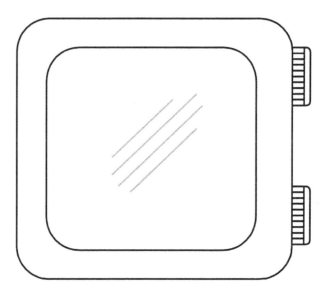

FIGURE 8.4 An example of a scroll checklist. The knobs on the right are turned to reveal various checklist items.

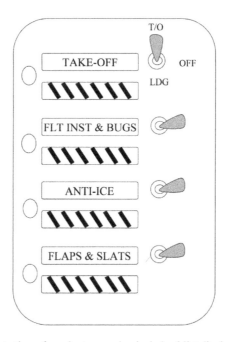

FIGURE 8.5 An illustration of an electro-mechanical checklist display.

System (ECAM) provides the critical take-off and landing checklists, while less critical checklists may be paper-based. In addition, the ECAM system incorporates a feedback loop, whereby the computer checks the status of items prior to 'clearing' the computer screen.

Like other interfaces, there is a considerable lack of standardisation between checklist design. This is due primarily to differences in operational settings, the critical nature of the environments, and the organisational culture. Enforcing the use of checklists in the absence of an appropriate organisational context and culture can be problematic and potentially lead to a deterioration in operational safety and efficiency (Catchpole & Russ, 2015).

8.5.3 Difficulties with Checklist Completion

In many environments, operational demands can emerge that prevent the timely and systematic application of checklists. For example, rapidly changing situations such as occur during emergencies reduce the time available and cause distractions for operators. Therefore, critical checklists should be designed to take into account environments that are likely to impose demands on operators, including distractions and time-constraints.

For operators, there is a need to appreciate the importance of checklists and ensure that the environment facilitates their accurate and timely completion. This appreciation is best instilled through an appropriate organisational culture and reinforced through training (Fourcade, Blache, Grenier, Bourgain, & Minvielle, 2012). For more experienced operators, the risk is a reliance on memory that obviates the requirement for a physical or electronic checklist.

Checklist chunking is a memory strategy employed by more experienced practitioners whereby items are recalled as 'chunks', rather than item by item (Cooke, 1990). The result is a reliance on short-term memory to complete the sequence, rather than referring to each checklist item in isolation. This type of strategy is likely to increase the likelihood of omissions unless a systematic strategy is engaged that supports the sequential process of checking that underpins checklist design.

The concept of 'flow' in the context of checklist completion is a systematic, sequential process of inspection engaged by experienced operators to evaluate the condition of system components (Burian, Clebone, Dismukes, & Ruskin, 2018). It is a process that is typically supported by the subsequent completion of the checklist and conforms to the physical characteristics of the interface. For example, in an aircraft cockpit, a pilot might begin the assessment of the condition of the aircraft at the overhead panel and then progress systematically from right to left examining the condition of each instrument or control as the process continues.

While the flow technique does not necessarily obviate the impact of distractions or time-constraints, together with the completion of the checklist, it offers a defence where an item overlooked during the initial inspection might be identified when the checklist is completed. In a multi-crew environment, a further level of defence can be provided through an independent inspection and the application of a 'challenge and response' protocol.

9 Workload and Attention

9.1 ATTENTION

Attention refers to a cognitive process that directs information-processing resources towards particular stimuli. The extent to which resources are allocated to stimuli depends on a number of factors including the salience of the task, the motivation to complete the task, and task difficulty. This differential allocation of attentional resources is most evident when comparing the performance of novice and expert operators during complex tasks. For example, novices will tend to devote a considerable proportion of their attentional resources to the performance of the task, often to the exclusion of other activities (Schriver, Morrow, Wickens, & Talleur, 2008; Shanteau, 1988). Experts, in contrast, tend to be capable of distributing their attention across a number of tasks (such as driving a car and conversing) (Pammer, Raineri, Beanland, Bell, & Borzycki, 2018).

9.2 THE SIGNIFICANCE OF ATTENTION

Attention is generally regarded as a limited resource that facilitates the accurate and efficient assessment of situations. Where inconsistencies are noted between the environment and expectations, attentional resources are allocated to resolve the discrepancy. Therefore, it provides the basis for diagnostic and problem-solving skills.

The impact of attention on operator performance is generally characterised by the extent to which salient, task-related information is neglected or overlooked during the performance of activities. This manifests in lapses or slips, and these human errors account for the majority of organisational accidents and incidents in which humans are involved (Feyer, Williamson, & Cairns, 1997; Runciman, Sellen, Webb, Williamson, Currie, Morgan, & Russle, 1993)

At an organisational level, encouraging attention to critical information, particularly where the information is repetitious, can be problematic, since it inevitably leads to expectations and what might be described as 'selective laziness'. This is evident in the application of rules-of-thumb or heuristics that are applied in lieu of a conscious and deliberate process of reasoning (Trouche, Johansson, Hall, & Mercier, 2016).

9.3 ATTENTION STRATEGIES

There are four major strategies through which attention may be allocated to a particular task: selective attention, focused attention, 'divided' attention, and sustained attention.

9.3.1 SELECTIVE ATTENTION

Selective attention is based on the number of channels through which information is expected to be received. For example, in a large control room, there are likely to be a number of displays, each of which is changing on a minute-by-minute basis with implications for the safety and security of the system. Clearly, it is not possible to attend simultaneously to all of the displays, and therefore, at any given time, the operator must select those displays to which to attend. This process of active sampling from multiple channels of information is referred to as *selective attention* (Johnston & Dark, 1986).

According to Moray (1981), there are difficulties associated with a selective approach to data sampling. For example, the nature of the short-term memory means that increasing the number of channels increases the probability that information will be overlooked or forgotten. This will generally occur through the substitution of new information at a rate beyond which the information can be transferred to long-term memory.

Where multiple channels are required to be sampled, the rate at which this interaction occurs will depend on the frequency with which the information changes per unit time. An instrument that changes frequently (such as a speedometer) will be sampled more frequently than an instrument that changes slowly (such as the fuel indicator). The result, however, is that operators may sample less frequently those instruments where changes occur on a less regular basis (Preciado, Munneke, & Theeuwes, 2017).

External factors, including goals and emotional arousal, have an impact on selective attention by occupying short-term memory and reducing the residual resources that are available to be directed towards stimuli (Finucane, 2011). Referred to as *cognitive narrowing*, it is characterised by the sampling of one or two critical pieces of information to the exclusion of other information (Harmon-Jones, Gable, & Price, 2012; Moray, 1982). Under normal conditions, sampling what is perceived to be relevant information, to the exclusion of less relevant information, is an adaptive mechanism that reduces demands on information processing. However, in high-tempo or non-normal conditions, where information needs to be derived from a range of sources, valuable information may be overlooked.

The tendency towards cognitive narrowing in high-tempo situations was illustrated following a fire that broke out in the Channel Tunnel on November 18, 1996, where rail controllers activated exhaust fans but failed to set the angle of the blades, so that they spun ineffectively for a number of minutes before the error was detected (Santos-Reyes & Beard, 2017). Vehicles were being carried by train through the tunnel, and fire broke out in one of the container vehicles. The fire took hold quickly, producing a combination of flames and toxic fumes. Understandably, the rail controllers were directing their attention to the location of the train and expeditious evacuation of passengers.

In designing systems to account for the demands for selective attention, Sanders and McCormick (1993) suggest that:

- As few channels as possible (this may mean increasing the signal rate on a single channel) should be used to reduce the likelihood that critical information will be overlooked;
- Information should be provided, either through training or design, as to the relative importance of the various channels so that attention can be directed appropriately;
- Operators should be taught to manage the impact of anxiety, particularly in high-tempo and/or unfamiliar situations;
- Provision should be made for effective scanning, either through training or through the judicious location of targets;
- Where information must be derived from multiple sources, these should be co-located to facilitate the acquisition and integration of information; and that
- Auditory channels should not mask one another.

9.3.2 Focused Attention

Like selective attention, focused attention involves the allocation of attentional resources to a limited number of channels, to the exclusion of extraneous information (Monheit & Johnston, 1994). The main difference between selective attention and focused attention lies in the extent to which the allocation of attentional resources is intentional. In the case of selective attention, resources are intentionally directed towards a stimulus. For example, approaching a pedestrian crossing, a driver might attend to the sidewalk on the assumption that pedestrians in close proximity to the crossing may present a hazard. This is a reasoned assumption and one that the driver would be able to explain.

Focused attention occurs when attention is drawn unintentionally towards a stimulus (Wijers, Okita, Mulder, Mulder, Lorist, Poiesz, & Scheffers, 1987). This occurs most frequently in response to warnings and alarms where the stimuli are presented in a form different from the ambient environment (Wickens, 2000). An auditory stimulus might be louder than the ambient environment, and/or embody a higher pitch or unique tone, where a visual stimulus might be brighter than the ambient environment, displayed in a distinctive hue, and/or at a unique frequency.

The challenge in managing focused attention lies in designing systems that draw attention in appropriate situations to appropriate stimuli. The presentation of stimuli injudiciously may draw attention from otherwise important information. This is especially evident in advertising, where the intention is to draw the attention of the customer towards a product or service. The difficulty occurs where attention is drawn from a task-critical activity such as maintaining a safe distance between vehicles while operating at high speed.

Sanders and McCormick (1993) suggest that focused attention is most appropriately drawn by ensuring that:

- Information is presented in a form as distinct as possible from the channel to which the operator would normally be attending;
- Competing sources of information are separated in physical and/or auditory 'space';
- The number of competing channels is reduced to ensure that the target is perceptible at the appropriate time; and that
- The channel of interest is sufficiently distinct to facilitate both the perception and interpretation of information.

9.3.3 'Divided' Attention

When attention is described as 'divided', there can be an assumption that attention is allocated simultaneously and proportionately to different tasks to enable what is described as 'multitasking' (Douglas, Raban, Walter, & Westbrook, 2017). A common analogy involves driving a motor vehicle while carrying out a conversation with a passenger, where it appears that the tasks are being undertaken simultaneously. However, despite the appearance of multitasking, there is considerable evidence to indicate that the process involves rapidly switching attentional resources from one task to the other (Skaugset, Farrell, Carney, Wolff, Santen, Perry, & Cico, 2016).

Although attentional resources are not allocated simultaneously to different tasks, they each consume resources so that switching from one task to another necessarily releases attentional resources (Wickens & Kramer, 1985). Therefore, it's possible that engaging in a number of demanding tasks will consume attentional resources to a point where the residual resources are insufficient to sustain performance on one or more of the tasks. This is the point at which operators will describe themselves as 'overloaded', and where they may disengage from one or more tasks through a process of task-shedding (Small, Wiggins, & Loveday, 2014).

Task-shedding or 'load-shedding' may be a conscious, intentional process where it is clear that the demands of additional tasks are likely to impede other tasks. For example, in a busy medical emergency room, a physician may delegate duties to other medical staff, thereby enabling attentional resources to be directed towards activities that are considered a higher priority. This capability requires both an understanding of the demands of the various activities and the foresight to devolve tasks that are likely to impose demands.

For less experienced practitioners, there may be neither the understanding of the relative priority associated with different tasks that need to be completed, nor the attentional resources necessary to exercise oversight over the range of tasks that might need to be completed (Koh, Park, & Wickens, 2014). In this case, the process of task-shedding may be nonconscious and may occur as an inevitable consequence of the demands imposed by the attentional resources drawn to additional tasks. Typically, activities associated with preceding tasks are forgotten and are not completed nor the relevant information interpreted, resulting in a lapse.

The differences between experienced and inexperienced practitioners are particularly evident where a number of different tasks need to be completed in rapid sequence. Where experienced practitioners are likely to complete a range of tasks rapidly and accurately, the performance of inexperienced practitioners tends to be less accurate and less efficient (Bliss, Harden, & Dischinger, 2013). These differences in performance are likely to be due to efficiencies in memory, with experienced practitioners' engaging in what might be described as 'automatic processing' and inexperienced practitioners relegated to 'controlled processing' (Shiffrin & Schneider, 1977).

Automatic processing involves the activation of learned associations from memory, where a stimulus 'automatically' triggers an associated response. For example, with experience, drivers develop a rapid and accurate association between the activation of a brake light in a preceding vehicle and its deceleration, enabling braking action. Amongst less experienced drivers, the association is likely to be less well established in memory, resulting in a delay in the braking response.

While automatic processing occurs nonconsciously, controlled processing is a conscious process that is activated in the absence of learned associations in memory. It engages declarative knowledge or 'knowledge of facts' that are resident in declarative memory (Anderson, 1993). For less experienced practitioners for whom controlled processing tends to be most evident, drawing declarative knowledge into working memory in an effort to resolve an unfamiliar situation imposes demands on attentional resources that can impede accuracy and/or response latency.

Since tasks impose different demands on operators, depending on their experience, establishing the impact on performance will enable an assessment of the appropriateness of the allocation of tasks to different operators. In each case, the goal is to optimise residual attention while maintaining performance, accounting for the imposition of additional tasks. The distinction is illustrated in Figure 9.1

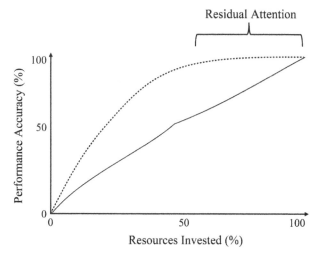

FIGURE 9.1 An illustration of the relationship between attentional resources invested and performance accuracy. (Adapted from Wickens, 1987.)

where increases in performance accuracy on Task A correspond to a broadly linear association with the proportion of attentional resources invested so that, at 90 per cent accuracy, almost 100 per cent of the attentional resources have been devoted to the task. For Task B, performance is maximised by investing only 60 per cent of the available attentional resources. Therefore, in comparison to Task A, Task B is preferable as it imposes fewer attentional demands on operators in achieving optimal performance.

This type of approach assumes a single attention resource from which performance is derived. While it explains the process through which performance diminishes during multiple tasks, it does not explain observations where operators appear to perform tasks concurrently when the demands of the task require distinct input modalities. Multiple Resources Theory offers an explanation where attentional resources are presumed to be distributed across four dimensions that characterise the nature of a task:

- Stages
- Input Modalities
- Processing Codes
- Responses

The stage at which information is processed is an important element that determines the extent to which tasks will conflict. For example, a simple tracking task is less likely to conflict with a mathematical task, since the former is processed at a more peripheral stage of information processing than the latter (Wickens & Kessel, 1980).

The process through which information is acquired also limits the potential conflict between sources of information, since auditory stimuli require cognitive resources that supposedly differ from visual stimuli (Isreal, 1980). Consistent with this perspective, spatial and verbal tasks are presumed to require distinct attentional resources (Wickens, 2008). Finally, Sanders and McCormick (1993) suggest that auditory responses can be initiated coincidentally with verbal responses so that a person can call out a number and note it physically.

9.4 SUSTAINED ATTENTION AND VIGILANCE

In many working environments, advances in technology have resulted in the transition of operators from a role as 'direct controller' to one of 'system monitor'. This transition is based on an implicit assumption that operators have the capacity to sustain attention for extended periods in the absence of interaction. In fact, amongst human observers, there is evidence to suggest that the probability of detecting a signal diminishes considerably following extended exposure to a particular task (Matthews, Warm, Reinerman-Jones, Langheim, Washburn, & Tripp, 2010; Rees, Wiggins, Helton, & Loveday, 2017).

Giambra and Quilter (1987) suggest that the probability of detecting a signal diminishes from approximately 90 per cent at three minutes into the task, to 62 per cent at 40 minutes into the task. Subsequent performance remains relatively stable up to an hour into the task. Therefore, it appears that human operators are ill-equipped to sustain attention to passive tasks over an extended period of time. This has important implications for the development of advanced technology systems.

The deterioration in performance that occurs over an extended period of observation is referred to as the vigilance decrement as it is typically evident as an increase in response latency to changes in a system state that require an intervention (Grier, Warm, Dember, Matthews, Galinsky, Szalma, & Parasuraman, 2003). The period of observation is referred to as a vigil, since there is an assumption that the operator is actively attending to the environment (Helton & Warm, 2008).

To assess the impact of vigilance decrements in advanced technology contexts, Molloy and Parasuraman (1996) examined the impact of task complexity and the time on task on the detection of a single failure associated with an automated system. The results indicated that, in general, the monitoring process was less impacted by the vigilance decrement when the operator was actively involved in the control of the system. Where the operator was a passive observer, approximately 45 per cent of participants noted the system failure as it occurred. The failure occurred between 20 and 30 minutes into the task and suggests that, as a rule-of-thumb, passive monitoring should be limited to a maximum of 30 minutes.

There are a number of different explanations for the vigilance decrement, including mindlessness theory (Thomson, Besner, & Smilek, 2015). In this explanation, the target of the vigil lacks sufficient stimulation to draw the attention of the observer. As a result, attentional resources are attracted towards other stimuli such that, when a change does occur, there is a period of re-engagement with the task that inevitably increases response latency and the potential for a lapse (Helton & Warm, 2008).

In contrast to explanations proffered by mindlessness theory, attentional resource theory is based on the proposition that sustaining attention to a task is demanding and consumes attentional resources to a point where attentional resources are no longer available (Thomson, Besner, & Smilek, 2016). To recover these resources, attention is directed away from the vigil, thereby increasing the probability of a lapse and an increase in response latency where changes occur to the system state (Thomson et al., 2015).

The malleable attention explanation of the vigilance decrement is based on the assumption that the capacity for attention shifts with the demands of a task. Therefore, if the demands of a task diminish, then the attention resources available also diminish (Young & Stanton, 2002). Consistent with attentional resource theory, sustaining attention is assumed to consume resources, although the point at which attentional resources are exhausted, and therefore, the deterioration in performance occurs, is likely to differ.

Irrespective of the explanation of the vigilance decrement, the recovery of attentional resources requires a break from the vigil, the duration of which will depend on the nature of the task, the duration of the vigil, and the experience of the operator. There is some evidence to suggest that a low-demand activity during the break is optimal in ensuring the recovery of attentional resources (Helton & Russel, 2015, 2017). While there has been some suggestion that experiences with nature-related stimuli have a positive impact on the recovery of attentional resources, the impact appears to be conflated with arousal and emotional resources (Finkbeiner, Russell, & Helton, 2016).

In addition to the management of the vigil, it is also possible to alter the environment so that, when changes in the system state do occur, they are sufficiently salient to draw the attention of operators. For auditory warnings, this might include the loudness, tempo, tone, or the nature of the auditory stimulus, taking account of competing auditory and visual stimuli that might coincide with the warning (Haas, & Edworthy, 1996; Perry, Stevens, Wiggins, & Howell, 2007). Variations in visual stimuli might include luminance, the positioning of a stimulus in a central location, and/or frequent changes in illumination.

As a part of this general approach towards aiding signal detection amongst operators, Milburn and Mertens (1997) examined the ergonomic requirements for altering mechanisms for targets displayed on an air traffic control console. The study was developed in response to observations that blinking targets are more 'attention-getting' than steady targets. It was assumed that a blinking target would be easier to locate and therefore, to identify. The blink rate and duration were held constant while the luminance was varied systematically from a standard of 51.4 Cd/m^2 (candles per square metre). While the non-blinking target resulted in an accuracy rate of 99.69 per cent, this increased to 100 per cent when the luminance of the blink rate was varied between 93.75 and 75 per cent of the standard. This illustrates both the importance of testing different levels of stimuli in drawing the attention of operators, and testing assumptions as to the types of stimuli that will draw attention, particularly under different environmental conditions.

9.5 WORKLOAD

Workload is defined as the frequency and/or number of tasks to be completed, together with the conditions under which the tasks are to be completed (Gawron, Schflett, & Miller, 1988). These conditions include the time and resources available. The load that is imposed during the completion of these tasks can impact the physical and/or mental condition of the operator and, at extreme levels, can impede performance (Sweller & Chandler, 1991).

In applied industrial contexts, the distinction needs to be drawn between the demands necessary to complete a task successfully and the resources invested by the operator in the completion of the task. The latter is referred to as 'cognitive load' and it is external to the task to be completed. The level of cognitive load is determined by the effort invested by the operator, and the nature of the information available to the operator (Sweller & Paas, 2017).

For more experienced operators who possess a repertoire of experiences, the effort invested in the successful performance of a task may be relatively low (Kalyuga & Sweller, 2018). However, amongst less experienced operators, the effort necessary to undertake a task successfully may be much higher, since the information will need to be processed in short-term memory. Increasing demands on short-term memory corresponds to increases in cognitive load (Sweller, 2016).

In the context of the assessment of human performance, workload is referred to as 'task load' and incorporates a combination of constructs, including mental load and physical load (Hancock & Matthews, 2019). For some tasks that are more cognitively demanding, the physical load may be relatively low, while the cognitive or mental load may be higher, culminating in a higher task load.

9.5.1 ATTENTION AND WORKLOAD

In complex operational environments, increases in perceptions of mental workload do not necessarily result in a systematic reduction in operator performance (Zeier, 1994). This suggests that the relationship between mental workload and performance is non-linear much like the relationship between attentional resources and performance. As workload increases, attentional resources are directed towards the performance of the task to maintain performance (see Figure 9.2). As workload increases, the proportion of attentional resources available will decrease, although residual attention may remain to cope effectively with further increases in the demands of the task (Shaw, Rietschel, Hendershot, Pruziner, Miller, Hatfield, & Gentili, 2018). Importantly, changes in mental workload may not immediately be evident in changes in the performance of operators.

The dissociation between attention and workload is captured by Yeh and Wickens (1988) who noted differences in perceptions of workload and associated performance, especially at higher levels of perceived workload. They concluded that, despite a systematic decrease in performance, the perception of workload reaches a stable asymptote, and does not continue to increase as might be expected. This suggests that, although attentional resources may remain, the operator perceives that the demands of the task exceed the resources available and has reached a state of 'overload'. The relationship is depicted in Figure 9.3.

The dissociation between perceived workload and performance might be considered a useful adaptive mechanism, since the operator perceives a limit that constrains the allocation of additional resources even though these resources might

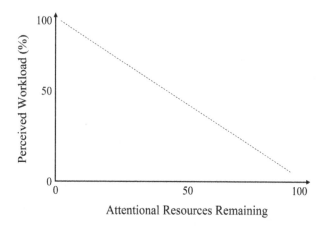

FIGURE 9.2 An illustration of the negative linear relationship which is presumed to exist between the perceived level of workload and the proportion of attentional resources remaining.

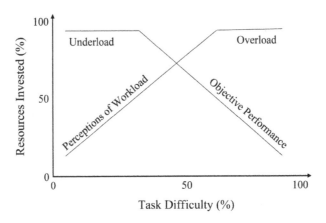

FIGURE 9.3 An illustration of the conceptual relationship between the supply of attentional resources and task-related performance. (Adapted from Yeh and Wickens, 1988, p. 113.)

be available. The residual resources are then available should additional demands emerge that might be critical to resolve in safeguarding the system.

9.6 MEASURING WORKLOAD

Since changes in workload do not necessarily coincide with changes in operational performance, quantifying the workload imposed by a task requires inferences to be drawn from a range of sources, including psychophysiological measures, subjective ratings, and secondary tasks. Typically, different sources will be employed to triangulate the outcomes and ensure an accurate assessment of the workload involved.

9.6.1 Psychophysiological Measures

A number of psychophysiological measures of workload have been employed to measure workload including heart rate, heart rate variability, electro-encephalography, electro-cardiography, electro-myography, and functional near infrared spectroscopy (Hancock, 1987; Liu, Ayaz, & Shewokis, 2017; Sturman et al., 2019; Verwey & Veltman, 1996). The major advantage associated with the application of psychophysiological measures of workload is the objectivity afforded by the measures. There is also evidence to suggest that physiological measures such as heart rate variability are predictive of the overall level of mental effort. For example, Tattersall and Hockey (1995) examined the relationship between heart rate variability and the mental load imposed on maintenance engineers during a three-hour task that involved the diagnosis, detection, and correction of a series of maintenance-related faults. The results indicated that those tasks that imposed a higher level of mental effort were associated with a suppression in the variability of the heart rate.

Despite the effectiveness of psychophysiological measures of workload, there are a number of difficulties associated with the application of these measures including individual variability in the data, artefacts, and the amount of raw data that can be generated. Wearing psychophysiological measures can also be intrusive and may distract operators from their primary role, impacting their operational performance. Therefore, the utility of psychophysiological measures will depend on the nature of the operational environment and the context in which the measures are employed (Charles & Nixon, 2019).

9.6.2 Subjective Ratings

A less intrusive alternative to psychophysiological measures is afforded by subjective measures of workload including the National Aeronautics and Space Administration (NASA) Task Load Index (TLX), subjective ratings of workload, and the Subjective Workload Assessment Test (SWAT) (Galy, Paxion, & Berthleon, 2018; Gawron et al., 1988; Yan, Wei, & Tran, 2019).

The NASA-TLX is based on six, subjective dimensions that are presumed to reflect the overall, weighted perception of workload. These dimensions include: mental effort, physical demand, temporal (time) demand, perceived performance, effort, and the level of frustration (Gawron et al., 1988). Although the NASA-TLX has been applied extensively as a subjective measure of workload, there has been debate as to the effectiveness of the strategy in accurately predicting overall perceptions of workload. Hendy, Liao, and Milgram (1997) note, for example, that single dimensions of workload, such as assessments of perceived 'effort', often outperform multidimensional tools such as the TLX. Galy et al. (2018) share a similar perspective, suggesting that the value in the NASA-TLX lies in considering the ratings for each individual dimension, rather than generating a global score across the six dimensions.

Like the TLX, the SWAT is a multidimensional scale designed to produce an interval scale of the level of subjective workload. The SWAT comprises a series of cards on which are listed a series of statements of the perceived level of workload (Farmer, 1988). These are sorted by the participant and the sequence is entered into a computer program that analyses the level of workload in the time load, the mental load, and the psychological stress load imposed by the task (Rubio, Diaz, Martin, & Puente, 2004).

9.6.3 Secondary Tasks

To overcome a number of the difficulties associated with the application of subjective measures of workload, Gopher and Braune (1984) advocate the application of secondary measures to determine the impact of a primary task on performance and thereby infer the workload involved. The application of secondary tasks is based on the proposition that a primary task occupies a proportion of the attentional resources available. The imposition of a second task should add to the load of the primary task and further reduce the attentional resources available. By increasing the workload involved in the performance of the primary task, a reduction in the performance of the secondary might be expected, provided that the accumulated load exceeds the attentional resources available (See Figure 9.4).

Generally, secondary tasks must be unrelated to the primary task to ensure that the attentional resources used are cumulative. As a result, a number of domain-independent

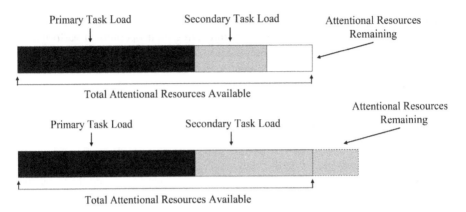

FIGURE 9.4 An illustration of the relationship between the primary and secondary task loads, and the attentional resources available.

secondary tasks have been developed including key tapping, mental arithmetic, and reaction time tasks. The application of secondary tasks is illustrated by Bortolussi, Hart, and Shively (1989) in their assessment of the workload imposed on pilots during a landing approach under instruments. Each pilot flew a normal instrument landing system approach and landing, during which 21 additional flight tasks were introduced, thereby imposing different levels of workload. The secondary tasks employed included a choice reaction time task and a time-production task that involved a response triggered by a tone.

The secondary tasks were either synchronised with specific flight tasks, or were presented randomly throughout the flight (Bortolussi et al., 1989). The results indicated that performance on the secondary tasks was sensitive to changes in the workload imposed during the flight. In this case, one of the advantages associated with the application of secondary tasks was the extent to which they could be synchronised with specific events during what would otherwise be a continuous activity. This is particularly useful in determining the impact of tasks that occur over a period of time.

Clearly, the main difficulty associated with the application of secondary tasks is the extent to which they may impact performance on the primary task. Normally, a participant is instructed to concentrate on performing the primary task successfully, rather than attempting to manage both the primary and secondary tasks (Scerbo, Britt, & Stefanidis, 2017). While this strategy may overcome the problem to a certain extent, there remains the potential for conflict between primary and secondary tasks.

The Detection Response Task (DRT) has emerged as a valid and reliable measure of workload, particularly in the context of driving. The procedure involves the presentation of a stimulus, usually unrelated to the primary task, but sufficiently engaging to ensure that it is capable of being perceived. In the driving context, this might include a vibration or a light presented at random intervals. The participant is required to respond to the stimulus, with the latency recorded. Changes in response latency are associated with changes in cognitive load where a higher load is associated with longer response latencies and a lower cognitive load is associated with shorter response latencies (Strayer, Turrill, Cooper, Coleman, Medeiros-Ward, & Biondi, 2015).

The DRT is recommended by the International Standards Organisation (2016) as a valid and reliable measure of cognitive load in continuous tasks, such as driving. Levels of sensitivity and validity are evident in other domains (e.g. DeLeeuw & Mayer, 2008) where cognitive load impacts performance, and its application is likely to be more prevalent in applied contexts in the future.

10 Situational Awareness

10.1 SITUATIONAL AWARENESS

Situational Awareness (SA) is a construct that has been very difficult for researchers and practitioners to define. However, it is a term that is often used in applied contexts to explain the consequences of system failures. Typically, these failures involve a breakdown in the process of acquiring and processing task-related information where valuable cues are either overlooked (lapse) or misinterpreted (mistake). To that end, effective SA is especially dependent on the initial stages of information processing where information is acquired and examined, and according to which subsequent decisions are made.

In addition to its application as an explanation of system failures, SA has also been used to explain differences in performance at elite levels. For example, during the Korean War, 10 per cent of United Nations pilots accounted for the loss of 50 per cent of North Korean and Chinese aircraft during air-to-air combat (Torrance, Rush, Kohn, & Doughty, 1957). These statistics are consistent with outcomes in World War I and World War II, and it suggests that a relatively small proportion of pilots possess exceptional skills in air-to-air combat. One of these skills appears to be the capacity to anticipate the behaviour of an enemy pilot, thereby enabling a protagonist to manoeuvre successfully into an offensive position.

SA is generally considered more than simply an awareness of the particular aspects of a situation. It involves higher-level cognitive structures that oversee the process of information acquisition and management (Endsley, 1995a; Durso & Sethumadhavan, 2008). In effect, operators with high levels of SA are able to accurately anticipate the next series of events. This anticipation enables the timely application of strategies that will either prevent an occurrence or enable opportunities to be advanced (Stanton, Chambers, & Piggott, 2001).

SA is a construct prevalent in the military context where history is replete with examples of commanders with presumably greater SA being able to neutralise and even defeat enemies of much greater size (Stephenson, 2010). The advantage afforded by SA lies in the capacity to deploy assets that will inflict significant damage or prevent an enemy from pressing an advantage. It a process that is now aided by technology, although it retains elements of information integration and interpretation, and this capacity is usually derived from operational experience (Carretta, Perry, & Ree, 1996; Stanton, Salmon, Walker, Salas, & Hancock, 2017).

Conceptually, SA is divided into three stages, the first of which is a perceptual stage where task-related features or cues are acquired and interpreted. This is referred to as Level 1 SA and provides the foundation for subsequent stages. Having acquired task-related information as part of Level 1 SA, the relevance of information must be assessed and the implications considered for the future operation of the system. This is a process like that which occurs when a driver refers to the fuel indicator in a motor vehicle and considers the importance of this information for the future operation of the vehicle. On a long drive with few opportunities to acquire fuel, the amount of fuel remaining in the tank will have implications that differ from those associated with a short drive and/or where there are multiple opportunities to acquire fuel. This process of integrating information and comprehending its importance in the context of an operational task characterises Level 2 SA.

Once task-related information is acquired, integrated, and its significance determined, an assessment is made concerning the future state of the system. In the context of fuel for a motor vehicle, an estimation will be made as to whether the vehicle has sufficient range to reach the destination, assuming that there are no opportunities to acquire fuel en route. This capacity for anticipation or projection is a particularly important part of situational awareness as it enables interventions to forestall negative outcomes. For a motor vehicle driver, this Level 3 SA provides the basis for a decision as to whether or not to refuel the vehicle prior to a drive. For a marine captain berthing a vessel, it provides the basis to intervene to correct a poor approach angle and reorientate the vessel well before it would otherwise collide with the dock.

This three-stage conceptualisation of SA is based on the principle that the relevance of specific information is highly dependent on the context within which it is provided (Walker, Stanton, Kazi, & Salmon, 2009). However, it also depends on the skills and capabilities of the operator to recognise those features of the environment that are associated with a specific object or event. For example, for marine captains to take account of drift in an approach to a berth, they must first recognise the direction and strength of the wind gusts, together with the angle of approach. With experience, captains will be able to finely tune their approach to account for a wide range of weather conditions. The difficulty lies with operators who lack a breadth of experience and who may, therefore, lack the SA necessary to respond to changes in the conditions that may have a significant impact on the performance of the vessel.

10.2 LEVEL 1 SITUATIONAL AWARENESS

At the initial stages of information processing, effective SA requires a capacity to identify and interpret salient features or cues from a number of different features that might characterise the operational environment (Friedman, Leedom, & Howell, 1991). For example, the capacity to identify the genesis of a system failure in a nuclear reactor might involve the recognition of a small rise in the core temperature in the absence of other indications on the instrument panel. Having identified a feature, it becomes necessary to interpret the information in the operation of the system.

Part of the process of identifying salient cues involves being able to recognise those situations that are not consistent with expectations (Falkland & Wiggins, 2019). However, the capacity to develop expectations is a product of experience, and a lack of task-related experience amongst operators can often mean that valuable information pertaining to the task remains unrecognised (Stanton et al., 2001). While advanced technology can be used to overcome some of the limitations associated with a lack of experience, difficulties can arise when the information provided is erroneous and/or is subject to misinterpretation.

To create efficiencies in both the number of displays and their utilisation, many advanced technology systems employ a menu-based interface that enables the selection of one or more modes of operation. However, in the absence of clear indicators as to the mode selected, operators can unwittingly specify an incorrect or inadvisable mode of operation given the conditions. An illustration of insidious nature of 'mode errors' can be drawn from the crash of Air Inter Flight 148 on approach to Strasbourg Airport in France in 1992. The Airbus A320 aircraft was en route from Lyon at night and in low cloud and snow.

The approach to Strasbourg Airport requires a precise approach angle due to the surrounding terrain and this is normally set by the pilots at 3.3 degrees descent to the horizontal, also referred to as the Flight Path Angle. However, using different modes associated with the same system, it is possible to set approach paths that are based on different metrics. One of these metrics is the rate of descent or the Vertical Speed Mode which the pilots had used prior to the approach to Strasbourg (Bureau Enquêtes Accidents, 1992). In commencing their approach, they failed to change the mode from Vertical Speed to Flight Path Angle, setting a rate of 3,3 which, in the Vertical Speed Mode, translated to a 3,300 feet rate of descent which was much higher than would have occurred with a 3.3 degrees descent angle. Importantly, the only indication in the cockpit that the pilots had selected an incorrect mode was a comma, rather than a full stop displayed between the digits on the panel. A '3.3' would have indicated that Flight Path Angle had been selected, while a '3,3' indicated that Vertical Speed had been selected.

The rate of descent selected resulted in a descent angle that was much greater than the 3.3 degrees approach that was necessary to successfully navigate the high terrain surrounding Strasbourg. Normally, pilots would be advised of their proximity to terrain by the Ground Proximity Warning System. In this case however, the pilots received no warning and the aircraft crashed into a ridge line in the Vosges Mountains, approximately 40 kilometres from Strasbourg Airport. Amongst the passengers and crew, 87 perished, while a further nine survived with serious injuries.

The accident involving Air Inter Flight 148 is an example where operators are said to be functioning 'out of the loop'. The 'out of the loop' syndrome occurs when an automated system performs functions that are not anticipated by the operator. This tends to be the most common type of error that occurs as a result of interactions with advanced technology as the cues necessary to interpret the operation of the system remain obscured and/or are absent. Therefore, it prevents an operator from acquiring Level 1 SA.

In addition to system-related limitations that might prevent the acquisition of Level 1 SA, there are also operator-related issues that may impact the acquisition of task-related information. Part of the difficulty appears to lie in both the accuracy and the reliability of systems, to the extent that operators become complacent regarding the potential system failures that may occur.

From an information-processing perspective, the likelihood that a system will perform functions that are unanticipated by the operator is related to the inherent behaviour of the automated system and the factors that impact the operator. Where a system is relatively unreliable, operators tend to maintain a relatively high level of vigilance, thereby reducing the reaction time in response to an unexpected change in the system state (Hitchcock, Warm, Matthews, Dember, Shear, Tripp, Mayleben, & Parasuraman, 2003). However, where a system is relatively reliable, operators may develop a level of trust in the system, the consequence of which may be an increase in the reaction time in response to an unexpected change in the system state.

The interrelationship that exists between contemporary operators and advanced technology systems underscores the requirement that considerable effort has to be directed towards the design of interfaces that enable the acquisition of task-related information quickly and accurately. Issues associated with trust and the development and maintenance of an accurate mental representation are equally important issues that have the potential to impact human performance in complex environments. Moreover, they reflect problems associated with the system that are more fundamental than might be implied by referring simply to training and design.

Irrespective of issues such as design and training, advanced technology itself has implications for SA, especially in failure detection and diagnosis. For example, Molloy and Parasuraman (1996) provide evidence to suggest that a lack of direct involvement in the performance of a task increases the time required to establish control of a system in the event of a failure. Therefore, the difficulty associated with advanced technology may arise, in part, due to the lack of active involvement in the performance of a task. In the absence of such involvement, the cues arising from changes that occur within the operational environment may no longer be evident.

10.3 LEVEL 2 SITUATIONAL AWARENESS

Rather than simply being aware of events that are occurring, SA also involves the interpretation and comprehension of the information arising from the environment, to the extent that meaning is derived concerning the nature of the system (Level 2 SA).

The skills necessary to derive an accurate interpretation depend on a number of features, including the previous experience of the operator and the nature of the representation of the domain in long-term memory. It is only by understanding the interaction between the various features that constitute the environment that relatively disparate pieces of information can be integrated to form a coherent understanding of the current state of the system (Jones & Endsley, 1996).

Ultimately, the accurate interpretation of the information arising from the operational environment depends on the development and maintenance of a mental model. A mental model is a representation in the mind of the structure and/or operation of a system (Mogford, 1997). Mental models are developed largely through experience and active interaction with the environment (Kieras & Bovair, 1984).

An inaccurate representation of the system may lead to difficulties in operating performance, particularly under conditions of high workload and/or where the time available to acquire and integrate information is constrained. Important information pertinent to a problem may be overlooked or disregarded as unimportant if an operator is unable to integrate this information into an accurate mental model of operation of the system.

An example of the consequences of inaccurate mental model can be drawn from an aircraft accident involving an Eastern Air Lines Flight 401, a Lockheed L-1011 Tristar that crashed on December 29, 1972, 18.7 miles west-northwest of Miami International Airport in Florida (National Transportation Safety Board, 1973). On approach into Miami, the flight crew had lowered the undercarriage, but one of the lights which normally indicate that the undercarriage has safely extended failed to illuminate. They subsequently conducted a missed approach and the aircraft was directed to a holding pattern over the uninhabited Florida Everglades where the crew could determine whether the warning was spurious or reflected a genuine failure.

Circling at 2,000ft, the flight crew, consisting of the captain, first officer, and flight engineer, directed their attention towards resolving the problem. The captain and first officer removed and examined the light globe, while the flight engineer made his way to the avionics bay to establish visually, whether or not the undercarriage had extended.

As is typical in a holding pattern, the aircraft was set to autopilot to ensure that the aircraft maintained a specified altitude and heading. Like other systems onboard aircraft, the autopilot in the L-1011 Tristar consisted of a number of modes of operation. Normally, in the situation faced by the pilots, the altitude mode would be selected so that the aircraft maintains a specified height above the ground. However, in the case of Flight 401, the autopilot selected was the Control Wheel Steering (CWS) mode that referred to a specified pitch at which the aircraft had been set. The pitch determines the angle of the aircraft to horizontal, so that a setting above the horizontal would normally cause the aircraft to climb, while a setting below the horizontal would normally cause the aircraft to descend.

In the CWS mode, the pitch is maintained at the point where the pilot releases pressure on the control column. In turning to speak to the flight engineer, the captain inadvertently placed pressure on the control column, resetting the pitch of the aircraft to cause a slow, imperceptible descent. It was only 10 seconds prior to the collision with terrain that one of the pilots noted that the aircraft had descended well below their intended altitude. Of the 163 passengers and 13 crew members aboard, 94 passengers and five crew members received fatal injuries. Two survivors died later as a result of their injuries.

Since the autopilot mode had been set inadvertently, the flight crew had developed a mental model to the effect that the aircraft was stabilised at a safe altitude and that this, in effect, would justify directing their attention towards the undercarriage warning light. This misunderstanding meant that none of the flight crew were directing their attention towards maintaining a watch over the aircraft. Further, an audible chime that was associated with the departure from assigned altitude appeared to have been ignored, possibly because it conflicted with the mental model under which the pilots were operating. It reflects the strength of mental models in directing the behaviour of operators, even when there may be features in the environment that might conflict with the prevailing model.

In dynamic, high-risk situations, one of the advantages associated with an accurate mental model is the capability to integrate information from a range of sources to derive meaning and enable the accurate and timely comprehension of the situation, consistent with Level 2 SA. For example, developing mental models that enable the differentiation of the common cold from meningitis is particularly important for medical practitioners, especially where many of the symptoms are shared. Meningitis is a potentially fatal condition which has a rapid onset and progression. However, there are specific features associated with meningitis that enable it to be differentiated from the common cold and that, if detected quickly, can increase survivability and recovery. The difficulty lies in integrating these symptoms to form a judgement when the rate at which meningitis affects the general population is vastly lower than the rate at which the common cold is contracted (Kostopoulou, Delaney, & Munro, 2008).

10.4 LEVEL 3 SITUATIONAL AWARENESS

Level 3 situational awareness builds on Levels 1 and 2 and involves anticipating the future state of the system. To do this accurately and efficiently, appropriate remedial strategies can be initiated that ensure that the system remains operating within prescribed limits. This is particularly advantageous for systems that are highly dynamic and where operating for extended periods beyond prescribed limits can lead to system failures.

The importance of Level 3 SA is evident in electricity control where power lines are designed to function at a specified voltage, beyond which the line is likely to fail, referred to as a 'trip'. Excessive voltages can occur when one part of the electrical network fails, and the residual voltage is directed towards other power lines. The increase in voltage may overload these lines, causing them to trip, which places further demands on the electrical network in a cascade, so that eventually the entire system may fail.

The two primary solutions in reducing the demands on a fragile electricity network are to reduce the supply of electricity by reducing generation, and/or to reduce demand by shutting off areas of the network. Reducing generation can be time-consuming, particularly in the case of large, thermal power stations (e.g. coal, gas). Reducing demand for electricity is comparatively faster, but it requires a swift response from the network controller.

On August 14, 2003, the Northeastern United States experience a widespread blackout that was triggered initially by a software failure that prevented alarms that would otherwise have alerted controllers in the FirstEnergy control room that a number of powerlines in the network were overloaded (Anderson, Santos, & Haimes, 2007). The excessive load on the lines caused them to expand and sag onto foliage, resulting in an increase in the electricity current, to which automatic systems responded by disconnecting the line from the network. This increased the electricity load on other parts or the network, together with adjacent networks.

The network controllers at FirstEnergy were advised that a line that they shared with American Electric Power had tripped and reactivated. However, the advice was disregarded by the controllers at FirstEnergy since they had not received any alarms to this effect. The resulting failures cascaded across the Northeastern United States and Canada within minutes, with an estimated 55,000,000 customers affected. The case demonstrates the importance of technology in enabling Level 3 SA in complex, contemporary systems.

Successful Level 3 SA is an outcome of an accurate and reliable mental model where experience has determined likely courses of action, together with the impact of various interventions. The capability to anticipate accurately the impact of future events enables strategies to be devised and implemented that will minimise the potential impact of changes in the system state. For example, a motor vehicle driver may observe that a number of vehicles fail to observe a stop sign at a particular intersection, potentially conflicting with through traffic. This experience enables the construction of a mental model to the effect that, on approaching the intersection as through traffic, there is a possibility that vehicles approaching on the cross-street will neglect to observe the stop sign. For the driver, the consequence is to anticipate and thereby exercise caution on approaching the intersection, consistent with Level 3 SA.

Given the reliance on mental models, the capacity to anticipate the consequences of events is particularly difficult for less experienced operators. Even for operators with a breadth of experience, never having dealt with a specific situation can be particularly problematic in anticipating the consequences and thereby determining the appropriate intervention. It highlights the importance of scenario-based training where operators are actively involved in anticipating and responding to changes in the system state.

'What-if' scenarios are particularly useful in developing the capacity for Level 3 SA, as it encourages operators to consider events that may be highly unlikely during the normal course of operations (Zagonel, Rohrbaugh, Richardson, & Andersen, 2004). These hypothetical scenarios build the framework for mental models, potentially enabling the generalisation of skills to adapt to different conditions.

Where possible, simulation is an important tool in developing Level 3 SA, as it allows operators to 'try out' different responses to changes in the system state. Importantly, the feedback provided offers an opportunity to identify gaps in knowledge and capability, together with a repertoire of successful interventions where appropriate (Couto, Barreto, Marcon, Mafra, & Accorsi, 2018). With experience, interventions can become more finely tuned to enable operators to better differentiate when specific responses may be more or less appropriate under the circumstances. This capacity for precision in recognising nuanced changes in the system state and responding accurately and efficiently is generally regarded as a characteristic of expertise (Savelsburgh, Van der Kam, William, & Ward, 2005).

10.5 TEAM SITUATIONAL AWARENESS

For many operational environments, the successful management of complex tasks requires the coordination of multiple operators. Inevitably, this coordination requires that operators possess a similar mental model of the operation to ensure that responses are appropriately sequenced. This degree of consistency in the nature of the mental models engaged provides the foundation for joint situational awareness.

Consistent with approaches to situational awareness at the level of the individual operator, joint situational awareness involves the perception and integration of task-related features to form an understanding of the system state and the projection of the future state of the system. However, a multi-crew environment imposes an added degree of complexity insofar as different features may be perceived differently by different operators. Integrating the information derived from these features requires effective and efficient communication to generate an understanding of the system state.

Having established the state of the system, communication is again necessary to ensure that appropriate responses are initiated, particularly where different operators are required to implement different responses. Therefore, unlike SA at the level of the individual operator, the development of joint SA depends both on the skills necessary to recognise and respond to features that are likely to impact the operational environment, and the skills necessary to communicate this information accurately and efficiently to other members of the crew.

In constructing joint SA, there may also be a requirement to recognise the need to acquire and distribute information outside the immediate environment. For example, analyses of aircraft in-flight emergencies suggest that operators who perform more effectively tend to both acquire and share information beyond their immediate context as a means of ensuring joint SA (Chute & Wiener, 1996). In effect, it constitutes a process of distributed cognition whereby information-processing capabilities are shared amongst different operators or systems.

In a distributed operational context, individual operators will have different levels of SA at different times during the performance of a task. Therefore, the SA of a group may not necessarily be superior to the SA of an individual at any given time. It is not necessarily a cumulative effect that might be evident in superior performance. Effective joint SA is largely dependent on the successful management of information through:

- Communication
- Leadership
- Group Dynamics (Green, Parry, Oeppen, Plint, Dale, & Brennan, 2017)

These group processes are underpinned by consistent and regular training that encourages interaction and the communication of information, standard operating procedures that ensure that information is communicated in an appropriate form and at the appropriate time, and the management of organisational culture and climate, all of which impact the rate and quality of communication in group settings (Sorensen & Stanton, 2016). An adherence to these principles provides the foundation for the development of joint SA that is fundamental for accurate and timely responses in complex, multi-operator settings.

Technological support for the development of joint SA is typically provided by displays that are both consistent and accurate, so that their interpretation contributes to a joint mental model of the state of the system (Feibush, Gagvani, & Williams,

2000). Consistent representations of the system are particularly important where operators are dispersed geographically, since they offer a comparable point of reference that obviates the requirement for extensive communication. This enables more targeted communication directed towards the management of the system, potentially improving performance, and recovering from or preventing errors, thereby minimising disruption.

The role of joint SA in preventing and recovering from errors is illustrated in the context of medical practice where complex pathologies can mask life-threatening conditions. For example, Rothschild, Hurley, Landrigan, Cronin, Martell-Waldrop, et al. (2006) report a case where a patient being treated for a heart condition became increasingly agitated and displayed delusional behaviour. A sedative was administered, although it was ineffective in managing the behaviour. A nurse working with the physician noted that the patient had been prescribed a lidocaine infusion for the preceding two days, one of the side effects of which is delusions. Communicating this explanation to the physician, the lidocaine infusion was concluded and the psychotic behaviour ceased.

10.6 ASSESSING SITUATIONAL AWARENESS

Situational awareness is generally regarded as a capability that is acquired through exposure to an operational context. It is not possible to possess high levels of SA in the absence of experience, since it depends on knowledge of the various features that reflect the operation of the system. For example, a novice driver may have little knowledge of the importance of the revolutions per minute (RPM) indicator in a motor vehicle, despite having been a passenger in a motor vehicle for many years. While a high RPM may be audible, its meaning in the context of the safe functioning of the vehicle will likely be unclear until it is associated in memory with the need to change gears.

While experience is a necessary prerequisite for the development of SA, there is evidence to suggest that, within a context, some operators develop the capacity for SA at a much faster rate than others. There is also evidence to suggest that, despite extensive experience, some operators fail to develop the capacity for SA. This suggests that there are individual differences in both the rate and capacity to acquire SA within a context.

It is important to note that the capacity for SA appears to be context-dependent so that high levels of SA in one context do not necessarily translate to high levels of SA in another context. This is probably due to an inherent capability to acquire and integrate context-related information. For example, a surgeon may be particularly adept in recognising the features that signify a deteriorating patient, but may have difficulty recognising a deceptive tennis shot from an opponent. While both situations rely on the interpretation of features, differences in experience and the nature of the environment likely explain the differences in performance.

Since SA is context-dependent and reliant on a degree of exposure, attempts to assess SA have generally involved measures of task-related behaviour from which levels of SA are inferred. The most common tool in the assessment of SA is referred to as the Situational Awareness Global Assessment Test (SAGAT). It is normally administered in the context of a simulation or operational task, and requires operators to respond to a series of probes that are intended to establish the level of SA at which operators are functioning (Endsley, 1995b).

Since SA is a process comprising three stages, the probes are designed to evaluate the extent to which information is perceived (Level 1), integrated and interpreted accurately (Level 2), or whether this information enables the anticipation of the future state of the system. It is important to note that performance at later stages of SA depends on success at the preceding stages. For example, it is not possible to anticipate the future state of the system (Level 3) without having first perceived the changes that may have occurred (Level 1) and interpreted the significance of the changes accurately (Level 2).

The SAGAT probes can be completed concurrently with the operational task or at the completion of the task. However, completing the SAGAT at the completion of task is reliant on the memory of the operator, since some of the probes may refer to activities that occurred early in the sequence of activities. The completion of the SAGAT concurrently risks imposing additional demands on cognitive resources, unless the activity is paused to allow the administration of the probes. This is best achieved in a simulated context that allows an operation to be paused and any visible or audible features to be masked.

Where the SAGAT is designed as tool to be completed by operators engaged in the performance of a task, the Situational Awareness Rating Scale (SARS) and the Situational Awareness Rating Technique (SART) are subjective assessments of SA designed to be completed by observers, typically using a Likert scale (Salmon, Stanton, Walker, & Green, 2006; Waag & Houck, 1994). Since the SARS and the SART rely on the judgement of the observer, items are typically anchored to observable behaviour (Saus, Johnsen, Eid, & Thayer, 2012).

Operator ratings scales are also available, including the Low-Event Task Subjective Situation Awareness (LETSSA) scale (Rose, Bearman, & Dorrian, 2018). In this case, operators are asked to self-rate the extent to which they agree or disagree with a series of statements related to the task. For example, in a rail context, drivers might be asked to respond to a statement such as: 'I saw/heard the vigilance alarm in time to respond before incurring a brake penalty'. Like the SART and the SARS, ratings are recorded using a Likert scale.

Various editions of situational awareness 'self' and observer-rating scales have been developed for specific domains, including control rooms (Situation Awareness Control Room Inventory – SACRI) (Hogg, Folles, Strand-Volden, & Torralba, 1995), air traffic control (Situation Present Assessment Method – SPAM) (Durso, Bleckley, & Dattell, 2006), and the military (Mission Awareness Rating Scale – MARS) (Matthews, Eid, Johnsen, & Boe, 2011). While these types of scales offer ease of application, their reliability and validity tend to differ, and they have rarely been evaluated beyond the context for which they were originally designed.

While the majority of assessment tools for SA have been developed for individual operators, there is a significant interest in tools that are capable of assessing team or joint SA. A number of approaches have been advocated, including verbal protocol analysis (Rose, Bearman, Naweed, & Dorrian, 2019) and a strategy that compares the similarities and differences in the mental models described by individual operators (Westli, Johnsen, Eid, Rasten, & Brattebø, 2010). The greater the differences in the mental models described, the poorer the SA, and the poorer the performance of the team.

Salmon et al. (2006) undertook a comparative analysis of a range of measures of situational awareness and assessed their applicability in team-based settings. Amongst the measures examined were the SAGAT, SART, SACRI, MARS, together with physiological measures such as eye tracking. It was concluded that existing measures of SA were insufficient to capture differences in performance at the team level. Consequently, it was recommended that, in assessing SA in the context of teams, a range of tools should be employed.

11 Human Factors and Decision-Making

11.1 MODELS OF DECISION-MAKING

Models of decision-making can generally be divided into two classes according to their origin and purpose. Descriptive decision-making models are those that are derived from empirical research, and that are used to explain the process through which decisions are made within various environments. Prescriptive models of decision-making are those that prescribe a particular process that is intended to improve decision-making performance.

The distinction between descriptive and prescriptive models of decision-making reflects different approaches to the analysis of the decision-making process. Nevertheless, both approaches are intended to identify those strategies necessary to improve decision-making within an operational context.

11.2 DESCRIPTIVE DECISION-MAKING

Descriptive models of decision-making can be further divided into those that are more intuitive, and those which are more analytical in the cognitive processes employed (Kleinmuntz, 1990). Possibly the most intuitive type of decision strategy is one where an operator selects the first alternative that is considered appropriate for the situation. This type of sequential or serial approach to decision-making is characteristic of more experienced practitioners faced with time-critical situations and is referred to as System 1 reasoning (Hendry & Hodges, 2019). Decision-makers who are faced with a less time-critical, but possibly higher risk situation (e.g. purchasing a house) are more likely to adopt a rational or analytical model where different alternatives are considered systematically (Zakay & Wooler, 1984). This type of analytical approach to decision-making is referred to as System 2 reasoning and tends to be more time-consuming but may be less prone to error, depending on the experience of practitioners (Zulkosky, White, Price, & Pretz, 2016).

Intuitive and analytical decision-making models likely exist along a metaphorical, dual-task continuum, rather than as two dichotomous approaches. The strategies that operators employ often depend on the situation and the experience of the person involved. Therefore, the goal in improving decision-making skills is to match the decision-making strategy with the appropriate conditions under which the decision is being made.

One of the most significant, descriptive models of analytical decision strategies is a model in which the formulation of a decision is influenced by the Subjective Expected Utility (SEU) of the outcomes (Navarro-Martinez, Loomes, Isoni, Butler, & Alaoui, 2018). Expected utility is the 'worth' or 'value' that a person ascribes to a particular option or outcome (Svenson, 1979). From an analytical perspective, this value is calculated as the product of the value of the satisfaction of the goal and the probability of the outcome. Therefore, if the desire for the goal is high but the probability of the outcome is low, there is a lower resultant expected utility than if both the value of the goal and the probability of the outcome are high. At a rational level, the relative values of these outcomes determine which options are selected.

It should be noted that subjective expected utility requires some form of numerical value to be assigned to an outcome. In addition, this type of approach is based on the assumption that effective decisions depend on a deliberate and considered analysis of the options available.

The analytical approach contrasts with the strategies employed during intuitive or naturalistic decision-making, where options are selected on the basis of familiarity and previous experience (Falzer, 2018; Orasanu, 1995a). In many cases, decisions are made without any conscious consideration. For example, changing gears in a motor vehicle with manual transmission is often performed nonconsciously by experienced drivers (Crampton & Adams, 1994). Therefore, it is very difficult for experienced operators to verbalise the basis on which the decision is made to change gears.

The appropriateness of the intuitive approach to decision-making is often dependent on a detailed understanding of the cues and the consequences associated with the particular task (Anderson, 1983; Johnston & Morrison, 2016). For example, experienced decision-makers typically possess a detailed and functional understanding of the interrelationships between components within the environment (Charness, 1991; Okoli, Weller, & Watt, 2016). Therefore, they possess mental models that can be drawn upon in responding to operational situations (see Chapter 10).

Naturalistic decision-making generally involves a serial evaluation of the options available. For example, Klein and Klinger (1991) noted that experienced fire commanders would, in most cases, select 'the most typical' option available, based on the information that they had available at the time. However, the initial decision was often revised as more information was acquired. Therefore, experienced practitioners who are faced with a time-critical decision tend to search for the nearest of their experiences that resembles the pattern of cues that is evident in the situation (Laureiro-Martinez & Brusoni, 2018). Once a decision is formulated, it may be revised, depending on additional information.

It might be reasonably argued that naturalistic decisions that involve the serial evaluation of options are best described as 'choices', rather than decisions. A decision normally involves the consideration of multiple options, a degree of uncertainty concerning the appropriateness of the options available, and differences in the degree of risk associated with the options available.

TABLE 11.1
A Summary of the Features That Characterise Analytical and Intuitive Approaches to Decision-Making

- Analytical decision-making can be time-consuming
- Involves concurrent evaluations of the options available
- More appropriate for less familiar tasks
- Intuitive decision-making can be less time-consuming
- Involves the serial evaluation of the options available
- More appropriate for more familiar tasks

The distinction between analytical and intuitive strategies generally depends on the nature of the situation. A summary of the main characteristics of intuitive and analytical decision strategies is listed in Table 11.1.

11.3 DECISION ERRORS

One of the main advantages associated with the application of intuitive strategies over analytical strategies is that the former tend to require less cognitive effort and resources than the latter. Therefore, cognitive resources are available for the management of other tasks. However, it should be noted that the success of the intuitive approach lies in the capacity to draw on knowledge structures that already exist within memory. These knowledge structures are referred to as schemas, and various inferences can be made on this basis (Araújo, Hristovski, Seifert, Carvalho, & Davids, 2019; Branscombe, 1988). These inferences form the basis of heuristics that can be used subsequently to guide human behaviour.

A heuristic is a 'rule' or 'algorithm' that is developed based on knowledge from previous experiences. For example, through exposure, a driver may develop a heuristic to the effect that the illumination of the brake light in a preceding vehicle is associated with deceleration. Heuristics tend to facilitate more rapid responses to changes within the environment and enable the development and maintenance of skilled performance (Hafenbrädl, Waeger, Marewski, & Gigerenzer, 2016). Therefore, heuristics provide the basis for effective and efficient responses, particularly in time-critical environments (Means, Salas, Crandall, & Jacobs, 1993).

Despite their role as an aid for the decision-making process, heuristics are also one of the main reasons why humans make decision-related errors (Itri & Patel, 2018). An extensive research effort has determined a large number of specific information-processing strategies that often lead to errors in information processing and decision-making. These include the representativeness bias, the confirmation bias, the availability bias, the anchoring bias, and misconceptions of chance.

11.3.1 Representativeness Bias

Representative bias relates to the probability that one object resembles another, based on previous experience and is illustrated in the following example:

Steve is very shy and withdrawn, invariably helpful, but with little interest in people, or in the world of reality. A meek and tidy soul, he has a need for order and structure and has a passion for detail.

Which of the following do you think is Steve's occupation?

An airline pilot
A farmer
A salesman
A librarian
A physician

Faced with this problem, most respondents assume that Steve is a librarian (Tversky & Kahneman, 1983). However, if this problem is considered from a purely rational perspective, a quite different result emerges.

If it is assumed that there are 20,000 librarians, of whom 30 per cent (6,000) are 'meek and tidy souls', and there are 150,000 farmers, of whom 10 per cent (15,000) are 'meek and tidy souls', then a person selected at random would more likely be a farmer than a librarian. Even if the figures were such that the number of farmers was reduced, the number of 'meek and tidy' farmers would likely be greater than the number of librarians with similar characteristics. This example demonstrates how humans make decisions based on stereotypes.

As part of the heuristic associated with 'librarians', there is probably a typical or stereotypical librarian that is recalled by the respondent when responding to the question. However, the same principle might be evident in making judgements concerning drivers, physicians, nurses, airline captains, maintenance engineers, or any other type of profession.

Overcoming the impact of the representativeness bias depends on a knowledge of the base rates associated with a particular occurrence. Base rates are the probability that an object will occur naturally within the environment (Tversky & Kahneman, 1983). An illustration of the significance of base rates can be drawn from train driver selection and testing. For example, an interviewer may have been selecting train drivers for five years with an average of 50 drivers per year selected into the railways. Overall, 70 per cent could be defined as appropriate selections. However, a question remains as to the effectiveness of the interviews, since there is no knowledge of the base rates involved.

Although it is clear that 70 per cent of the drivers selected might have been appropriate, it remains unclear how many of the drivers who were not selected might otherwise have been outstanding operators, had they been selected. In the absence of this information, there is no standardised criterion against which to evaluate the performance of the interviewer.

In the broader community, the representativeness bias has been used successfully by politicians as a means of consolidating support. For example, Adolf Hitler used a common misconception of European Jewry to gather support and consolidate power (Fleming, 1984). The ethnic cleansing that has occurred in both Central Europe and Africa also relates to stereotyping, leading to fear and, ultimately, to warfare (Keane, 1996).

Human Factors and Decision-Making

11.3.2 CONFIRMATION BIAS

Like the representativeness bias, confirmation bias also plays an important role in the decision-making process. This bias relates to the assumption that humans will reduce the cognitive effort involved in a decision, either to improve the response latency or to manage another task concurrently (Fischoff & Beyth-Marom, 1983; Talluri, Urai, Tsetsos, Usher, & Donner, 2018). Generally, it involves a person seeking information to confirm an expectation or assumption and rejecting information that conflicts with an expectation.

An example of confirmation bias can be drawn from the situation where an operator has a strong motivation for a particular activity or diagnosis. Under these circumstances, the operator is more likely to seek out that information which confirms a decision, and reject, ignore, or explain away information that conflicts with the decision (Levin, Schneider, & Gaeth, 1998).

In the medical environment, confirmation bias can often arise where physicians receive a prior report or case notes that may pre-empt a particular condition (Croskerry, 2002). In a case described by Pines (2006), a treating physician in an emergency department is provided a case report by an emergency department nurse that summarises previous admissions and an examination performed the preceding evening by another attending physician. Quite reasonably, and with other cases requiring attention at the time, Pines (2006) suggests that the treating physician may be inclined to accept any initial impressions or pre-existing assumptions concerning the case, particularly if these are triggered by trusted associates.

Recognising confirmation bias can be particularly difficult, since there is a perception that the information available is being perceived and interpreted accurately. In the medical context, the solution to confirmation bias is a differential diagnosis where different conditions are identified that might share symptoms and are then discounted systematically through additional testing (Hussain & Oestreicher, 2018).

Keijzers, Fatovich, Egerton-Warburton, Cullen, Scott, Glasziou, and Croskerry (2018) propose a process of 'slowing down' the diagnostic process, a strategy described as clinical inertia. It allows for a review of the cognitive basis of the diagnostic strategy using metacognition or 'thinking about the process of thinking'. The goal is to encourage the consideration of alternative possibilities, thereby potentially minimising the tendency towards confirmation bias (Althubaiti, 2016).

11.3.3 AVAILABILITY BIAS

Consistent with other biases, the availability bias is based on a relatively simple heuristic that, under normal circumstances, yields accurate outcomes. It engages the principle that the easier that examples can be brought to mind, the greater the likelihood that they are representative of the probability of cases in reality (Tversky and Kahneman, 1973). For example, Mamede, van Gog, van den Berge, Rikers, van Saase, van Guldener, & Schmidt (2010) demonstrated that medical resident physicians were more likely to make diagnostic errors where the cases that they considered were similar to ones that they had considered previously. In each case, they were presumed to be drawing on memory and, on this basis, overestimating the likelihood of a particular condition.

Since the availability bias impacts perceptions of likelihood, previous experience can also result in underestimations of particular cases, where operators overlook rare cases. For example, some airport security screeners may never experience a situation where they are required to detect a genuine firearm deliberately being brought illicitly into a secure area. Therefore, any attempts to build a repertoire of experiences through training constitute analogies that are intended to represent 'reality'. The difficulty with this approach is that it potentially acts as a distraction that may not necessarily reflect the conditions that will occur in those especially rare situations where detection is critical (Goodwin & Wright, 2010).

While challenging, Goodwin and Wright (2010) suggest that preparing for especially rare events requires scenario planning where a range of conditions is considered, irrespective of the estimated likelihood. This is an approach consistent with 'what-if' scenario planning that enables some degree of anticipation to recognise the confluence of factors should they emerge.

11.3.4 Anchoring Bias

The anchoring bias is based on the assumption that conclusions are usually drawn from initial observations (Slovic, Griffen, & Tversky, 1990). For example, respondents who are asked to quickly estimate the value of the following multiplication:

$$8 \times 7 \times 6 \times 5 \times 4 \times 3 \times 2 \times 1$$

generally arrive at an estimate greater than if the sequence of numbers is reversed as in:

$$1 \times 2 \times 3 \times 4 \times 5 \times 6 \times 7 \times 8$$

The difference in the estimates appears to be 'anchored' to the information that was received during the initial process of encoding (Fischoff, 1975).

The anchoring bias has important implications for a wide range of contexts where information is received progressively in the form of reports or briefings. For example, the Value of a Statistical Life (VSL) is typically anchored to market estimates from the United States. This appears to have resulted in a bias in the publication of VSL data, since both authors and journals appear to anchor their estimates of VSL to US data. The result is a bias towards overestimating the level of VSL that is considered appropriate internationally (Viscusi & Masterman, 2017).

11.3.5 Misconceptions of Chance

In gambling, one of the most common decision errors that occurs is due to the misconception of chance (Plous, 1993). For example, when respondents are asked which of the following two sequences of heads and tails is more likely to occur, the majority select (a) over (b).

(a) H-T-H-T-T-H
(b) H-H-H-H-H-H

In fact, neither of these sequences is more likely to occur. In each case, the probability of flipping a head or a tail is 1 in 2 or 50 per cent. Since each flip of the coin is independent of the others, the probability of throwing a sequence of heads or tails will remain exactly the same.

Within the operational context, misconceptions of chance are most evident in statements from operators such as 'my luck has to change'. This statement reflects an expectation that events in the past are associated with events in the future. However, there may be no association, and the probability of an undesirable outcome may remain consistent, irrespective of previous events.

11.4 PRESCRIPTIVE DECISION-MAKING

Where descriptive models of decision-making purport to explain the process through which individuals formulate decisions, prescriptive models of decision-making are designed to facilitate or guide the process of decision-making, particularly amongst inexperienced practitioners. Unlike descriptive decision-models, prescriptive models of decision-making are generally analytical approaches. This is due primarily to a number of assumptions that have been made by prescriptive theorists including that:

1. Most decision-makers use intuitive strategies;
2. Intuitive strategies are subject to biases and are therefore unreliable;
3. Analytical strategies tend to be more accurate than intuitive strategies; and that
4. Analytical strategies can be prescribed as part of a decision tree or linear sequence (Hammond, Hamm, Grassia, & Pearson, 1987).

On the basis of these assumptions, the primary goal of prescriptive decision-making is to ensure that biases are avoided, and that decisions are based on rational and valid criteria. To this end, many decision-making models have employed mathematical constructs as the basis for decision-making. This attempt to quantify decision values or decision elements is designed to ensure that information is acquired and evaluated against a rational criterion that minimises the impact of biases.

11.4.1 THE DECIDE MODEL

The DECIDE model was originally developed within the fire-fighting domain and adapted subsequently for application in other domains such as aviation (Guo, 2008; Jensen, 1997; Krieger, 2014). It is based on the acronym:

- D Detect a change in the environment
- E Estimate the significance of the change
- C Choose an appropriate goal
- I Identify the action to achieve the goal
- D Do the actions
- E Evaluate the efficacy of the decision

The DECIDE model is one of a number of models that have emerged within various domains, all of which possess similar qualities. These include GRADE (Gather, Review, Analyse, Decide, Evaluate) and SADIE (Share, Analyse, Decide, Implement, Evaluate). Unlike other analytical models, these are not based on mathematical constructs (Guo, 2008). This particular class of prescriptive strategies is based on the assumption that humans can make rational decisions on the condition that they possess all of the relevant information (Krieger, 2014).

Despite their pervasiveness, the DECIDE model is one of few of these models to have been tested empirically. Three pilots were trained using the DECIDE model and their decision processes were compared during a simulated task against participants who had not undergone the training (Jensen, 1997). The results suggest that the DECIDE model may have some potential as a means improving performance in the period immediately following training.

11.4.2 Bayes' Theorem

Most prescriptive models of decision-making are based on the notion of rationality. To this end, the Bayesian approach to decision-making involves the development of an understanding of probabilities during the decision-making process.

Many decision-makers find it considerably difficult to estimate accurately the relative probability of events. For example, Eddy (1982) asked a series of doctors to assume that they had just examined a lump on a woman's breast and had ordered a mammogram. They were advised that the probability of malignancy in this case is approximately 1 in 100, while the mammogram has an accuracy of 80 per cent for malignant tumours and 90 per cent for benign tumours.

Eddy's (1982) main interest was in doctors' perception of the probability of malignancy given a positive test. The results indicated that 95 per cent of respondents considered the probability of malignancy at approximately 75 per cent. However, from a Bayesian perspective, a different outcome emerges.

Bayes' Theorem is a formula that calculates the probability of an outcome as a proportion of the total possible outcomes (Fischoff & Beyth-Marom, 1983). In the case of Eddy (1982), the probability of cancer given a positive test result is the probability of an accurate test result, calculated as a proportion of the total probability of cancer, divided by the total probability of an accurate test (either malignant or benign), where:

p(Cancer|Positive) is the probability of cancer given a positive result
p(Positive|Cancer) is the probability of a positive result which is malignant
p(Positive|Benign) is the probability of a positive result which is actually benign
p(Cancer) is the overall probability of malignancy
p(Benign) is the overall probability of non-malignancy

At the time, it was known that the overall possibility of malignancy was 1 per cent (of the total population) while the corresponding probability that a person did not have cancer was 99 per cent.

Substituting,

$$p(Cancer \mid Positive) = \frac{(.80)(.01)}{(.80)(.01) + (.10)(.99)} = \frac{.008}{.107} = .075 = 7.5\%$$

where:

$p^{(Positive|Cancer)} = 80\% = .80$
$p^{(Positive|Benign)} = 10\%^1 = .10$
$p^{(Cancer)} = 1$ in 100 or .01
$p^{(Benign)} = 99$ in 100 or .99

In this case, the overall probability of cancer with a positive test result is 7.5 per cent, a great deal less than the 75 per cent probability estimated by doctors. This outcome demonstrates the difficulty that humans experience in integrating probabilities and risks.

11.4.3 FRAMING AND RISK ASSESSMENT

Risk assessment is an essential component of contemporary approaches to decision-making. For example, potential investors in an organisation will want to know the risks involved in an investment and the capital gain that is expected to occur. The balance between the investment and the odds of a return will govern the final decision.

For many decision-makers, the notion of risk is difficult to quantify accurately. For example, Tversky and Kahneman (1981) sought the relative preferences for the following alternatives:

(a) A sure gain of $240
(b) A 25 per cent chance to gain $1,000, and a 75 per cent chance to gain nothing

They noted that the majority of respondents selected alternative (a) with a guaranteed return. However, a simple calculation indicates that the overall Expected Value (EV) of the decision is:

$$EV = (.25)(\$1,000) + (.75)(\$0) = \$250$$

Therefore, the expected value of alternative (b) is greater than the value of alternative (a).

When the alternatives were framed in losses, such as:

(a) A sure loss of $750
(b) An 80 per cent chance of losing $1,000, and a 20 per cent chance of losing $0

most respondents selected alternative (b) (Tversky & Kahneman, 1981). In this case, the expected value is calculated

$$EV = (.80)(\$1,000) + (.20)(\$0) = \$800$$

indicating that the expected value for alternative (b) is greater than for alternative (a).

Tversky and Kahneman (1981) refer to this apparent paradox as the *framing effect*, since they suspected that the outcome was due to the way in which the problems were framed: either as gains or losses. In situations where risk-related outcomes are referred to as gains (e.g. optimising the health and well-being of employees), humans tend to be risk averse. However, where risks are referred to as losses (e.g. failure to meet production targets), humans tend to be risk-seeking.

Framing has important implications for safety initiatives, since the types of safety messages can influence the behaviour in which recipients are likely to engage. For example, in communicating the effects of alcohol consumption during pregnancy, Yu, Ahern, Connolly-Ahern, & Shen (2010) demonstrated that framing the related statistics in the context of gains improved the perceived efficacy in adopting strategies that were intended to reduce Foetal Alcohol Spectrum Disorder. Similarly, Nan, Xie, and Madden (2012) demonstrated that, amongst older adults with a lower perception of the efficacy of the H1N1 vaccine, a loss-framed message improved their willingness to receive the vaccine.

Despite the apparent universality of the framing effect (e.g. Mayhorn, Fisk, & Whittle, 2002), there is evidence to suggest that there are individual differences in the effectiveness of framing. For example, Simon, Fagley, & Hallerman (2004) demonstrated that framing effects tend to be less effective where recipients demonstrate a relatively higher need for cognition and a higher depth of processing. Differences in framing effects have also been established on the basis of sex and vocational background (Cullis, Jones, & Lewis, 2006).

11.4.4 Hazardous Attitudes

In applied industrial environments, successful decision-making is generally thought to require both the rational capabilities necessary to integrate and synthesise information, and the motivation necessary to implement these strategies. The motivational component of decision-making can be conceptualised according to six hazardous thoughts or attitudes (Buch & Diehl, 1984; Telfer, 1988) (see Table 11.2).

The validity of these concepts has been investigated in a number of environments including the aeromedical (Adams & Thompson, 1987) and flight training domains (Lester, Diehl, & Buch, 1985), and in orthopaedic surgery (Bruinsma, Becker, Guitton, Kadzielski, & Ring, 2015). In most cases, training has involved practitioners

TABLE 11.2
The Five Hazardous Attitudes and Their Associated Descriptors

Macho	'I'll show you – I'm the best'
Invulnerability	'It's never going to happen to me'
Anti-Authority	'No one can tell me what to do'
Impulsivity	'Do something – Quickly'
Resignation	'What's the use?'
Deference	'It's someone else's problem'

TABLE 11.3
The Five Hazardous Attitudes and Their Associated Antidotes

Macho	'Taking chances is foolish'
Invulnerability	'It could happen to me'
Anti-Authority	'Follow the rules, they are usually right'
Impulsivity	'Not so fast. Think first'
Resignation	'I'm not helpless, I can make a difference'
Deference	'It's everyone's responsibility'

being taught to recognise situations where hazardous attitudes are presumed to have played a part, and subsequently instigating relevant antidotes (see Table 11.3).

Based on a psychometric analysis, Hunter (2005) suggests the application of a Likert scale as the basis to assess a propensity for a particular hazardous attitude. Developed in the aviation context and referred to as the Aviation Safety Attitude Scale (ASAS), it shows good psychometric properties and can be readily adapted to different contexts.

11.4.5 COMBATING THE INFLUENCE OF BIASES

In response to the difficulties faced by decision-makers, Plous (1993) has developed a useful set of recommendations that are designed to minimise the impact of biases and encourage effective and efficient decision-making. These can be applied, irrespective of the operational environment and require decision-makers to:

- Pay attention to base rates;
- Recall that chance is not self-correcting; and
- Avoid misinterpreting regression to the mean.

11.5 DECISION-MAKING IN COMPLEX ENVIRONMENTS

On the basis of a model of system complexity proposed by Woods (1988), there are effectively two mechanisms that are available to reduce the potential for a system failure:

1. Reduce the potential for a failure amongst system components; and/or
2. Reduce the impact of the failure of a component on the system, by containing the consequences.

The development of effective and efficient decision-making strategies reflects an attempt to reduce the potential for a failure amongst system components. From an information-processing perspective, effective decision-making requires: (1) that task-relevant information is perceived accurately and at the appropriate time; (2) that this information is processed accurately and efficiently; and (3) that the response developed is implemented appropriately. This conceptual model of decision-making is applicable at both an individual and an organisational level, and can be approached from either a training perspective or from the perspective of system design.

In complex environments, the decision-making process is necessarily more complicated due to the number of issues that need to be considered and processed. For example, task-related information within a complex environment might need to be derived from a range of sources and might be presented in a range of forms that must be synthesised, and the information integrated to form a coherent and accurate understanding of the nature of the situation. Therefore, effective decision-making within a complex organisational environment depends on both the accurate and reliable acquisition of task-related information, together with an organisational structure that enables the communication of this information for subsequent analysis and interpretation.

Where an organisational structure is inappropriate or inefficient, the potential for a system failure tends to increase. However, effective decision-making in a complex environment does not necessarily depend on a highly structured system. A comparison conducted in the military environment revealed that high-reliability organisations tended to be associated with a more flexible organisational structure (Roberts, Stout, & Halpern, 1994). Therefore, it is the appropriateness of an organisational structure, rather than the characteristics of the structure itself, that is likely to be more important for decision-making in complex environments.

An example of the relationship between organisational structure and the effectiveness of complex decision-making can be drawn from an incident in July 1988 involving the USS *Vincennes*, in which Iranian Airlines Flight 655, an Airbus A300 aircraft, was targeted and destroyed inadvertently over the Persian Gulf, resulting in 290 fatalities (Rouse, Cannon-Bowers, & Salas, 1992). Although the aircraft was climbing through 12,500ft, the radar return was interpreted by the crew on board the Vincennes as an F-14 Tomcat military aircraft on descent through 9,000ft, and preparing for an attack.

The main difficulty faced by the crew of the *Vincennes* was the requirement for a rapid response that involved the acquisition, integration, and communication of an array of information to the appropriate decision-makers within the hierarchy (Meshkati, 1991; Bisbey, Reyes, Traylor, & Salas, 2019). In this case, the organisational structure can only be effective if it has the capacity to respond at a rate that is faster than the rate at which the events are changing within the operational environment. The situation involving the *Vincennes* and the Iranian airliner lasted for a total of seven minutes, during which the environment changed constantly.

Although there were a number of design-induced failures implicated in the destruction of the A-300, the hierarchical structure of the chain of command aboard the *Vincennes* also contributed to the final outcome. Brehmer (1988) suggests that a hierarchical organisational structure may be incapable of responding appropriately to the rapid changes that occur in operational environments involving advanced technology. Large amounts of information may be obtained at a rapid rate, and the information is not necessarily available in a sequential form. Further, different pieces of information may be more or less available at different levels within a hierarchy, thereby increasing the potential for information to be overlooked or processed incorrectly.

Human Factors and Decision-Making 135

The capability of an organisation to adapt to changes in the demands associated with a task is likely to improve decision-making performance by ensuring that decision-makers are provided with all of the information necessary to assess the situation (Lanceley, Savage, Menon, & Jacobs, 2008). Nevertheless, decision-makers are still faced with the requirement to process the information and integrate the data to formulate an understanding of the nature of the environment.

11.6 DECISION-MAKING AND RELIABILITY ANALYSIS

Reliability analysis is a process that yields a quantitative value of the probability of an occurrence, given a range of diverse factors. However, an occurrence can be perceived as 'positive' or 'negative' under the circumstances and can be assigned an associated 'value'. This process of combining the probability of an event with the value of an outcome is generally referred to as *decision analysis*, and can be used to determine the most appropriate option, given a range of possibilities within a complex decision-making environment (Oliveira, Lopes, & Bana e Costa, 2018).

In most cases, decision analysis is used where the outcomes of a decision can be quantified accurately. For example, the decision to purchase a new train might be determined on the basis of the number of passengers, and might be offset against the purchase costs and the operational expenses. Therefore, a quantitative assessment of the outcome can be made relatively accurately, and can be compared to the projected outcome associated with each alternative strategy.

Where human behaviour is concerned, the costs associated with a particular outcome can be more difficult to quantify. For example, the choice between two motor vehicles might be made on the basis of visual appeal, rather than simply the monetary costs involved. In this case, visual appeal has an inherent subjective value to the decision-maker that must be taken into account during the process of decision analysis.

Svenson (1979) suggests that subjective assessments of decisions can be taken into account by rating each attribute of a decision outcome in terms of its attractiveness to the decision-maker. The level of attractiveness can be quantified using a standard metric, such a ranked scale equivalent to the utility to the decision-maker.

Combining the utility of a decision outcome with the probability of its occurrence is presumed to explain the behaviour of decision-makers faced with decisions under uncertainty (Kahn & Baron, 1995). According to *prospect theory*, the value of a decision option can be explained as the product of the subjective value of the outcome and weighting ascribed on the basis of the perceived probability of the outcome (Tversky & Fox, 1995). This relationship can be summarised as:

$$\text{Decision Outcome Value} = w(p)v(x)$$

where w refers to the attractiveness of the prospect, p refers to the given probability, and v refers to the subjective utility of the outcome (x).

Although prospect theory is normally used to explain or describe the process of individual decision-making, the general principles can be applied in more complex decision-making scenarios that might occur within an organisational context. The advantage associated with this process is that it provides a quantifiable value of the outcome of a potential decision. For example, it might be determined that the probability of entering incorrect data into one type (Type A) of a medical intravenous (IV) pump is 0.01 (1 in 100 entries), while the probability of an error in another medical intravenous pump (Type B) is 0.09 (9 in 100 entries). However, the cost of purchasing (A) might be $250,000 where (B) might cost $190,000.

Prospect theory suggests that the weighting ascribed to the probability of an event will differ, depending upon whether the event is interpreted as a 'gain' or a 'loss' and whether the probability of the event is relatively low or relatively high (Tversky & Fox, 1995). The value of the decision weighting is based on the general principle that the greater the level of certainty associated with an event, the greater the value of the decision weighting. However, the association can be described more accurately as a non-linear relationship where the perceived values between extreme (high and low) levels of probability tend to differ from the perceived weighting between moderate levels of probability. As a result, a shift in the probability of an event from 0.8 to 0.9 or 0 to 0.1 is typically perceived as greater than a shift from 0.6 to 0.7 (Tversky & Fox, 1995).

In the choice between the error probabilities associated with the use of the two IV pumps, the probability values might be ascribed different weightings, given that a difference from 0.01 to 0.09 could be interpreted as a significant shift in reliability. Using a normalised decision weighting scale from 0.0 to 1.0, the decision-maker is asked to assign a relative weighting value to each of the probabilities. In some cases, a decision-maker may assign an equal value where the difference between the levels of probability is perceived as so small as to be relatively insignificant.

Given that the values represent the probability of a 'loss', a relatively greater decision weight is ascribed to higher levels of probability (greater certainty associated with the event). For example, a nominal weighting of 0.01 might be assigned to the probability value of IV pump (A), while a weighting of 0.05 might be assigned to the probability value of IV pump (B) to indicate the relative significance of the difference between the probabilities. The values in the formula can be substituted such that:

$$\text{Decision Outcome Value (FMC A)} = 0.01 \times -\$250{,}000 = -\$2{,}500$$

and

$$\text{Decision Outcomes Value (FMC B)} = 0.05 \times -\$190{,}000 = -\$9{,}500$$

As a result of this analysis, the subjective values for the different models of IV pumps suggest that (A) has the greatest utility, despite the fact that initial cost is greater than for (B). Consequently, it provides a basis for decision-making, since there is a clear cost-benefit that is associated with IV pump (A).

11.7 DECISION-MAKING UNDER UNCERTAINTY

In complex, dynamic environments, there are often factors involved that cannot be associated with a discrete value, either financially or otherwise. As a result, nominal values may be assigned as a means of quantifying the effects of these factors (Wu, Yan, Wang, & Soares, 2017). However, where these values differ from the values that occur in reality, the effects may be underestimated or overestimated during a decision analysis.

Where there is a level of uncertainty associated with the discrete value of factors involved in a decision analysis, there are a number of strategies that can be employed to increase the overall accuracy of the decision outcome. The first of these strategies requires that the value of each factor be held constant at either a low or high level. The effects on the decision outcome can be compared, so that the relative influence of each factor can be determined. The result is an indication of the relative effect of changes associated with one factor on the subsequent decision outcome (Mann & Ball, 1994).

As an example of the management of uncertainty, a situation might be considered involving a decision between two competing training strategies, the precise training outcomes of which are subject to uncertainty. Other factors that may influence the decision could include the cost of the course, the appropriateness of the training philosophy, and the length of the course. In this case, factors such as the length of the course and the costs involved are subject to relatively less uncertainty than the potential outcomes. Therefore, discrete values can normally be ascribed to these factors and a model can be developed to determine the decision outcome.

In the case of the training outcomes, this factor is also included in the model. However, the value of the training outcomes is ascribed a low value initially, to determine the effect on the decision outcome. For example, if the training outcomes for both courses were relatively poor, it is necessary to determine whether the values for the other factors would offset the poor training outcomes for either or both courses. In effect, this represents the 'worse case' scenario.

In the subsequent analysis, the value of the training outcomes in both cases might be ascribed a relatively high value to determine the effect on the decision outcome. This represents the 'best case' scenario, and a comparison is made between the potential decision outcomes for both training courses. By comparing the decision outcomes for the 'worse case' scenario with the outcomes for the 'best case' scenario, it is possible to determine the option which is likely to yield the greatest utility for the decision. More importantly, the impact of uncertainty has been taken into account.

Where more complex decisions are concerned, the number of uncertain factors involved may increase, to the extent that most of the factors associated with the decision analysis may be subjected to some level of uncertainty. However, Morgan and Henrion (1990) suggest that, in most cases, the number of factors that have a significant impact on the decision outcome is likely to be limited to a relatively small number of key elements, despite the apparent complexity of the decision-making environment. For example, the relative influence of the style of carpeting in a restaurant is probably likely to have a minimal impact on a passenger's decision to visit the restaurant. As a result, an assessment might be made as to the relative influence of a specific factor, to the point where some factors might be excluded from the decision analysis.

Clearly, excluding a factor from a decision analysis can be problematic in the absence of safeguards to protect against errors that may occur. Nevertheless, the importance of uncertainty is the extent to which it influences the decision, and not simply the decision outcome. This distinction can be illustrated by a choice between the purchase of two houses. Although there may be some uncertainty concerning aspects of the neighbourhood, this uncertainty may not be directly relevant to the decision-making process, since the decision specifically relates to the nature and quality of the house. Consequently, the first step in reducing the number of uncertain factors that need to be taken into account as part of the decision-making process is to ensure that the factors actually relate to the decision under consideration, and are not simply related to the decision-making environment.

NOTE

1 Given that the probability of correctly identifying a benign tumour is 90 per cent (100–90 = 10 per cent).

12 Fatigue

12.1 FATIGUE IN PRACTICE

In an increasingly connected work environment, fatigue has emerged as a significant risk, both in the context of human performance, and as a psychosocial factor related to worker health and well-being. Fatigue is the reaction of the body to extended exertion, physically and/or mentally, and is evident in a reduction in alertness, a lack of attention to critical stimuli, poor decision-making, and a response latency that is slower than is typically the case (Lerman, Eskin, Flower, George, Gerson, Hartenbaum, Hursh, & Moore-Ede, 2012). Extreme levels of fatigue can be accompanied by a strong desire for sleep.

The recovery from fatigue is normally a period of rest or sedentary behaviour that is typically associated with sleep. In the case of chronic sleep deprivation, a series of sleep episodes may be required to enable recovery. However, the period and rate of recovery is highly variable, as is the impact of fatigue on human performance (Williamson & Friswell, 2013). The effects of fatigue also appear to be impacted by the time of day.

12.2 CIRCADIAN RHYTHMS

Circadian rhythms are also known as 'body rhythms' and relate to the cycle of both psychological and physiological dimensions within the body (Pilcher & Morris, 2020). The impact of circadian rhythms is most noticeable following trans-meridian flight, or amongst shift workers where the physiological and psychological responses of the body are not synchronised with the environment.

Those features within the environment that are presumed to indicate psychological and physiological responses are referred to as *zeitgebers*. These include features such the time of meals, waking times, and sleeping times. Amongst operational personnel, shift workers are likely to be most susceptible to circadian dysrythmia due to timing and lack of consistency of duty shifts. This combination of factors may lead to an increase in the potential for error and an overall reduction in human performance.

In general, both psychological and physiological human functions form a sinusoidal curve across time (usually a 24–5-hour cycle). This effect is evident in responses including concentrations of hormones, levels of arousal, and levels of motivation.

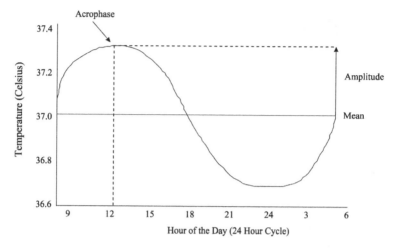

FIGURE 12.1 A conceptual diagram of a circadian rhythmic function illustrating the acrophase, amplitude, mean, and the period of body temperature response. (Adapted from Samel and Wegmann, 1988, p. 405.)

From a mathematical perspective, Samel and Wegmann (1988) have defined the sine curve on the basis of four distinct elements: the daily mean; the acrophase (the upper limit), the amplitude of overall deviation, and the period of time over which the effect takes place. These are depicted graphically in Figure 12.1.

As is evident from Figure 12.1, the most significant deterioration in human performance typically occurs between midnight and 0600 hours. However, the impact of fatigue on human performance is not necessarily limited to this period. For example, in an analysis of incidents reported to the Aviation Safety Reporting System (ASRS) in the United States, approximately 21 per cent were related to fatigue (Lyman & Orlady, 1980). Similarly, Åkerstedt (2000) estimates that up to 20 per cent of accidents in transportation are directly attributable to the impact of fatigue.

Despite the concern associated with fatigue in the contemporary workplace, the impact on human performance has been the target of investigation for some time. As early as 1943, experiments were being conducted to determine the effect of fatigue on human error (Gibson & Harrison, 1984). However, researchers have yet to identify a strategy that prevents the impact of fatigue reliably and effectively within complex and demanding work environments.

From a system safety perspective, the management of fatigue and circadian rhythms has generated significant interest in recent years, primarily due to the experience of military crews during the First and Second Gulf Wars. In the case of bomber crews, flights were initiated from the United States and personnel would be operational for considerable periods. Throughout the Gulf War, a number of pharmacological agents were tested with some limited success. These agents generally focused on improvements in sleeping patterns, since this was the most common physiological symptom associated with circadian dysrhythmia (Russo, 2007).

12.3 SLEEP DISTURBANCE AND STAGES

The main characteristics of sleep disturbance following circadian dysrhythmia include initial difficulty falling asleep, spontaneous waking during the night, and an accumulated deficit of sleep arising from early awakenings (McKenna, Reiss, & Martin, 2017). Sleeping patterns can be identified using the electroencephalograph (EEG) measuring differences in electrical potential between two or more sites on the skull (Dyro, 1989; Fisk, Tam, Brown, Vyazovskiy, Bannerman, & Peirson, 2018). This non-invasive procedure produces characteristic frequencies of electrical activity. On the basis of these frequencies, five stages of sleep have been identified and are listed in Table 12.1.

To improve the quality of sleep, it is generally regarded as desirable to reach stages 3, 4, and/or REM sleep, particularly during recovery from sleep deprivation (Hobson, 1989). Typically, it takes approximately 40 minutes to reach these stages on entering sleep.

12.4 FATIGUE AND HUMAN PERFORMANCE

Allied to the area of circadian rhythms is the growing problem of personnel operating for extended periods in the role of monitor. During these periods, operators likely engage in micro-sleeps (Beh & McLaughlin, 1997). Consequently, there is some concern that the lack of direct involvement in the performance of the activity may be associated with a reduction in the capacity to respond quickly and effectively to non-normal or emergency situations (Boyle, Tippin, Paul, & Rizzo, 2008).

From an operational perspective, the impact of fatigue was illustrated in the crash of Colgan Air Flight 3407, a Bombardier Q400, on February 12, 2009. The aircraft was on descent to Buffalo, New York, having departed Newark, New Jersey. Light snow and fog resulted in ice forming on the leading edge of the wings, increasing drag, and causing a reduction in the airspeed of the aircraft to a point where the stall warning activated. In responding to an impending stall, pilots are taught to lower the attitude of the aircraft and then to increase power. However, in the case of the Flight 3407, the captain pulled back on the yoke, and subsequently applied power, The result

TABLE 12.1
The Five Stages of Sleep and Their Corresponding Frequencies

Stage	Characteristic	Duration
Stage 0	Wakefulness	N/A
Stage 1	Sleep is entered	1–10 minutes
Stage 2	Sleep is entered	Approx. 10 minutes
Stage 3	Deep sleep	Approx. 5 minutes
Stage 4	Deep sleep	5–60 minutes
REM	Dreaming	1–40 minutes

was a further 'pitch-up' of the aircraft, further restricting the margin for recovery (NSTB, 2010).

In addition to a stall warning, the Q400 is equipped with a tactile 'stick-shaker' which causes a vibration of the control column reflecting the buffeting that occurs on the leading edge of wing during the stall and is intended to cue pilots. The captain of Flight 3407 overrode the system and continued to pull back on the control column, eventually causing the aircraft to crash.

The subsequent investigation into the crash of Flight 3407 revealed that both the captain and the first officer had commuted to the Colgan Air duty base at Newark. In the case of the first officer, this had involved a transcontinental, overnight flight from Seattle, seated in the jumpseat of two aircraft operated by Federal Express. Similarly, the captain had commuted from Tampa, Florida. Amongst flight crew at Colgan Air at the time, commuting to the Newark duty base was not unusual.

Prior to the accident flight, the flight crew were scheduled to depart on their first flight at 1.00 pm. However, due to the weather conditions, a number of flights were cancelled, and the flight crew eventually departed Newark at 9.18 pm. While it is difficult to establish conclusively, the combination of changes in circadian rhythms associated with transcontinental flight, disrupted sleep patterns, and extended periods of duty-time likely impeded the pilots' capacity to recognise and prepare for the deteriorating conditions and the impact of icing. Similarly, the response to the warnings was impeded by the application of ill-timed and imprecise actions. These responses are broadly consistent with the effects of fatigue on operational performance (see Table 12.2).

TABLE 12.2
The Effects of Fatigue and Sleep Loss on Flight-Crew Performance

Performance Category	Effects
Reaction Time:	Increased timing errors
	Less smooth control
	Require enhanced stimuli
Attention:	Overlook/ misplace tasks
	Preoccupation with single tasks
	Reduced audio-visual scan
	Less aware of poor performance
Memory:	Inaccurate recall
	Forget peripheral tasks
	Revert to 'old' habits
Mood:	Less likely to converse
	Distracted by discomfort
	Irritability
	'Don't Care' Attitude

Source: From Graeber (1988, p. 338).

12.5 SHIFT WORK

Shift work has become much more pervasive in the contemporary workplace due, in part, to the expectations of consumers and the demands of globalisation. As a result, there have been significant efforts invested in understanding the implications of shift work, especially in the context of human performance. These efforts generally involve either analyses of sleep-wake data or estimates of patterns of sleep that are derived from work rosters (Roets & Christiaens, 2019). The outcome is a model of the fatigue-related risks associated with a schedule of shifts.

Models of fatigue-related risk have enabled the development of risk indices that can be applied in practice to compare the potential impact of different strategies that are intended to manage fatigue in the workplace. The challenge associated with these models has been in taking into account the range of variables, in addition to the shift-schedule itself, that are likely to impact human performance. For example, demands can occur during non-work periods that impede recovery and may impact performance during a subsequent shift.

In a more detailed consideration of the impact of work-rest schedules on sleeping patterns, Cruz and Della Rocco (1995) noted that different rosters resulted in distinct sleeping patterns. A so-called '2-2-1' roster involves two afternoon shifts, followed directly by two morning shifts, followed by a night shift. This was examined against a '2-1-2' roster that involved two afternoon shifts, followed by one midday shift, followed by two early morning shifts. The results revealed a relatively greater Total Sleep Time (TST) for the '2-2-1' shift compared to the '2-1-2' shift over a five-day period.

Although there was no significant difference evident between the two schedules in overall levels of sleepiness, Cruz and Della Rocco (1995) noted that the '2-2-1' schedule had a greater number of extreme sleepiness ratings than the '2-1-2' schedule. This was particularly evident during the drive home after the evening shift on the '2-2-1' schedule. However, it should be noted that these data were collected over a five-day period and the long-term impact of these types of schedules remains unclear.

Despite the challenges in establishing the impact of different work schedules on workplace safety, Garde, Begtrup, Bjorvatn, Bonde, Hansen, et al. (2020) recommend that night shifts in particular should be limited to no more than three consecutive nights, that intervals between shifts should be greater than 11 hours, and that shifts should be limited to no more than nine hours. While there are some work environments that may enable adjustments to these recommendations without risks to workplace performance, they remain a useful guide for operations within single time zones.

12.6 ALERTNESS-MAINTENANCE AND SUSTAINED ATTENTION

Semi-automated process control environments are becoming increasingly ubiquitous as organisations strive to reduce labour costs and minimise the potential for human error. However, the development of these working environments has come at a cost to human performance. The unpredictable nature of process control environments such as rail control, air traffic control, and power system control is such that operators are required to direct and maintain their attention to tasks for long periods, during which there may be little requirement for intervention (Metzger & Parasuraman, 2001; Vicente, Roth, & Mumaw, 2001).

In process control environments, the role of the operator is to minimise disruptions to continuous operations by recognising and responding quickly to changes in the system state. This requires a continuous process of assessment so that changes in the system state can be recognised at a stage that will enable the efficient and effective implementation of an intervention (Helton & Warm, 2008).

Despite the demands on operators to sustain their attention towards automated tasks over time, empirical evidence suggests that humans find it difficult to continue to direct their attention to passive stimuli over extended periods (Warm, Parasuraman, & Matthews, 2008). Typically, the result is a gradual dissociation from the task and a corresponding increase in response latency when changes necessitate a response to the system. This dissociation also tends to result in extreme variations in actual and perceived workload as operators transit between monitoring tasks and active control over the system (Wiener, Curry, & Faustina, 1984).

Sustained attention to a task, or 'vigilance', has been the subject of a significant research effort to develop an understanding of the underlying cognitive features of the construct and thereby identify appropriate management strategies within the operational context. One of the outcomes of this research effort has been the distinction between vigilance and alertness-maintenance strategies. Where vigilance is directed towards specific stimuli, alertness is associated with a generalised state, similar to arousal (Oken, Salinksy, & Elsas, 2006). Therefore, alertness has been considered a preparatory state without which vigilance in detecting changes to specific features within the environment would not be possible.

Evidence to support the distinction between sustained attention and alertness can be drawn from the apparent association between sustained attention and mental workload. Specifically, increases in workload are associated with reductions in attentional resources, thereby impeding the capacity for vigilance (Grier, Warm, Dember, Matthews, Galinsky, Szalma, & Parasuraman, 2003).

The rate at which vigilance degrades in the absence of engagement with a task increases with increases in both levels of fatigue and the time on shift (Parasuraman, Molloy, & Singh, 1993). Since increases in fatigue are associated with reductions in generalised alertness, it might be surmised that vigilance and alertness represent separate, although related, constructs. This has implications for remedial strategies, since vigilance-related interventions may be less effective in the absence of interventions that also contribute to improvements in alertness.

At a systemic level, vigilance-management tasks tend to involve relatively targeted strategies such as the use of alarms and the design of interfaces to improve the acquisition and integration of task-related information (Neigal, Claypoole, & Szalma, 2019). By contrast, alertness-management strategies tend to be more generalised in focus and involve strategies such as workforce scheduling, and/or the use of pharmacological interventions (Bowden & Ragsdale, 2018). However, despite the utility of initiatives at the systemic level, the nature of complex process control systems means that there will remain a reliance on individual strategies, particularly in terms of the application of alertness-maintenance strategies.

12.7 INDIVIDUAL DIFFERENCES AND ALERTNESS-MAINTENANCE (EXPERTISE)

Individual differences in the capacity to sustain attention to task may be explained by fundamental differences in operators' underlying cognitive capacity. However, these differences may also be explained by operators' capacity to implement so-called alertness-maintenance tasks. Alertness-maintenance tasks comprise activities that are related to the primary activity but which enable an operator to briefly disengage from the primary task (Oron-Gilad, Ronen, & Shinar, 2008). They are typically less demanding than the primary task and thereby reduce the demands on cognitive load. This enables the application of additional, residual cognitive resources towards re-engagement with the primary task.

In remaining cognisant of changes in the system state and thereby reducing the cognitive demands associated with system-related interventions, operators engage a number of strategies, most of which are designed to maintain a generalised level of alertness. For example, drivers in low arousal environments adopt alertness maintenance strategies such as listening to music or driving with the window down (Pylkkönen, Sihvola, Hyvärinen, Puttonen, Hublin, & Sallinen, 2015). Similarly, pilots of long-haul commercial aircraft engage in conversations between crew members and the generation of hypothetical scenarios (Sallinen, Sihvola, Puttonen, Ketola, Tuori, Härmä, Kecklund, & Åkerstedt, 2017). Although these strategies might be considered short-term solutions to reductions in alertness, two important observations can be made on the basis of this evidence. First, it appears that operators have the capacity to recognise a potential loss in their own state of alertness; and second, it suggests that operators will adopt alertness maintenance techniques that are closely associated with the primary task.

The utilisation of alertness-maintenance tasks has been reported in a range of operational domains including aviation, motor vehicle operations (Verwey & Zaidel, 1999), and rail operations (Coplen & Sussmen, 2000). However, what is not clear is whether alertness-maintenance tasks are used in practice, whether they are employed by experienced practitioners, and/or whether their application is more evident with increasing levels of fatigue or time on shift.

12.8 FATIGUE MANAGEMENT SYSTEMS (FMS)

To account for individual and organisational differences in the incidence and impact of fatigue, a number of jurisdictions have authorised the development and implementation of Fatigue Management Systems, providing some flexibility to organisations and industries in their approach to the management of fatigue (Cabon, Deharvengt, Grau, Maille, Berechet, & Mollard, 2012). Drawn from approaches to risk management and safety management systems, FMS are based on the principle that different activities and time on duty impact operators differently, and impose different organisational risks (Sawatzky, 2017). These risks need to be identified, measured, and monitored to enable the implementation of restorative or other mitigating strategies as necessary (Phillips, Kecklund, Anund, & Sallinen, 2017).

Like SMS, the development of an FMS begins with the identification of threats or hazards that might impact levels of fatigue. These include excessive physical or monitoring activities, extended periods of overtime, short periods between shifts, and/ or breaks that fail to enable recovery. These hazards can be entered into software that will determine overall levels of risk and assist with establishing the case for remedial activities. This might include restructuring shifts, reducing particular types of shifts to account for changes in circadian rhythms, and/or ensuring that periods between shifts allow for recovery (Williamson & Friswell, 2013).

While identifying fatigue-related hazards provides the opportunity to develop an appropriate architecture to minimise the impact of fatigue on operational performance, the principles may need to be adjusted where there are organisational or operational changes that impact the effectiveness of the FMS (Steege, Pinekenstein, Rainbow, & Knudsen, 2017). Catalysts for changes to an FMS can emerge from pulse surveys or other measurement strategies where subjective perceptions of fatigue increase and are sustained beyond a prescribed threshold. Other catalysts might include incidents where fatigue is identified as a significant factor.

There are a number of scales that purport to assess fatigue, including the Karolinska Sleepiness Scale (Åkerstedt & Gillberg, 1990), the Epworth Sleepiness Scale (Johns, 1991), and the Swedish Occupational Fatigue Inventory (SOFI) (Åhsberg, & Fürst, 2001). Although they are subjective tools, monitoring responses over time and taking into account different activities and periods of assessment can provide a reasonably clear, evidence-based overview of perceptions of fatigue. It may also reveal individual differences in perceptions of fatigue, the outcomes of which can be employed to institute interventions where necessary, at the level of individual operators. This may be necessary where particular operators experience life changes that impact their performance, enabling a tailored approach to the management of fatigue that maintains the safety of the system while ensuring the health and well-being of employees.

Part IV

Group Processes and Human Factors

13 Groups and Teams

13.1 THE NATURE OF GROUPS

Groups are an integral part of human existence, and most people have had the experience of being a group member. However, the quality of this experience can vary from one that is particularly rewarding to one that can be frustrating and from which very little is achieved. Therefore, ensuring that a group is successful has benefits, both in achieving immediate outcomes and over the longer term, where performance can be sustained through group identity, membership, and commitment.

In the context of problem-solving, the advantage of a group over an individual is often presumed to lie in the information-processing capability afforded through the additional ideas, different methods, and levels of creativity that are proffered by additional group members (Hüffmeier, Zerres, Freund, Backhaus, Trotschel et al., 2019). However, there is also a pervasive view that, amongst high-performing teams, the outcomes are greater than the sum of their parts. This suggests that the outcomes are not simply an additive function, but that involvement in a high-performing team transforms the behaviour and capabilities of the team, a notion referred to as 'synergy' (Hackman, 1993; Ratcheva, 2008).

Synergy is a construct that is difficult to assess, but is evident in the way in which a team negotiates a problem and develops solutions. Communication tends to be more efficient, the behaviour of team members is more predictable, and team members tend to exhibit a higher level of joint situational awareness (Westli et al., 2010). Therefore, synergy is evident as much in the process as it is in the outcomes of team-related activity.

Synergistic processing suggests that there is potentially more associated with the team process than simply a collection of individuals. For example, amongst professional athletics teams, those that display greater synergy also show a greater dependence on teamwork and interpersonal relationships and can succeed in the absence of the best players (Arashin, 2010; Jones, 1974). This capability was highlighted by the performance of the Boston Celtics basketball team who, from 1957 to 1969, won the United States National Basketball Association 11 times, with no single player amongst the highest three scorers in the Association (Cohen, 1982)

13.2 GROUP DEVELOPMENT VERSUS TEAM DEVELOPMENT

A team is a group that displays unique characteristics. As a special 'group', a team will still display group characteristics (structure, norms, roles, interaction) and be influenced by group processes. However, the quality of the performance of a team is of critical importance. The notion that there is a difference between a group and team is governed by the premise that 'a collection of capable individuals does not always produce a capable group. Mature adults often form an immature working team' (Lippitt, 1969, p. 101).

Over time, individuals develop an understanding of the nature and capabilities of members of a group, eventually enabling the transformation into a team (Drach-Zahavy & Somech, 2001; Tyson, 1989). With further interaction, a team will develop what might be described as 'spirit', or what Forsyth (1983) described as 'unity, cohesiveness, comraderie, esprit de corps' (p. 449). Organisations, including the military, develop tools and strategies to foster the development of this spirit, including shared experiences and symbols of group identity (Boermans, Kamphuis, Delahaij, Ven den Berg, & Euwema, 2014).

Fostering the development of team spirit is presumed to engender a prioritisation of the common or team goal over any individual outcome. It is also presumed to encourage sustained commitment and cooperation in response to adversity, and enhance creativity (Ratzmann, Pesch, Bouncken, & Climent, 2018). As a consequence, commercial organisations invest considerably in team-building activities, often outside the workplace, that are intended to foster team spirit.

A meta-analysis targeting the outcomes of team-building exercises in organisations suggests that these types of intervention can have an impact on team performance depending on the features of teams that are targeted. For example, team-building exercises that target the clarification of roles in teams tend to be associated with improvements in objective metrics of performance. However, the effects are less clear for team-building interventions that target interpersonal relationships (Salas, Rozell, Mullen, & Driskell, 1999). The evidence suggests that team-building interventions are positively associated with subjective perceptions of team behaviour and interactions that might serve as a precursor to objective improvements in team performance (Klein, DiazGranados, Sala, Le, Burke, Lyons, & Goodwin, 2009).

13.3 PHASES OF TEAM DEVELOPMENT

While team-building interventions can foster productive and efficient teams, a period of time and opportunities for interaction are also necessary to acquire, consolidate, and practise the skills necessary for team performance (Lippitt, 1969). This process of interaction appears to progress through a series of phases that will eventually enable the transition from a collection of individuals to a team. Tuckman (1965) conceptualised these phases as:

- Forming
- Storming
- Norming
- Performing

In the Forming stage of team development, there is both an orientation to the task for which the team is being formed, and an understanding of any dependencies on which performance is likely to depend. This might include the activities of other groups or teams, or it may involve a requirement for additional support in the form of further skill-sets or technical support.

Importantly, the precursor to the Forming stage of team development is a recognition that each member of the group brings individual capabilities and skills, a degree of motivation to engage in group-based activities, and team skills that are both generic and task-oriented. These individual differences in capability and motivation are an important determinant of the subsequent phases of team development.

In addition to team development, individual differences in team capabilities and skills also impact performance once the team is established. Encapsulated in the Input-Process-Output (IPO) model of team performance, 'inputs' constitute the composition, organisational support, and resources that impact 'outputs' through the 'processes' that the team engages during problem-solving (Cohen & Bailey, 1997). These processes include communication, coordination, and role and goal clarity associated with the task.

Once formed, Tuckman (1965) proposes that groups progress through a 'storming' phase during which a degree of group conflict occurs as members of the group clarify and consolidate their roles. This conflict reflects what Silva, Cunha, Clegg, Neves, Rego, and Rodriguez (2014) describe as a paradox of selfless egoism that characterises the clash between individualism and the establishment of a common goal. It leads to the establishment of norms of behaviour that facilitate group cohesion and allow information to be communicated efficiently and accurately, thereby contributing to joint situational awareness.

Finally, having resolved conflicts, established modes of communication, and developed an understanding of the capabilities of individual team members, the team is in a position to engage with the problem. Further efficiencies may emerge as part of this process, including an increased capability to account for changes in task demands.

Successful teams will develop the capacity to adapt and change their strategies, depending on the nature of the problem being encountered (Salas, Cooke & Rosen, 2008). For example, successful sports teams will adapt their pattern of play in response to different opponents. Even within the context of a game, changes will emerge that reflect the ebb and flow of the activities. The successful team is one that quickly identifies the need to change strategies and efficiently draws on additional capabilities to implement an appropriate response. Importantly, these capabilities may not have been at the forefront initially, emerging only as the potential threat became evident.

Extending Tuckman's (1965) four-phase model of team development, Morgan, Glickerman, Wooward, Blaiwes, and Salas (1986) developed the Team Evolution and Maturation (TEAM) model to better identify the phase at which a group had progressed in the emergence of a team. This is particularly useful in the context of team training, since it allows the introduction of additional resources that might facilitate the transition through the phases (Morgan, Salas, & Glickman, 1993; Salas, Morgan & Glickman, 1987). For example, norming might be encouraged through team-building exercises administered once the process of storming has concluded.

The TEAM model has two activity tracks that are completed prior to successful performance as a team (see Figure 13.1). The 'operational team skills training' track includes task-orientated skills that members of a team must possess to complete a task. In addition, these skills include members acquiring information, clarifying goals, and understanding task requirements. The 'generic team skills training' track is the developmental phase through which a collection of individuals progress before they display team-oriented behaviours. Specific behaviours can be identified in the different phases of team development, including cooperation, coordination, adaptability, communication, and giving and receiving feedback.

Phases similar to the TEAM model have been suggested by other researchers, including Woodcock (1989) (see Table 13.1). In the different approaches, groups are expected to progress through several phases of development before they eventually emerge as a team. These stages are broadly categorised as an 'aggregate' or simply a collection of individuals, a 'group', where some interpersonal strategies have emerged, finally, a 'team'. Among the proponents of a phased-based approach to team development, there is a general expectation that different groups will progress systematically through the phases of team development, that progression will occur at different rates depending on the nature of the team, and that not all groups will transition to team-based performance.

Consistent with a three-phase approach to team development, Buchholz and Roth (1987) suggest that there is a lack of common purpose and shared responsibility when a group of individuals is asked initially to form a team (see Figure 13.2). As interaction continues, members start to develop a group identity, clarify their purpose, and establish norms for working together. This is the phase at which they begin behaving as a group with a common purpose and a sense of shared responsibility. As the members of a group share both the responsibilities and the rewards of the group, and members commit to the group's purpose, the group transitions into a team.

Buchholz and Roth (1987) examined a diversity of team applications and interviewed managers and workers to determine their peak experiences in working as a team. As a result, they were able to identify specific attributes that were distinctive of high-performing teams. These comprised:

- Participative leadership
- Shared responsibility
- Alignment of purpose
- High communication
- Future focused
- Focused on task
- The use of creative talent and rapid responses

For each attribute of a high-performance team, Buchholz and Roth (1987) identified different characteristic behaviours within each phase of team development. For example, a team that is aligned to its purpose is characterised by all the members being personally committed to fulfilling their task, and making reference to their purpose when making decisions. However, a collection of individuals is characterised by members being unclear about the group's purpose or task.

Groups and Teams

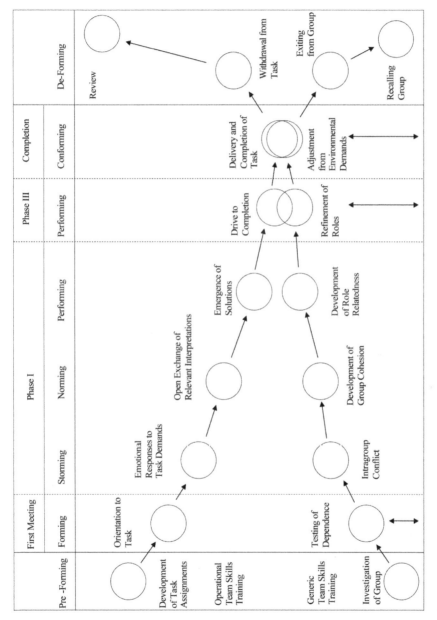

FIGURE 13.1 The Team Evolution and Maturation Model. (Adapted from Morgan et al., 1986.)

TABLE 13.1
Characteristics of an Aggregate, Group, and Team

Author	Aggregate	Group	Team
Tyson (1989)	Low interaction Lack of social structure Unlikely to meet again in the same combination	Members interact Members have an awareness of group identity Members understand the values, roles, & norms of the group Members have a common task Members have established communication patterns Group has clear goals	Characterised by effective work procedures and high productivity Cohesion Satisfaction among members Mature role structure
Woodcock (1989)	Unclear objectives No group involvement in decision-making Display poor listening skills Weaknesses cover up No 'rocking the boat'	Wider options considered Concern for others Personal feelings raised Greater listening Leader makes most decisions Agreed procedures	Risky issues debated Methodical working High flexibility Appropriate leadership Maximum use of energy and ability of members Essential principles and social aspects considered Needs of members met
Buchholz & Roth (1987)	Individual-centred Individual goals rather than group goals Members do not share responsibility Avoid change Do not deal with conflict	Leader-centred Members develop a group identity Define roles Clarify group's purpose Establish norms for working together Leader provides direction, delegates tasks	Purpose-centred Respond rapidly to opportunities and events Share responsibilities and rewards Members understand and are committed to their purpose

Groups and Teams

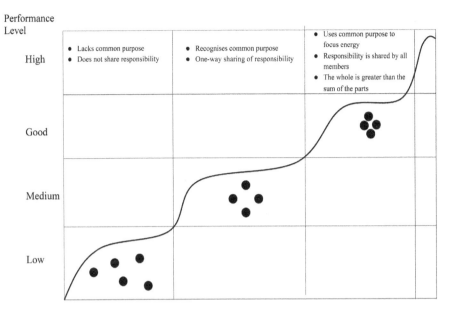

FIGURE 13.2 A conceptual diagram of team development and performance. (Adapted from Buchholz and Roth, 1987.)

13.4 FACILITATING GROUP PERFORMANCE

Working collaboratively as part of a team requires a range of skills and capabilities, including leadership, followership, communication, and coordination. Exercising these skills and capabilities depends on the nature of task, the nature of the team, and the context within which the team is functioning. As a consequence, it can be very difficult to ascribe the ideal characteristics of a team that will enable it to function efficiently and effectively in a specific situation.

There are individual differences in both the inclination to participate as part of a team and the skills and capabilities necessary to function as a team member. Some of these differences can be managed through repeated interaction where one member of the team might engage in activities that support the capability of another team member. For example, in software development, the coordination of expertise is critical for the successful and efficient management of problems, and working consistently to support another team member will provide an understanding of both the capabilities and the limitations of members of the team (Faraj & Sproull, 2000).

Where some teams are well-established and function together over an extended period (e.g. sporting teams), other teams are ad-hoc, and are only brought together on occasion to participate in specific activities. For example, medical practitioners might constitute ad-hoc teams in an emergency department, while pilots will form ad-hoc teams on the flightdeck of aircraft. Irrespective of the context, there is a need to support the communication and coordination strategies engaged by members of ad-hoc teams to enable and sustain team performance.

For ad-hoc teams in particular, Standard Operating Procedures, including standardised communication strategies, provide a reference point for communication and coordination. Standard Operating Procedures (SOPs) comprise strategies or procedures that are applied consistently in response to a system event(s) (Kontogiannis, Leva, & Balfe, 2017). For example, in response to a patient with chest pains, medical practitioners will request, as an SOP, the results of an electrocardiogram (ECG). Similarly, pilots aboard marine vessels will, as an SOP, assume command of the vessel from the captain. In these contexts, the SOPs are broadly accepted and this facilitates communication and coordination. It constitutes a 'process' in the IPO model of team performance obviating the requirement for detailed instructions or an explanation as a precursor to a behaviour or request.

In addition to SOPs, the behaviour of ad-hoc teams is likely to be supported by other processes, including standard language or phraseology. For example, in highly technical environments, terminology might be employed that relates to a particular device or procedure, the meaning of which is recognised by operators within a context. In the military context, the term 'sitrep' is used to denote a *situation report*. The abbreviation is used to both improve the efficiency of communication and avoid miscommunication. Importantly, it is a term that will be recognised by military personnel, irrespective of their position or unit.

The application of SOPs is based on the assumption that operators within a context possess the technical skills necessary to contribute successfully to a team. Thereafter, the management of a team is assumed to be supported by non-technical skills, including communication and coordination (Gordon, Fell, Box, Farrell, & Stewart, 2017). In combination, these non-technical skills, particularly in the context of team and group settings, has come to be referred to as *Resource Management* (see Chapter 16 for a comprehensive review)

The interest in resource management in the context of teams emerged largely through system failures where it was clear that resources were available to assist the operators, but for a lack of non-technical skills (Bacon, McCoy, & Henshaw, 2020). Poor leadership, a lack of communication and coordination, and/or the inability to develop joint situational awareness have all contributed to a number of serious accidents, despite the skills of the individual operators involved (Havinga, De Boer, Rae, & Dekker, 2017).

The typical response to what might be regarded as poor resource management has been the development of workshops, much like team-building workshops. Ideally, the development of a resource management workshop begins with the establishment of a committee that canvasses a number of issues including:

- The level of support from management;
- The culture of the organisation; and
- The current attitudes within the organisation towards resource management.

The steering committee should also be tasked with the assessment of the current state of resource management within a team and/or the organisation more broadly. This assessment has important implications for the development of resource management workshops, since some strategies may be more appropriate than others given the pre-existing attitudes and behaviours within the organisation (Gross, Rusin, Kiesewetter, Zottmann, Fischer, Prückner, & Zech, 2019). Among the difficulties that are likely to confront the steering committee include the wholesale rejection of the initiative by 'management' or 'operators' or both. Therefore, the steering committee plays a crucial role in 'marketing' the intervention at all levels within the organisation.

A critical aspect of the process of training course development involves the identification of the characteristics of the population for whom the course is designed. For example, a resource management training course that might be intended for flight attendants will likely differ from a course that is designed for maintenance engineers. This enables the initiative to be tailored to the specific needs of the participants, taking into account the gaps that need to be addressed.

Where necessary, the needs of the participants can be assessed through a needs analysis or task analysis that identifies gaps in performance (Truta, Boeriu, Copotoiu, Petrisor, Turucz, Vatau, & Lazarovici, 2018). These analyses might be initiated where a team fails to progress as expected through the various phases of team development. Interviews with members of the team can establish the type and level of organisational support required, including the need for training interventions. An important principle is to ensure that, as much as is possible, the interventions meet the needs of prospective participants in a training initiative.

Although the ultimate goal of resource management training is behavioural change reflected through improvements in team performance, it is unlikely that behaviour change will be realised in the short term unless significant resources are invested to ensure the acquisition and application of skills. At the initial stages, resource management initiatives are often designed to create an awareness of the issues. This provides the foundation for skill-based initiatives.

Consistent with training initiatives more broadly, the learning objectives for resource management initiatives should always be designed to meet the needs of potential participants. They also need to be subjected to evaluation and be achievable within an appropriate time frame considering the capabilities of participants.

Although the precise topics for resource management training initiatives will differ depending upon the needs and capabilities of participants, some of the issues considered may include:

- Communication
- Leadership
- Decision-making
- Team Management
- Stress Management
- Information Processing
- Situational Awareness
- Personality

In addition to knowledge, resource management training initiatives offer the opportunity to consider the mechanics of team processes. Therefore, all of the members of a prospective team need to be present so that the dynamics of actual team processes can be reproduced. In participating in a training initiative,

> it may soon appear ... that there are serious gaps in the skills possessed by members of the team, or in the roles that they are allotted. Not infrequently, teams are made up of people whose skills and backgrounds are altogether too similar.
>
> (McCowen & McCowen, 1989, p. 108)

Through practice and feedback, members of a prospective team have the opportunity to acquire skills including active listening, interpersonal communication, and coordination. However, to ensure the transfer of these skills to the operational context, there is a need to ensure that the environment within which the resource management skills are acquired and applied closely matches the operational environment.

In medicine and transportation, this realism is often provided through simulation. For airline operations, Line-Oriented Simulation (LOS) represents one of the most useful mechanisms to develop resource management skills. It creates a highly realistic context within which members of a team can practise their resource management skills as members of a team. It has the capacity to mimic both the environmental conditions under which responses might be required and the technical equipment that might be available. Feedback can be provided quickly and accurately with the opportunity to practise skills until a level of competence is acquired.

Despite the attraction of LOS, there are a number of pragmatic difficulties associated with its application, including the cost of trainers, access to simulation, and the additional training time that might be required for participants. While these costs can be ameliorated through the use of part-task training devices (Holzman, Cooper, Gaba, Philip, Small, & Feinstem, 1995), they do not necessarily provide the level of fidelity necessary for the transfer of training outcomes.

In addition to access to high-fidelity simulation, the quality of the feedback provided depends on the skills and capabilities of facilitators. One of the challenges associated with the application of resource management skills is that they need to be inferred through behavioural markers. In the aviation context, Smith and Hanebuth (1995) suggest that some of the behavioural markers associated with the effective application of resource management include:

- Briefing crew appropriately
- Adapting to differences between crew members
- Coping effectively with a stressful situation
- Distributing tasks to maximise the efficiency of the operation
- Undertaking a self-critique and
- Demonstrating an appropriate level of assertion

Establishing consistent responses amongst assessors can be challenging and may be moderated through a process of calibration. However, since feedback is provided on the basis of these assessments, it constitutes a particularly important issue for the development of high-performance teams.

13.5 TEAM PERFORMANCE

Effective team performance is typically the outcome of a dynamic process that includes features of the team, and features of the context within which the team is functioning. These can be broadly conceptualised as factors that enhance or hinder team effectiveness and the dependencies on which these are based (see Table 13.2). Salas, Dickinson, Converse, & Tannenbaum (1992) summarise the major factors that affect the performance of a team as task and work characteristics, individual, team, and environmental factors.

The Integrated Model of Team Performance and Training (Tannenbaum, Beard, & Salas, 1992) integrates Tuckman (1965) and Morgan et al.'s (1986) perspectives and emphasises the organisational context of a team as a factor that either facilitates or hinders the team process and, eventually, team performance (see Figure 13.3). The model also suggests that team performance is a dynamic process, resulting from the interchange of information and resources between members. Finally, the model conceptualises team performance and training as changing continuously, with the input of feedback and additional training introduced into the loop.

TABLE 13.2
A Summary of Team Performance Models

Author and Model	Factors That Enhance or Hinder Team Effectiveness	Team Effectiveness Variables
Hackman (1983) Normative Model	Capability of team members to work together over time Satisfaction of team members Acceptability of task outcomes by those who demand or receive them Resources allocated to the team	Level of effort exerted by team members Amount of knowledge & skill members apply to task Appropriateness of task performance strategies
Gladstein (1984) Task Group Effectiveness Model	Open communication, supportiveness, active leadership, experience and training Group structure (clear roles and goals, and specific work norms)	Group process (open communication, discussion of strategies, weighing individual inputs) Group task demands (task complexity, environmental uncertainty, level of team interdependence) Group composition (group size, structure, resources available, organisational structure) which also affect group process

Source: Adapted from Salas et al. (1992, pp. 5–15).

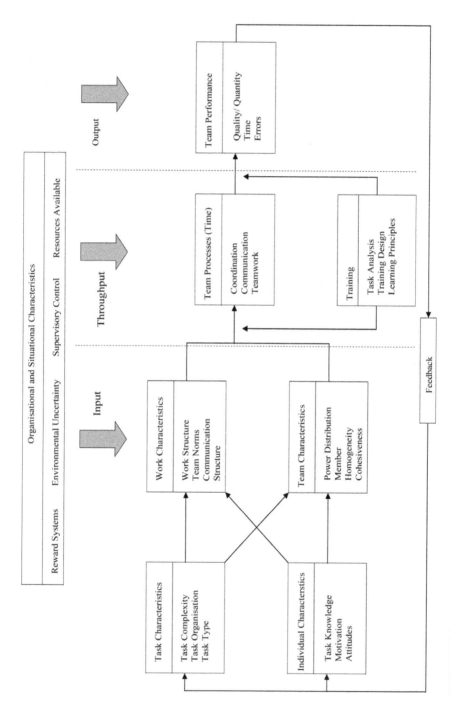

FIGURE 13.3 Integrated model of team performance and training. (Adapted from Salas et al., 1992.)

13.6 CRITICAL TEAM BEHAVIOURS

Morgan et al. (1986) identified critical teamwork behaviours that related to the success or failure of a team in a naval environment. The identification of seven critical teamwork dimensions (giving suggestions or criticisms, cooperation, communication, team spirit and morale, adaptability, coordination, and acceptance of suggestions or criticism) and examples of behaviours within each dimension led to the development of the Critical Team Behaviours Form (CTBF).

The CTBF has been validated in related contexts (Glickerman, Zimmer, Montero, Guerette, Campbell, et al., 1987; Oser, McCallum, Salas, & Morgan, 1989), and the outcomes allowed the identification of similar dimensions of teamwork in other team environments (Helmreich, Wilhelm, Gregorich, & Chidester, 1990). The behavioural data observed from 13 naval tactical teams suggested that the members of successful teams:

- Provide and ask for assistance when necessary;
- Regularly monitor the team's performance; and
- Encourage team spirit between team members.

These data provided an insight into the behavioural differences observed for these teams and suggest several behaviours that discriminate more effective from less effective teams in applied contexts.

14 Human Factors and Leadership

14.1 LEADERSHIP IN PRACTICE

Like other industrial settings, contemporary multi-crewed aircraft have developed to a point where coordination between crew members is a central component for the effective management of information and resources. However, inappropriate aircrew coordination has previously been implicated in approximately 68 per cent of accidents involving commercial aircraft (Shaud, 1989). A notable example occurred in Detroit in 1990 when a Boeing 727 collided with a Douglas DC-9 during heavy fog. Eight passengers died when the wing of the Boeing, under take-off power, collided with the main fuselage of the DC-9 (NTSB, 1991a). The subsequent investigation by the NTSB concluded that the primary cause of the accident was a lack of crew coordination that resulted in the DC-9 inadvertently taxiing onto an active runway.

In commenting on the lack of crew coordination, the inquiry into the collision between the Boeing 727 and the DC-9 specifically noted that, during the events immediately preceding the accident, the captain had 'tacitly relinquished his command role of the aircraft' (NTSB, 1991a, p. 35). It was also noted that the first officer had 'failed to follow repeated instructions from the captain' (NTSB, 1991a, p. 35). Implicit amongst these findings was the proposition that the captain's lack of appropriate leadership resulted in the breakdown of communication and coordination which ultimately led to the collision.

The extent to which this type of accident characterises the problem of leadership more broadly is difficult to determine. However, in an analysis of 50 accidents between 2007 and 2017 involving Controlled Flight into Terrain (CFIT), Kelly and Efthymiou (2019) recorded leadership as a factor in at least 56 per cent of occurrences. Beyond aviation, poor leadership has been attributed to a number of significant failures including in finance (Falk & Blaylock, 2012), energy production (Broadribb, 2015), and politics (O'Brien, 2010).

Although there are a range of definitions of the term *leadership*, it tends to involve a process of social influence whereby a person directs or facilitates members of a group towards a common goal (Chun, Cho, & Sosik, 2016). Consistent with this perspective, Etzioni (1965) suggests that the principal component of leadership, as distinct from power or authority, is the capacity to facilitate a voluntary change in the preferences and behaviour of others. Consequently, there is a need to clearly differentiate leadership behaviour in which the power and authority is vested in the office, from leadership as purely an influencing or motivating stimulus that is independent of power and authority.

Initial approaches to the study of organisational leadership have tended to characterise effective leaders on the basis of personality traits such as introversion-extroversion, dominance, personal adjustment, and/or emotional control (Bryman, 1986). For example, Chidester (1990) conducted a series of experiments with airline flight crews and noted that positive, task-oriented captains led crews that made fewer errors and were rated by a recently retired, experienced airline captain as consistently more effective in coordinating their activities. By contrast, captains with lower motivation led crews that recorded a greater frequency of errors and tended to be rated as less effective in coordinating their activities.

Although Chidester (1990) provided some support for the relationship between leadership traits and operational performance, it was not possible to infer a causal relationship. This follows an extensive review of trait-based approaches to leadership across a number of domains where consistent, causal relationships could not be established (Stodgill, 1974). It suggests, as Stodgill (1974) noted, that 'the qualities, characteristics and skills required in a leader are determined to a large extent by the demands of the situation in which he [she] has to function as a leader' (p. 65).

14.2 CONTINGENCY APPROACHES TO LEADERSHIP

The lack of evidence for a universally superior style of leadership has provided the impetus for contingency models of leadership where the effectiveness of groups is presumed to depend on the appropriate interaction between a leader and followers, and the extent to which an approach to leadership meets the demands of an environment or context (Bryman, 1986; Fiedler, 1978). For example, Fiedler (1978) developed a theory of leadership where group performance is purportedly contingent on the relationship between the leader's motivational structure and his/her degree of situational control and influence. More specifically, the theory is based on the assumption that those leaders who are principally motivated by the task requirements are more effective in situations where their capacity for control is high or relatively low. In contrast, leaders whose motivation is directed towards the development of positive group relations tend to be more effective in situations of moderate control and influence (Fiedler, 1978; Foushee & Helmreich, 1988).

The fundamental assumption associated with this model of leadership is that a leader's profile can be defined according to his/her motivation. Moreover, Fiedler (1978) maintains that this structure can be inferred through the summed scores on an 18-point, bipolar adjective scale where individuals are asked to describe their Least Preferred Co-worker (LPC). The resultant score is considered a reflection of the leader's attitude towards a comparably poor co-worker. Relatively low LPC scores (i.e. below 57) are indicative of a leader who considers task completion of primary importance and fails to note any positive attributes in poor co-workers. Fiedler (1967) identified these individuals as *Task Motivated* leaders. In contrast, *Relationship Motivated* leaders describe their LPC in more favourable terms, and are therefore,

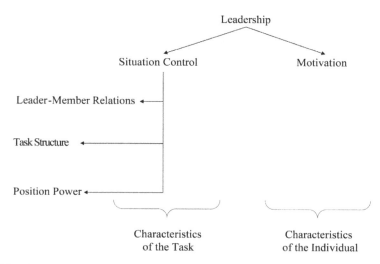

FIGURE 14.1 A diagram of the components of Fiedler's (1978) contingency model of leadership.

less concerned with task completion and more interested in the positive qualities of the individual worker.

The second component that characterises Fiedler's (1978) leadership model concerns the degree to which the situation provides the leader with control and influence (see Figure 14.1). A quantitative indication of situational control can be derived from the combined scores on the *Leader–Member Relations, Task Structure,* and *Position Power* subscales (Fiedler, 1978). The *Leader–Member Relations* scale requires leaders to evaluate the relationship between themselves and members of their own and subordinate groups. This is intended to provide an indication of the extent to which the leader considers that he/she has the support and loyalty of group members. Fiedler (1978) regards leader–member relations as the most important of the three situational dimensions, since a leader has greater certainty that the group members will comply with his/her decisions.

The *Task Structure* subscale examines the level of situational control and influence that is inherent in the nature of a task. Leaders whose operating instructions are explicitly defined by an organisational structure are usually assured of compliance from subordinates. By contrast, leadership control and influence are considerably diminished in organisations where tasks are ill-structured.

The final dimension of situation control is *Position Power*, and this subscale examines the extent to which a leader can enforce the compliance of his/her subordinates through reward or punishment. Fiedler (1978) recommends that the leader's superiors evaluate both the position power and task structure dimensions of situation control to avoid the inherent distortion that is prevalent in self-assessment.

Situation control is typically represented by an eight-point continuum on which high and low control situations occupy opposite ends. The position of any task can be determined on the basis of whether its scores fall above or below the median

TABLE 14.1
Leader Effectiveness Table Based on Fiedler's Contingency Model of Leadership

Favourableness	High							Low
Leader–Member Relations	Good				Poor			
Task Structure	Structured		Unstructured		Structured		Unstructured	
Position Power	Strong	Weak	Strong	Weak	Strong	Weak	Strong	Weak
	I	II	III	IV	V	VI	VII	VIII
Most Effective Leader Orientation	Task				Relationship			Task

on each of the three subscales (Fiedler, 1971, 1978). For example, individuals who score above the median on all three dimensions supposedly operate in high control situations, whereas those who consistently score below the mean are considered to function in low control situations. Intermediate groups are defined by various combinations of each. The eight groups are presented in detail in Table 14.1.

Situations that provide a leader with a relatively high level of control and influence are those where there is a high probability that his/her decisions and actions will have predictable outcomes. In these environments, low-LPC leaders are perceived as inconsiderate and more concerned with the task than with group members. This is largely due to the fact that low-LPC individuals provide strong leadership in situations where no more than minimal guidance is required. In contrast, high-LPC leaders provide minimal supervision, and are considered pleasant and relaxed in group interaction. The net result is a relatively greater collective performance.

Situations of moderate control are characterised by tasks that involve the expression of group members' opinions and attitudes. Under these conditions, high-LPC leaders are oriented towards solving the interpersonal problems of the group and thereby increasing productivity. According to Fiedler (1978), these leaders are rated as amiable and considerate and achieve a greater collective performance relative to low-LPC leaders. During situations of moderate control, low-LPC leaders tend to become more task-oriented and are likely to face an uncooperative working relationship within the team.

Finally, in situations of low control, high-LPC leaders are supposedly less effective than low-LPC leaders. This distinction becomes particularly apparent during stressful tasks and is due largely to the high-LPC leader's anxiety and subsequent withdrawal from the leadership role (Fiedler, 1978). By contrast, low-LPC leaders become extremely committed to task achievement, regardless of the consequences to member relations. In this case, collective performance is maintained through the perception that members will be punished for poor individual performance.

The validity of Fiedler's (1971, 1978) model of leadership has been the topic of considerable debate and has yet to be resolved (Oc, 2018). Nevertheless, an extensive validation study conducted by Chemers and Skrzypek (1972) has provided some support for the model in a military environment. High and low-LPC leaders were selected from a pool of United States military cadets whose scores fell at least one standard deviation above and below the median. The leader–member relations and position power dimensions were defined prior to the experiment in which groups were asked to perform one structured and one unstructured task. The experimental manipulations were such that each of the eight cells of situation control were equally represented.

According to Chemers and Skrzypek (1972), the results revealed a significant interaction between LPC score, leader acceptance, and task structure. This accounted for 28 per cent of the overall group performance variability. Therefore, the results were broadly consistent with Fiedler's (1967) assertion concerning the relationship between collective performance, LPC score, and situation control. Subsequent studies have provided further support for the universality of Fiedler's model in studies of school children (Hardy, 1971) and corporate managers (Blake & Mouton, 1978).

Despite its complexity, Fielder's (1978) conceptualisation of leadership is useful since it reflects a contingency model that distinguishes what might be regarded as 'relationship-oriented' behaviour from 'task-oriented' behaviour in the context of leadership. This is an approach adopted by Blake and Moulton (1978) who derived styles of leadership that could be could be broadly considered across two dimensions, one of which involves a concern for the welfare of subordinates, while the other involves a concern for task-oriented outcomes.

While there is an assumption that styles of leadership tend to emerge and remain stable over time, the implication arising from Blake and Moulton's (1982) Leadership Grid is that it should correspond to the demands of task and the characteristics and capabilities of subordinates. This is an approach consistent with the Situational Leadership Theory (SLT) developed by Hersey & Blanchard (1969) whereby the success of a particular leadership style in enabling performance depends on the capabilities and motivation of subordinates, referred to as *readiness*. As readiness increases, there is an assumption that effective leadership requires a relatively greater emphasis on the management of relationships (relationship-oriented) and less emphasis on structuring tasks (task-oriented) (Cairns, Hollenback, Preziosi, & Snow, 1998).

While the broad distinction between relationship-orientation and task-orientation has intuitive appeal in the context of leadership, the impact on subordinate performance has been difficult to establish empirically. However, capitalising on the broad distinction between relationship- and task-orientation, Bass and Avolio (1993) developed Transformational Leadership Theory, in which they distinguished transformational from transactional approaches to leadership. Transformational leadership tends to focus on the needs and capabilities of individual subordinates, engendering both a motivation to contribute and a challenge to pursue excellence. By contrast, transactional leadership involves the application of rewards that are contingent on performance and the management of deviance from the accepted norm. The absence of leadership is a style referred to as laissez-faire, and is characterised by a lack of engagement with subordinates (Bass, 1998, 1999).

In a comprehensive meta-analysis intended to investigate the validity of Transformational Leadership Theory, Judge and Piccolo (2004) noted that transformational leadership was positively associated with follower satisfaction, follower motivation, leader job performance, and leader effectiveness. However, similar relationships were evident for contingent reward, a component of transactional leadership. As expected, the laissez-faire approach to leadership was negatively associated with follower satisfaction and leader effectiveness.

Consistent with the outcomes of Judge and Piccolo (2004), Bass (1999) argues that transformational and transactional leadership are not mutually exclusive. Rather, features of both approaches can be employed as a strategy that impresses an appropriate organisational culture for a particular task or operational environment. Evidence to support Bass (1999) can be drawn from Clarke (2013) who noted that, in the context of safety behaviours, transformational leadership tends to be associated with increases in engagement with safety initiatives amongst subordinates, while transactional leadership tends to engender compliance with standards, rules, and regulations.

14.3 INEFFECTIVE LEADERSHIP

One of the clearest examples of laissez-faire leadership involved the crash of a Boeing B-52 bomber, Czar 52, in 1994. The aircraft was piloted by a senior officer who had been authorised to practise a series of manoeuvres in preparation for an air show (United States Air Force, 1995). On return to the airfield, the aircraft was instructed to conduct a go-around due to another aircraft located on the runway. The B-52 subsequently entered a tight turn at approximately 250ft above ground level (United States Air Force, 1995). However, as the aircraft passed through 90° bank, it stalled, descended rapidly, and collided with terrain.

The subsequent investigation into this accident revealed a catalogue of failures on the part of the pilot's senior officers to recognise and respond to the erratic and often dangerous behaviour displayed by the captain of Czar 52 during the period between 1991 and 1994. For example, one senior officer remarked of the captain of Czar 52 that he was 'as good an aviator as I have ever seen' (Kern, 1995, p. 9). However, according to Kern (1995), junior officers commented that 'I'm not going to fly with him, I think he's dangerous. He's going to kill someone someday and it's not going to be me' (p. 10).

Kern (1995) suggests that the differences evident between the senior and junior officers' perceptions of the captain of Czar 52 reflect two main failures amongst the management of Fairchild Air Force Base. In the first instance, the management tacitly supported an organisational climate where blatant violations of regulations were committed on a regular basis. In the second instance, the leadership failed to respond adequately when specific cases of violations had occurred.

On at least seven occasions prior to the accident, the captain of Czar 52 was observed violating standard operating procedures, and exceeding the normal operating procedures associated with the operation of the aircraft. More importantly, however, his behaviour was reflected in at least two instances where junior officers attempted to 'copy' his manoeuvres (Kern, 1995). Consequently, the impact of inappropriate leadership behaviour has an impact that reaches beyond a single sequence of events.

The behaviour of the captain of Czar 52 and the apparent admiration for his abilities amongst senior officers at Fairchild Air Force Base raises an important question concerning the organisational culture that was being encouraged at the time. While the captain of Czar 52 was recognised as possessing exceptional piloting abilities, the laissez-faire approach to leadership adopted by his senior leadership permitted a culture to emerge that conflicted with the goals and aspirations of the organisation.

14.4 LEADERSHIP AND SAFETY CULTURE

As the case of Czar 52 illustrates, approaches to leadership have a fundamental impact on the culture within an organisation which, in turn, has an impact on behaviour. This relationship is well-established but can be difficult to manage in practice due to the difficulty in both recognising and then correcting inappropriate cultural practices.

As a general construct, *culture* is an historically perpetuated system of shared norms, attitudes, rituals, symbols, and behaviours that define a person as being part of a group (Merritt & Helmreich, 1996). The identity that emerges as a part of a culture is associated with particular attitudes and behaviours. Over time, these attitudes and behaviours are perpetuated and reinforced, and become highly resistant to change.

To be considered a cultural group, members must define themselves as a unit (Andersen, Nørdam, Joensson, Kines, & Nielsen, 2018). Units may be based on nationality, ethnicity, sex, religion, language, organisation, geography, profession, and/or occupation. Even within an organisation, multiple units will exist, each with their own sets of rituals and symbols. Individuals may belong simultaneously to a range of cultural groupings and subgroupings with the relative strength of association changing, depending on the context.

Individuals who are part of a cultural group share systems of norms or acceptable ways of behaving, thinking, and communicating (Petitta, Probst, & Barbaranelli, 2017). These cultural norms are communicated to prospective new members through activities and rituals, where values and expectations are established through observational learning (Tear, Reader, Shorrock, & Kirwan, 2020). For example, a new employee might attend an induction that includes a summary of the history of the organisation, its mission, policies, and practices. This process of enculturation may also include cultural symbols including uniforms, badges, logos, and mottos.

From an organisational perspective, the goal of enculturation is to ensure that appropriate corporate values and philosophies are imparted (O'Kelley, 2019). However, their adoption in practice must be reinforced in the workplace through the behaviour of both colleagues and leaders. Behaviour from immediate colleagues or leaders that conflict with corporate values may impede or even prevent the adoption of corporate values (Petitta, Probst, Barbaranelli, & Ghezzi, 2017). Rather, new employees will tend to identify and thereby adopt the attitudes and values of proximal leaders and colleagues, even though they may conflict with corporate values.

Due to the disparities between different groups within an organisation, maintaining a universal culture can be difficult. Leaders at all levels of the organisation must reinforce a singular perspective, taking careful note where behaviour appears to be at odds with corporate values. Adherence to safe practices, including honesty and respectfulness, is a corporate value that most contemporary organisations will espouse. However, like other values, it needs to be reinforced through the attitudes and behaviour of all staff as a reflection of the organisational culture (Eljiz, Greenfield, Derrett, & Radmore, 2019).

Changes in organisational culture can occur over an extended period and almost imperceptibly unless there is a concerted effort to acquire data that are symptomatic of change (Alsalem, Bowie, & Morrison, 2018). For example, an increase in violations of standard practices and /or incidents might be indicative of an underlying change in organisational culture. Similarly, attitudes to safety can be assessed through measures of organisational climate.

Unlike organisational culture, organisational climate tends to be a more volatile construct and constitutes a litmus test of perceptions or attitudes (Casey, Griffin, Flatau Harrison, & Neal, 2017). Normally assessed using a Likert scale, responses might reflect more systemic issues that impact safety behaviour, including a lack of training, a lack of suitable tools, and/or a failure to address previous issues that have been reported (Alruqi, Hallowell, & Techera, 2018).

The role of leadership in fostering and maintaining a safety-oriented organisational culture lies in ensuring that appropriate strategies are in place to ensure that the appropriate values and expectations are communicated to new staff, and that appropriate comparative data are collected on a regular basis to ensure that any changes in organisational climate that might reflect changes in values and an adherence to a desired corporate culture are identified at the earliest opportunity (Lundell & Marcham, 2018). This enables the implementation of remedial strategies and, importantly, positions leaders as the champions of a culture that promotes an adherence to safety practices.

14.5 JUST CULTURE AND SAFETY LEADERSHIP

Allied to a culture of safety is an organisational environment where there is an acknowledgement of both the inevitably of error and a non-punitive organisational response in the event that errors are reported. Referred to as a 'Just Culture', it imbues a willingness to report events that might otherwise reflect poorly on the performance of an employee (Dekker & Breakey, 2016). It also adds to the weight of data that can be employed by organisational leaders as tools to identify areas of potential vulnerability that might emerge within the organisation.

14.6 PRINCIPLES OF EFFECTIVE HUMAN FACTORS LEADERSHIP

In general, it appears that supportive and open communication is one of the most important features associated with effective leadership, since it is the primary mechanism through which the process of influence takes place. However, from a pragmatic perspective, the process involves the application of a range of skills, including:

- Envisioning
- Modelling
- Influence
- Receptiveness
- Initiative and
- Adaptability.

Envisioning refers to the communication of a plan of action or idea as part of an operational process (Choi, 2006; Dunlap and Mangold, 1998). This promotes the development of an ethos or set of standards against which decisions and procedures can be evaluated in the absence of direct oversight by the leader. Ideally, this ethos should be supported through the behaviour of the leader and provide a model for subordinate crew members to emulate (Hu, Griffin, & Bertuleit, 2016). This influences the development of a positive organisational culture, which encourages and motivates adherence to accepted patterns of behaviour as part of group norms.

While modelling appropriate behaviour is an essential principle that underpins successful approaches to the management of safety, complementary strategies include engagement with employees on a regular basis and the communication of case examples that demonstrate the successful application of safety-related initiatives in practice. These activities serve to ensure that leaders are accessible to subordinates while conveying a level of trust in their skills and capabilities (Campbell, White, & Johnson, 2003). Arguably, this process establishes and maintains a rapport between leaders and subordinates that, in turn, encourages and reinforces a positive organisational culture.

Sustaining a level of rapport between leaders and subordinates requires that leaders remain receptive to suggestions and constructive criticism. While this receptiveness can be personally challenging, it reflects a degree of humility, and a willingness and capability amongst leaders to respond to the needs of subordinates (Walters & Diab, 2016). This assures subordinates that their views are valued, thereby encouraging further interaction, and this is especially useful in the context of safety-related information.

Responding to the different needs of subordinates demands a level of Emotional Intelligence (EI) since it requires active engagement with personal and interpersonal affective responses that then direct and guide subsequent interactions. Closely associated with transformational leadership, higher levels of EI amongst leaders tend to be associated with a positive safety climate, trust, and organisational commitment (Squires, Tourangeau, Spence Laschinger, & Doran, 2010). This provides an important foundation for the introduction of new, safety-related initiatives that depend on the engagement and commitment of employees.

Where safety-related initiatives are ineffective or impractical, a level of rapport between leaders and subordinates enables the timely communication of any concerns, and provides the opportunity to adapt an approach to improve organisational performance. Options to adapt an initiative might be drawn from subordinates or alternatively, a leader might draw on the experience of other organisations, the outcomes of empirical research, and/or best practice.

14.7 FOLLOWERSHIP

While contingency approaches to leadership acknowledge the relationship between leaders and followers, implicit in these models is an assumption that the quality of the relationship and therefore the commitment of followers, depends on the behaviour of the leader (Kelley, 2008). However, in the context of human factors initiatives, the skills and capabilities for effective followership are also important in ensuring that appropriate principles and practices are implemented purposefully. This proposition capitalises on the notion of followership as a role that demands active engagement in understanding the position of the leader in enacting the goals and priorities of the organisation (Baker, 2007). It also draws on the responsibility of followers to provide support for the leader by offering suggestions, and taking responsibility where appropriate (Follett, 1996).

Active followership has a broader impact on organisational performance by granting implicit authority to a leader (Hansen, 1987). The implication is that this perception can transcend immediate subordinates, thereby potentially extending the scope of influence to other teams or parts of the organisation. The advantage is that organisational change can be perpetuated in the absence of explicit engagement with a leader.

According to Kelley (2008), effective followership requires critical thinking and active participation in activities. This is especially important in the context of human factors initiatives where the introduction of a programme of organisational change can lead to unintended, negative consequences. From a system safety perspective, critical thinking and active participation on the part of followers will assist in identifying potential latent conditions, enabling changes or the timely development of defences.

In addition to its role in system safety, active engagement amongst followers has been associated with higher levels of job satisfaction and organisational commitment (Blanchard, Welbourne, Gilmore, & Bullock, 2009). Therefore, encouraging and enabling active engagement amongst followers is likely to have benefits for an organisation beyond the immediate implications for human factors. Levels of active engagement and critical thinking amongst followers can be assessed using the Kelley Followership Questionnaire (KFQ) (Kelley, 2008) or alternatively, a revised version of this questionnaire, the KFQ-Revised (Ligon, Stolz, Rowell, & Lewis, 2019)

14.8 PARTICIPATIVE LEADERSHIP

Participative leadership refers to a strategy where subordinates are involved in the decision-making process (Johns, 1988). The extent to which subordinates are involved tends to vary, depending on the nature of the situation. For example, at one extreme, a leader may seek the opinions of the subordinates prior to formulating a decision. However, at the other extreme, a leader may define limits of behaviour within which subordinates might function semi-autonomously (see Figure 14.2). It contrasts with directive leadership where subordinates have little or no influence on the decision-making process (Somech, 2005)

The concept of participative leadership is especially applicable in the context of human factors, since it encourages communication between leaders and subordinates without the loss of the authority of the leader. As Johns (1988) indicates, participative leadership is not an abdication of authority. Rather, it is a means of improving the group decision-making process, creating a level of motivation towards the group goal, and encouraging a level of acceptance for the decisions that are formulated (Huang, Iun, Liu, & Gong, 2010).

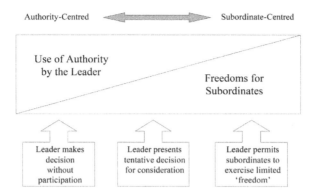

FIGURE 14.2 The relationship between authority-centred and subordinate centred forms of leadership. (Adapted from Johns, 1988.)

However, despite the advantages associated with participative leadership, it does require time to accomplish successfully, and some leaders may perceive the process as a loss of authority (Chen & Tjosvold, 2006; Somech, 2003). For subordinates, the relationship between participative leadership and task performance appears to be mediated by the level of trust in the leadership (Huang et al., 2010). Therefore, some subordinates may not be as receptive as others to a participative style.

15 Communication

15.1 THE NATURE OF COMMUNICATION

Communication is a process that is often taken for granted in organisational settings. However, it is one of the primary processes through which errors can occur. During their working day, most employees will spend a significant proportion of their time engaging in some form of communication. Generally, the types of communication will vary across four different modes (writing, speaking, reading, and listening). However, in the case of highly technical environments such as air traffic control and power system transmission, communication may also occur remotely through human–machine interfaces.

Communication is an interdependent process that involves the exchange of information (Cushing, 1994). Accurate and/or efficient communication can be a significant aid to people in technical roles, particularly in time-constrained and uncertain situations. Inaccurate and/or inefficient communication can seriously degrade performance and increase the frequency and severity of errors (Gibson, Megaw, Young, & Lowe, 2006).

Communication is important for a number of reasons, including:

- Coordination
- Articulating Goals
- Defining Roles
- Defining Group Norms
- Attention to a Task
- Motivation/Effort

- (in multi-role environments)
- (to establish joint goals)
- (leaders and followers)
- (procedures and practices)
- (situational awareness)
- (reinforcement)

The extent to which a message is clear and understood depends on a range of factors, including the characteristics of the sender (language, pronunciation, etc.), the characteristics of the receiver (hearing, etc.), and noise (Molesworth & Estival, 2015). Noise includes competing signals or signals that reduce the extent to which the message can be received (see Figure 15.1). Information is relayed via a channel and the sender may await feedback from the receiver to establish that information has been received.

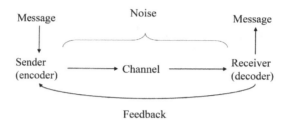

FIGURE 15.1 A diagram of the factors that influence the relationship.

15.2 COMMUNICATION ERRORS

Due to its reliance on radio communication and the detailed analysis of accidents and incidents, the aviation industry offers some useful examples of the types of communication errors that can occur in complex industrial environments, and the conditions under which they occur. For example, in an analysis of aviation incidents reported between 1988 and 1991, 36 per cent of occurrences involved communication errors (Prinzo, 1996). These included instances of a lack of communication, and miscommunication arising from the use of mitigating language and issues of expectancy (Foushee, 1982; Green, 1990).

An example of the lack of communication between flight crew can be drawn from the crash in 1987 of Flight 255, a Northwest Airlines McDonnell Douglas DC-9 (Super 82). The aircraft was departing Detroit Metropolitan Airport en route to Phoenix when the aircraft collided with a building and subsequently crashed onto a roadway approximately one mile from the threshold of the runway.

According to the National Transportation Safety Board (1988), the most significant factor involved in the crash was the failure of the flight crew to extend the flaps and leading-edge slats prior to take-off. The flaps and leading-edge slats extend the surface area of the wing, increasing the capacity of the aircraft to produce lift at a lower airspeed. It comprised an item on the pre-take-off checklist that, if not completed, would result in a significantly lower angle of climb than would normally be the case (see Figure 15.2). In the case of Flight 255, the inability to climb at the normal angle resulted in the collision with the building, following which the aircraft rotated through 90° before a second collision occurred.

A subsequent investigation of the Cockpit Voice Recorder (CVR) indicated that the captain failed to confirm the after-start, before take-off, and taxi checklists. This failure occurred due to a number of factors, including distractions, and the assumption that the responsibility for the checklists had been relegated to the first officer, despite the fact that this procedure would have contravened company policy (National Transportation Safety Board, 1988).

The crash involving Flight 255 highlights both the consequences of a lack of communication and the circumstances under which these failures occur, despite the availability of checklists. The pre-take-off phase of flight can be particularly demanding for flight crews, with a number of potential sources of distraction, including communication from cabin crew, air traffic control, and ground crews, navigating to the appropriate runway, and planning for the take-off and after-take-off phases of flight.

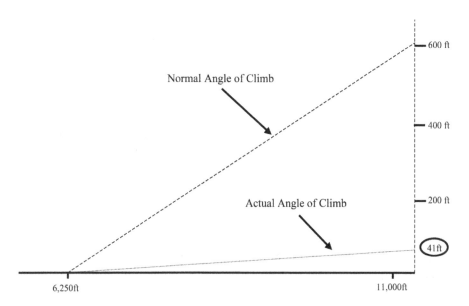

FIGURE 15.2 An illustration of the difference between the 'normal' angle of climb, and the angle of climb for Northwest 255.

Like aviation, potential sources of distraction occur in other contexts, including emergency medicine, transportation, and energy generation and transmission. To mitigate the opportunity for distractions, particularly during critical operations, standard operating procedures can be employed to restrict communication where it is unrelated to the performance of the task. In the aviation context, this restriction is imposed on operations below 10,000ft and is referred to as the 'sterile-cockpit' rule (Wu, Molesworth, & Estival, 2019). However, a similar strategy has been employed in healthcare during the administration of medication, whereby conversations are restricted, routine distractions are prevented, and non-relevant phone calls are avoided (Anthony, Wiencek, Bauer, Daly, & Anthony, 2010; Hohenhaus & Powell, 2008).

In addition to the administration of medication, restrictions to manage interruptions in healthcare have also been implemented in anaesthesia (Broom, Capek, Carachi, & Akeroyd, 2011) and surgery (Seager, Smith, Patel, Brunt, & Brennan, 2013). However, these restrictions need to be considered in the context of the organisational culture within which they are implemented. For example, a 'sterile' environment may impede communication where there is some ambiguity as to the relevance of information and/or where there are potential negative consequences associated with an error.

An illustration of the consequences of ambiguity amongst crew members is highlighted by the crash of an Air Florida Boeing 737 on January 13, 1982, shortly after take-off from Ronald Reagan Airport in Washington, DC. Flight 90 began the take-off roll with snow and ice having accumulated on the wings and what was suspected to be ice accretion in the compressor inlets of the two engines (National Transportation Safety Board, 1982).

The ice accretion in the compressor inlets resulted in a higher-than-normal reading on the Engine Pressure Ratio (EPR), creating an anomaly between the reading for the EPR and the physical position of the throttles. The first officer commented on the anomaly, and possibly expected that the captain had understood the problem and had made the decision to continue the take-off. In the event, the aircraft was significantly under-powered and it failed to become airborne, colliding with the 14th Street Bridge over the Potomac River.

The crash of Air Florida 90 has significant implications, both in terms of the characteristics of communication strategies, and in the first officer's expectation that his concerns had been communicated successfully to the captain. The captain had advanced the throttles to the point where the prescribed EPR was achieved and had then transferred control of the throttles to the first officer who was the pilot flying. It was clear from the cockpit voice recorder that the first officer was concerned with the performance of the aircraft during the take-off roll (see Table 15.1). However, in expressing his concern, the first officer used mitigating language. This type of strategy is often characterised by the use of qualifiers such as 'should', 'might', or 'seem' and tends to occur in environments where there is a distinction between levels of authority and/or where there is some ambiguity as to the accuracy of the information (Linde, 1988; Cushing, 1994).

The main difficulty associated with the use of mitigating language is that it is deliberately circumspect, and therefore, is subject to misinterpretation, leading to miscommunication (Linde, 1988). In the context of Air Florida 90, the situation was further compounded by the pilots' relative inexperience in operating in snow and icy conditions, and the time constraints imposed in having to recognise, communicate, and respond to the situation within a very short period.

The use of mitigating language is also evident in other environments, including healthcare, where there are differences in hierarchical relationships. For example, in an analysis of root-cause analyses conducted at hospitals in Denmark, Rabøl, Andersen, Østergaard, Bjørn, Lilja, & Mogensen (2011) noted that 52 per cent referred to communication errors, of which 86 per cent were associated with errors in handover and 23 per cent were associated with hesitancy to communicate concerns.

TABLE 15.1
An Extract of the Transcript from the Cockpit Voice Recorder Following the Crash of Air Florida 90

15:59:32 CAM-1 Okay, your throttles.
15:59:35 [SOUND OF ENGINE SPOOLUP]
15:59:49 CAM-1 Holler if you need the wipers.
15:59:51 CAM-1 It's spooled. Real cold, real cold.
15:59:58 CAM-2 God, look at that thing. That don't seem right, does it? Uh, that's not right.
16:00:09 CAM-1 Yes it is, there's eighty.
16:00:10 CAM-2 Naw, I don't think that's right. Ah, maybe it is.
16:00:21 CAM-1 Hundred and twenty.
16:00:23 CAM-2 I don't know
16:00:31 CAM-1 Vee-one. Easy, vee-two.

A number of remedial strategies have been advocated to address communication failures in healthcare, including graded assertiveness whereby concerns are raised initially through a 'hint'. In the absence of a response, the expression of concern escalates to an expression of a preferred course of action, a query, a shared suggestion, a statement, and finally, a command (Brindley & Reynolds, 2011). This also constitutes a 'call out' where members of a medical team are encouraged to raise issues of concern as they occur.

In addition to graded assertiveness, structured communication strategies have been advocated in healthcare to better ensure that the information communicated is sufficiently comprehensive, concise, and addresses the potential for omissions. Developed originally in the military context, the process begins with a statement that articulates the situation that is being addressed, the aim of which is to ensure that the communicator and recipient share the same context for communication. A summary of the relevant background follows to ensure that subsequent decisions are appropriately informed and that preceding conditions or constraints are taken into account. This culminates in an assessment of the situation and the articulation of recommendations. Summarised as SBAR (Situation, Background, Assessment, Recommendations), the effectiveness of the process has been tested in a range of contexts, the outcomes of which suggest that it improves communication by ensuring a shared mental model between the communicator and the receiver (Haig, Sutton, & Whittington, 2006). This appears to be particularly helpful where communication is necessary between professions (Kostoff, Burkhardt, Winter, & Shrader, 2016).

15.3 CAUSES OF COMMUNICATION ERRORS

At a fundamental level, communication errors are due to a mismatch between the expectations of a communicator and the interpretation of a receiver. In delivering information, the communicator has an expectation that the information has been conveyed accurately and has been interpreted accurately by the receiver. The receiver may have interpreted the information inaccurately, but remains under the impression that the information has been interpreted accurately. Unless detected, the consequences of this mismatch may be perpetuated, leading to operational errors.

One of the most challenging antecedents of miscommunication is organisational culture, particularly if it fails to encourage and support precise communication. Imprecise communication enables ambiguous terminology and phrasing, and is evident in the use of colloquial language, context-related jargon, and the failure to follow standard operating procedures.

An example of the implications of context-related jargon can be drawn from the circumstances preceding the collision between two freight trains near Gunter, Texas, on May 19, 2004 (National Transportation Safety Board, 2006). On the single-line track on which the trains were operating, it was necessary for one train to be positioned in a siding off the main railway line, after which the other train could pass. In this case, three trains were involved, two of which were heading north and needed to clear the main track so that a southbound train could pass safely.

At 5.05 pm, the engineer on the southbound train sought clarification from the rail controller as to the procedure necessary in passing the two northbound trains. He clarified with the controller 'were we heading in the hole there?', referring to a railway siding. The rail controller replied 'No, no, no, you're going to hold there till the Sherman Rock, excuse me, until the Sherman Switcher clears up at Dorchester. Then I'll take you guys on south to Prosper to meet the Sherman Rock Train.'

From the controller's perspective, the intention was that the southbound train would wait until both northbound trains had been routed into sidings before continuing past the Dorchester siding. Consistent with this expectation, at 5.39 pm, the southbound train received a track warrant authority that permitted the train to continue on the main track after the arrival in the siding of the second of the northbound trains.

To confirm that both trains had cleared the main line, the crew aboard the southbound train were expected to confirm by radio that they had sighted the second of the northbound trains prior to proceeding (National Transportation Safety Board, 2006). However, they mistook the first of the northbound trains for the second, and failed to the confirm the engine number, proceeding under the assumption that both northbound trains had been passed. In this case, the use of jargon and ambiguous phrasing in communicating with the crew of the southbound train possibly led to a false expectation that it would be safe to continue once they had received authority to proceed past the Dorchester siding.

In the aviation context, assumptions arising from ambiguous phrasing were also evident in the crash of an Avianca Boeing 747-238B on November 26, 1983. Flight 011 had departed Charles de Gaulle Airport at 10.25 pm and the pilots were attempting to conduct an Instrument Landing System (ILS) approach at Madrid Barajas International Airport, Spain (Gero, 1993). The pilot-in-command was unaware of his precise position and that he was no longer under radar coverage. The approach controller who had been in communication with the flight crew had handed over the aircraft to the tower controller without relaying any positional information, contrary to the standard procedure recommended by the International Civil Aviation Organisation (ICAO). Further, the controller had advised the flight crew that they were 'approaching' the navigation aid, but had not provided any precise distance. Presumably, the flight crew assumed that they remained under radar control and were therefore safe from obstacles. However, the aircraft descended below the Minimum Descent Altitude (MDA) in the area and collided with terrain approximately 7.5 miles from the airport.

A significant proportion of incidents involving either operational errors or pilot deviations have been associated with miscommunication between pilots and air traffic controllers (Prinzo, 1996) (see Table 15.2). According to Grayson and Billings (1981), miscommunication between pilots and air traffic controllers was cited as a significant factor in approximately 70 per cent of confidential reports to the Aviation Safety Reporting System (ASRS).

Miscommunication errors can be classified according to ten distinct characteristics, including:

- Misinterpretable statements
- Inaccurate statements
- Inaccuracies in content
- Incomplete content
- Ambiguous phraseology
- Untimely transmissions
- Garbled phraseology
- Absent – not sent
- Absent – equipment failure and
- Recipient not monitoring.

TABLE 15.2
Pilot/Controller Miscommunication as a Factor in Incidents during 1994

Incident	Total	% of Incidents Involving Communication
Operational Errors	476	26
Pilot Deviations	791	41
Near Midair Collisions	147	14

Source: Adapted from Prinzo (1996, p. 2).

These characteristics are applicable across most operational environments, and can be used to categorise the different types of miscommunication that are likely to be evident. Classifying different forms of miscommunication enables an assessment of common antecedents and the development of targeted interventions.

From an operational perspective, it is also possible to consider the types of operational incidents that are most likely to arise as a consequence of miscommunication. For example, in the aviation context, the four main classes of hazardous incident arising from miscommunication are:

- Deviations from assigned altitudes/flight levels
- Deviations in headings
- Failures to 'hold short' of the active runway and
- Deviations from airways routing (Monan, 1986)

In radiology, the consequences include:

- Inadequate patient handovers between clinical settings
- Inappropriate transfers of a patient
- Incorrect requests
- Delayed communication and the
- Communication of an incorrect diagnosis (Hannaford, Mandel, Crock, Buckley, Magrabi, Ong, Allen, & Schultz, 2013)

Establishing the consequences that are most likely to arise following miscommunication enables the development of defences that, in the event of miscommunication, will prevent an incident or accident. For example, checklists might be introduced that structure the types of information to be communicated, together with the sequence in which the information should be communicated (Segall, Bonifacio, Schroeder, Barbeito, Rogers, et al., 2012). Similarly, innovations of information technology may also be introduced that facilitate communication between different technical systems and support efficient communication across a range of different stakeholders (Singh, Naik, Rao, & Petersen, 2008).

While the development of organisational systems and structures can enable communication, their effectiveness depends on a capacity for synergy between the communicator and receiver. Similar ways of working, the use of standard terminology, and a common understanding of the system ensure that a consistent mental representation exists and creates a synergistic environment where accurate communication is achieved as efficiently as possible.

Balancing accuracy with efficiency in communication is necessary both to avoid error and to ensure that the system is not delayed unduly with repetition and clarification. However, determining the appropriate balance can be difficult in an operational context. In highly dynamic environments, where there is an incentive for brevity, communication failures can occur when assumptions are made concerning the perceived salience of information that lead subsequently to differences in expectations. For example, in 2014, General Motors (GM) was required to recall approximately 30 million vehicles due to the installation of a dangerous ignition switch, the problems about which had been known within GM for many years. Reports had been received where the ignition switch would cause the vehicle to stall and, importantly, would prevent the airbags from deploying following a collision. The problem appeared to be related to the way in which these reports were communicated within GM as issues related to 'customer convenience', rather than safety. In effect, reference to the issue in the context of convenience altered the perceived salience of the problem to the point where it was regarded primarily as a matter related to the preferences of customers (Eifler & Howard, 2018).

15.4 COMMUNICATION AND TEAM PERFORMANCE

According to Orasanu (1990), a strong relationship exists between the type of functional communication initiated between team members, and the subsequent performance of the team. For example, a comparison between high-performance and low-performance flight crews during a simulated flight, revealed differences in the way in which the two groups of pilots communicated. Table 15.3 illustrates the nature of this relationship where high-performance crews recorded a higher percentage of observational statements than their low-performance counterparts, when considered as a proportion of the total number of functional statements within a communication exchange.

Functional statements comprise all those statements that pertain to the function of the aircraft, and include:

Commands	A specific assignment of responsibility
Failure to Acknowledge	Does not evaluate previous speech act
Inquiries	Request for factual information
Observations	Recognising/noting a fact

Although causality cannot be inferred between the frequency of statements of observation and performance, it might be argued that the role of observational

TABLE 15.3
Percentage of the Total Number of Functional Statements for High Performance and Low-Performance Crews during a Simulated Abnormal Flight Task

Type of Statement	High-Performance (%)	Low-Performance (%)
Command	11.0	11.3
Failure to Acknowledge	3.5	6.0
Inquiry	3.0	0.9
Observation	23.8	15.9

Source: Adapted from Orasanu (1990).

statements in the context of high-performing flight crews lies in making salient information that might otherwise be overlooked by other team members. This is important in dynamic, high-workload environments where there are competing demands for attentional resources. The advantage afforded by team members in this case is the opportunity to observe and communicate relevant information at the appropriate time during an operational activity.

The categories of communication that were adopted by Orasanu (1990) were based, in part, on a speech act coding scheme developed by Kanki and Foushee (1989). The process involved the classification of utterances into distinct categories, the frequency and distribution of which could be examined subsequently to determine any differences or similarities between groups. This is a potentially useful technique in diagnosing and isolating difficulties associated with underperforming teams more generally, although it is not a process that should be adopted in the absence of a consideration of the context.

Linde (1988) suggests that the communication context will likely determine both the type and frequency of utterances amongst teams and importantly, will point towards the intention of the utterance. For example, in approaching heavy traffic in a motor vehicle, a front-seat passenger, observing the build-up of traffic, will normally communicate with the driver less frequently. Implicitly, the intention is to enable the driver to direct attentional resources to navigate what is perceived by the passenger as a high-demand situation. Therefore, less communication in this context is associated with a positive intention.

Amongst teams, the context for communication could also relate to the perceived hierarchy or deference to differences in perceived levels of expertise. For example, Linde (1988) considered both utterances and context in an analysis of mitigating statements that occurred during communication sequences preceding aircraft accidents. The results indicated that, the less direct a statement, the less likely that a superior would ratify the statement. Similarly, the more indirect a statement, the less likely that a recipient would continue the topic of conversation.

Differences in the receptiveness to statements have significant implications for communication strategies in the context of a hierarchy or where an organisational culture discourages forthright communication. It also confirms that mitigating language is a likely source of miscommunication. Importantly, this process of classifying utterances in context offers a technique that might be engaged to establish whether a team is more or less likely than others to engage in mitigating language, thereby enabling the development and introduction of appropriate interventions.

15.5 COMMUNICATION ACROSS TEAMS

One of the challenges in large industrial organisations involves ensuring that information is communicated throughout different parts of the business. The absence of an appreciation of the need to communicate relevant information beyond an isolated team inevitably leads to a situation where different parts of an organisation are functioning based on different assumptions, leading to inefficiencies and, in some cases, system failures.

In communicating across teams, the difficulty lies in knowing who constitutes a relevant stakeholder and what information needs to be communicated. For example, in the context of cybersecurity, phishing emails are often the first indicator of an impending cyberattack on an organisation. However, the recipients of these emails often fail to appreciate the importance of reporting phishing emails and once recognised, they are simply deleted. Reporting emails enables the relevant information technology groups within the organisation to assess whether phishing emails are isolated events or part of a strategic attack that requires an intervention at the organisational level.

Even within the same operational context, there are challenges in encouraging communication across teams, and this is particularly evident in the context of flight and cabin crew onboard aircraft. In a case involving the crash of Flight 1363, an Air Ontario Fokker F28 on March 10, 1989, it was evident that, immediately prior to the crash, the cabin crew were made aware of concerns regarding ice and snow accretion on the wings that would impede the performance of the aircraft on take-off (Chute & Wiener, 1996; Evans, Cardiff, & Sheps, 2006). However, they were reluctant to communicate their concerns to the pilots as the pilots prepared for take-off from Dryden, Ontario. As one flight attendant noted during the investigation following the accident:

> Well, we have – the pilots and the flight attendants have respect amongst one another as friends, but when it comes to working as a crew, we don't work as a crew. We work as two crews. You have the front-end crew and a back-end crew, and we are looked upon as serving coffee and lunch and things like that.
>
> (Moshansky, 1992, p. 1075)

It appears likely that, in the case of flight and cabin crew, the reluctance to communicate is due to an organisational culture that reinforces differences in job roles and responsibility. Deconstructing these barriers to communication can be challenging, but begins with an understanding of the perceptions of different stakeholders. Targeting flight and cabin crews, Chute and Wiener (1995) conducted one of the most detailed comparative analyses, and noted that 63 per cent of pilots and 68 per cent of flight attendants supported the notion of a single department for flight and cabin crew, thereby deconstructing an organisational barrier that would otherwise accentuate differences between the two work teams.

Despite the similarities in aspirations, differences between flight and cabin crew were evident in their reflections on the nature and frequency of the interaction that occurred, particularly in advance of the flight. Bearing in mind the ad-hoc nature of the teams that are assigned to flights, the pre-flight briefing is an opportunity for the aircraft captain to set any expectations and prepare the crew for any issues that are likely to arise. It constitutes an important opportunity to encourage communication that might impact the safety and/or security of the flight.

Amongst cabin crew at the time, 81 per cent rated the frequency of these pre-flight briefings as between 'never' and 'sometimes', while amongst flight crew, 71 per cent rated the frequency of these briefings as between 'sometimes' and 'frequently'. Chute and Wiener (1995) suggest that these differences in perception probably reflect differences in expectations concerning the nature and role of communication that occurs between flight and cabin crew. For example, a captain might consider a short introduction equivalent to 'briefing', where cabin crew might regard this as a casual or informal conversation.

Differences in perceptions across teams are likely to reflect differences in organisational culture. Therefore, changes need to be initiated at the organisational level if differences in culture, practice, and expectations are to be minimised across different organisational teams. This might include joint training initiatives and the development of joint goals during briefings where expectations can be established that encourage communication. A similar strategy is advocated in the surgical environment between surgeons and nursing staff, where the lead surgeon is encouraged to provide a briefing that establishes the basis for communication during the operation (Einav, Gopher, Kara, Ben-Yosef, Lawn, et al., 2010).

Despite joint initiatives and opportunities for familiarisation, it often remains difficult for team members to know when and how to communicate successfully across teams, particularly where one team might be engaged in a high-risk, high-demand phase of the operation. The uncertainty that exists needs to be managed through standard operating procedures, but nevertheless requires a degree of judgement as to the importance of the information.

To address the uncertainty associated with initiating communication in critical situations, Wadhera et al. (2010) have proposed a structured communication process, targeted towards specific events. For example, in the context of a cardiopulmonary bypass, events might include the initiation of the bypass and the removal of cross-clamps. In addition to standardised phraseology and the verbalisation of activities, it offers an opportunity for other members of the team to raise concerns. In effect, this 'normalises' the process of inquiry and removes a degree of uncertainty concerning the most appropriate period at which to raise issues. A similar approach is advocated in the aviation context, where cabin crew are encouraged to contact the flight crew:

1. Whenever there is a threat to the safety of the aircraft; and/or
2. Whenever there is an unexpected change in the condition of the aircraft.

15.6 SUCCESSFUL COMMUNICATION

Although considerable interest is often directed towards the analysis of instances of communication failure, a great deal of knowledge can be derived from examples that were resolved through 'effective' communication. According to Predmore (1991), the analysis of the communication strategies involved in incidents that were resolved successfully can provide instructive information pertaining to the use of communication during high-workload, time-constrained events.

One incident that is often cited as an example of the effective use of communication involved United Airlines Flight 811, a Boeing 747-122 that departed Los Angeles for Sydney, Australia, on February 24, 1989, with intermediate stops in Honolulu and Auckland. Following its departure from Honolulu, the aircraft was climbing through 22,000ft when the forward cargo door detached, impacting the fuselage and causing the cabin floor above to buckle, ejecting both the seats and nine passengers (National Transportation Safety Board, 1990a). The debris also resulted in the shutdown of two of the aircraft's four engines.

Following the loss of the cargo door, the captain immediately initiated a descent, and the aircraft was cleared for an approach to Honolulu. The loss of part of the fuselage created a number of changes in the aerodynamic characteristics of the aircraft, and the flight crew were required to modify the descent and approach accordingly. Despite these challenges, the aircraft landed safely.

An analysis of both the frequency and type of communication strategies initiated during the emergency suggested that, although the frequency of communication statements increased, the relative ratios of the types of communication statements remained relatively consistent (Predmore, 1991). However, during the latter stages of the flight, it appeared that observational statements became more frequent, possibly as a result of the need to direct a large amount of information towards the pilot who had control of the aircraft.

A comparison between the communication strategies prior to and subsequent to the emergency indicated that the rate of incomplete statements tended to increase. Predmore (1991) suggests that this change in the nature of communication is indicative of the high rate of exchange required to manage catastrophic, time-constrained events about which there is considerable uncertainty. However, it also represents an aspect of communication during emergencies or non-normal situations where valuable, task-related information might be lost or miscommunicated in the case of less experienced or less capable crews.

On July 19, 1989, only a few months after the events involving United Airlines Flight 811, United Airlines Flight 232 experienced a catastrophic engine failure during a flight from Denver, Colorado, to Chicago, Illinois. The aircraft was a Mc Donnell DC-10, recognisable for its three engines, one of which is located beneath the tail (Number 2), and two on either wing (Numbers 1 and 3).

In the case of the DC-10, control over the aircraft is normally exercised through a hydraulic, fluid-filled system, three of which are provided to ensure a degree of redundancy should one or more of the hydraulic systems fail. To ensure a further level of redundancy, the hydraulic systems are routed differently throughout the aircraft, coalescing only in one location, beneath the Number 2 engine.

Flying at 37,000ft, the fan blade housed within the Number 2 engine sheared out of its casing, and sliced through the one location on the DC-10 where all three hydraulic systems coalesced, rendering the hydraulic system unusable (National Transportation Safety Board, 1990b). The pilots resorted to controlling the aircraft using the engines on the wings where an increase in power to both engines simultaneously would cause the aircraft to pitch up, while withdrawing power resulted in the aircraft pitching down. Rudimentary lateral movements could be exercised by adjusting power on one of the two engines.

Like the Boeing 747-122 series aircraft, the DC-10 comprised a flight crew of two pilots and a flight engineer. However, on this occasion, a senior pilot from United Airlines was on board as a passenger and offered to assist the flight crew. It was this pilot who operated the throttles, controlling power to the engines. The remaining crew members were attempting to control the aircraft, communicate with air traffic control, and resolve the hydraulic systems by communicating with the United Airlines maintenance facility.

Soon after the engine failure, the flight crew sought advice from air traffic control as to the most appropriate location for an emergency landing. Sioux City Gateway Airport was selected, and communication was established between the aircraft and the local air traffic controllers. As the aircraft could only sustain turns to the right, the air traffic controllers at Sioux City Airport presented the flight crew with a number of options to land the aircraft, including on a highway. In the event, the crew managed to position the aircraft for a landing at Sioux City Airport, lining up for an approach on a closed Runway 22. In the final stages of the approach, the aircraft turned again to the right, at which point the right wingtip collided with the runway and the aircraft cartwheeled to a stop disintegrating in the process.

Although a wing-drop during the landing phase resulted in a significant loss of life, the efforts of the flight crew in reaching the airport, despite the condition of the aircraft, were recognised by the National Transportation Safety Board (1990b) as exemplary. Moreover, an analysis of the communication strategies employed by the crew of Flight 232 revealed that the distribution of tasks was more than likely a significant feature in the successful management of the situation. For example, the captain was willing to accept the additional support offered by the travelling pilot as a means of managing the tasks required. Further, the duties appeared to be delegated so that the flight engineer was more involved in damage assessment, while the captain and first officer were more engaged in establishing flight control (see Table 15.4).

In addition to the performance of the flight crew, the incidents involving Flight 811 and Flight 232 demonstrated the value of effective communication from an organisational perspective. United Airlines had not experienced a passenger fatality since 1979 and, consequently, there was some concern over the implementation of the crisis management plan (Doughty, 1993).

Doughty (1993) suggests that the communication aspects of the United Airlines crisis plan were ultimately dependent upon a number of key features including:

- Cooperation with media;
- Positioning the notion of 'safety' as the key issue and ensuring that mistakes would not be repeated;

TABLE 15.4
% of Functional Statements for Each Action Decision Sequence for the Captain, First Officer, and Flight Engineer of Flight 232

Type of Activity	Captain	First Officer	Flight Engineer
Flight Control	43	51	16
Damage Assessment	12	12	41
Problem Solution	9	6	29
Landing	26	22	9
Emergency Prep	9	8	4
Social	2	1	0

Source: Adapted from Predmore (1991).

- Avoiding speculation;
- Correcting erroneous media reports promptly;
- Cooperating with investigation and regulatory agencies;
- Enabling delegated team members to concentrate on tasks;
- Recognising and valuing non-company personnel who provide assistance;
- Releasing passenger and crews list as quickly as possible; and
- Keeping employees informed of the progress.

The application of these principles resulted in a relatively effective communication strategy, so that the public understood even intricate details associated with the events.

15.7 SEMANTIC AND PROSODIC ASPECTS OF COMMUNICATION

Communication conveys meaning most explicitly through 'what' is said, or what is referred to as its semantic quality. In many operational environments, the semantic quality of communicative statements is assured by adopting standard phraseology (Estival & Molesworth, 2011). This is particularly important in situations where operators are communicating in languages other than their native language and where there is a risk of miscommunication (Auton, Wiggins, Searle, & Rattanasone, 2017).

The risk to communication tends to increase where communication occurs in the absence of non-verbal cues (Catt, Miller, & Hindi, 2005). Non-verbal cues include facial expressions and body posture and can signal to a communicator that information has not been understood, thereby providing an opportunity for clarification. This opportunity for clarification reduces the likelihood of miscommunication, especially in those environments where standard phraseology is not the norm.

Beyond non-verbal, visual cues, communicators also respond to prosodic cues that are associated with a communicative statement, including intonation, pauses, and interjections (e.g. 'um' or 'uh'). For operators in high-risk, high-consequence environments, detecting changes in prosodic cues offers an additional opportunity to identify potential sources of miscommunication (Auton, Wiggins, Searle, Loveday, & Rattanasone, 2013). In the case of a receiver reading back a command that has just been issued, an increase in the intonation at the end of a sentence normally indicates a degree of uncertainty as to its meaning and the intended response. In English, a rising intonation is normally associated with a question. Therefore, the implication is something like: 'this is what I heard, have I heard it correctly?'.

During a read back, an increase in the number of pauses or the length of pauses suggests that the receiver is having difficulty recalling the information (Auton et al., 2017). This may have implications for a subsequent operation where it might be expected that the commands issued are implemented, possibly in a specific sequence.

Consistent with pauses, an increase in the frequency of interjections during a readback is typically associated with a lack of understanding and might be interpreted as an opportunity for further clarification (Auton et al., 2013). Together with other forms of prosodic cues, it offers a defence, particularly in those operational environments where communication occurs primarily through human–machine interfaces.

16 Resource Management

16.1 RESOURCE MANAGEMENT

Resource management is a construct that emerged within the aviation industry and is generally ascribed to two significant catalysts. The first was a series of aircraft accidents where it was evident that the failure was due to a lack of coordination between the flight crew, rather than any structural or mechanical failure. The second was a landmark research initiative conducted under the auspices of the National Aeronautics and Space Administration that revealed serious deficiencies in the performance of some flight crew, particularly in their ability to manage the resources available during a flight (Ruffell-Smith, 1979).

Ruffell-Smith (1979) engaged a full-mission simulator to examine a series of Boeing 747 crews during a demanding, but realistic, transatlantic scenario. The variability in the performance of crews, together with recent accidents that appeared to confirm the inability of some crews to manage non-normal conditions, led to the development of a training initiative, the intention of which was to develop the capacity of flight crews to manage the resources available to ensure the safe and expeditious operation of aircraft.

Initial attempts to address the problem of poor resource management drew on concepts relating to organisational psychology, including personality, styles of communication, negotiation, and team skills (Lauber, 1986). The approach to training also mirrored the approach that was evident in the business community and included workshops and activities that were largely knowledge-based and provided participants with a general awareness of the issues (Raby, 1987; Sams, 1987).

It was assumed that by raising awareness of the issues that might prevent the effective and efficient management of resources, these interventions would provide a foundation for the development of what was referred to at the time as effective Cockpit Resource Management (CRM). CRM was formally defined as 'the effective utilisation of all available resources – hardware, software, and liveware – to achieve safe, efficient flight operations' (Lauber, 1986, p. 9). Despite the initial emphasis on the cockpit environment, this definition remains relevant within the contemporary industrial environment, since a context is not ascribed.

16.2 AN ATTITUDE APPROACH TO RESOURCE MANAGEMENT

In addition to training strategies, initial interventions to address poor CRM focused on the development of an understanding of pilot personality and attitudes (Helmreich, 1987). This focus on pilot attitudes emerged from anecdotal evidence to suggest that the attitudes of some pilots towards other members of the crew were not necessarily conducive to the development of effective CRM. Foushee (1982), in particular, targeted the notion of the *right stuff* as an attitude which, at the time, was relatively pervasive throughout military aviation in the United States and elsewhere. The notion of the *right stuff* described the characteristics of aviators and astronauts as courageous individualists, who were single-mindedly driven towards the completion of a challenging task. They were epitomised by aviators such as Chuck Yeager, a fighter pilot during World War II, and the first person to break the sound barrier (Wolfe, 1983).

There was a general recognition that, for the flight crew of commercial aircraft, a capacity to collaborate with and involve other members of the crew, together with external resources, constituted an approach to flight management that was likely to be preferable to courageous individualism. However, the assumption that poor resource management related to inappropriate attitudes implied that changes were possible through appropriate and targeted interventions. It also provided an opportunity for the development of quantifiable assessment tools, thereby providing an evidence base for comparison.

During early CRM training programs, the analysis of attitudes formed the primary means of evaluating the outcomes of these initiatives (Gregorich, Helmreich, & Wilhelm, 1990). This led to the development of the Cockpit Management Attitudes Questionnaire (CMAQ) that comprised a series of 25 statements that were rated by respondents using a Likert scale (Helmreich & Wilhelm, 1991). A number of editions of the CMAQ were developed and the psychometric properties were established across a range of samples (Gregorich et al., 1990).

With the introduction of advanced technology aircraft, the statements that comprised the CMAQ were refined to include assessments of attitudes towards automation and safety more generally. Referred to as the Flight Management Attitudes Questionnaire (FMAQ), it emerged as an important and cost-effective tool that could be administered quickly and easily across large numbers of respondents. Like the CMAQ, the FMAQ was further revised, particularly for application outside the aviation context and this edition is referred to as the Safety Attitudes Questionnaire (SAQ) (Sexton, Helmreich, Neilands, Rowan, Vella, et al., 2006).

Analysis of the psychometric properties of the SAQ revealed a six-factor model that included teamwork climate, perceptions of management, job satisfaction, and safety climate. It is administered most regularly in medical settings and demonstrates good reliability. While cross-sectional studies suggest differences in attitudes between groups, establishing the validity of these measures tends to be challenging due to difficulties in assessing operator performance and safety-related outcomes.

Changes in attitudes following resource management initiatives constitute a tangible, quantifiable outcome reflecting changes in behaviour that are expected to occur in the operational environment (Helmreich, 1984; Helmreich & Wilhelm, 1991).

However, the direction of the relationship between attitudes and behaviour has yet to be established in practice. In some cases, it appears clear that attitudes influence behaviour, while in other cases, engagement in behaviour appears to provide the catalyst for changes in attitudes (Iversen, 2004; Rundmo & Hale, 2003).

In industrial environments, changes in performance are likely to emerge from involvement in training initiatives that target both attitude and behavioural change (O'Connor, Campbell, Newon, Melton, Salas, & Wilson, 2008). For example, amongst less experienced operators, exposure to cases that illustrate poor resource management may foster behaviours that encourage the contributions of team members. Amongst more experienced operators where behaviour is well-established, it may be necessary to model or encourage active participation in behaviours that are consistent with the effective management of resources, enabling an appreciation of the outcomes and thereby changing attitudes (Prince, Oser, Salas, & Woodruff, 1993). This approach appears to be most effective in a simulated learning environment where there are opportunities for review and the provision of feedback (Salas, Burke, Bowers, & Wilson, 2001).

16.3 RESOURCE MANAGEMENT REDEFINED

Resource Management first emerged as an initiative targeted towards the cockpit environment and the role of the flight crew in the management of aircraft. This approach was initiated largely in response to the perception at the time that errors amongst flight crew represented the most significant factor in commercial aircraft accidents (Billings & Reynard, 1984). However, the importance of other members of the aviation industry in precipitating effective resource management was readily apparent, especially in the relationship between the flight crew and Air Traffic Control (ATC). For example, Billings and Cheaney (1981) conducted a detailed analysis of communication errors arising from the confidential Aviation Safety Reporting System (ASRS) managed by the National Aeronautics and Space Administration (NASA), and noted that 73.3 per cent of reports between May 1978 and July 1980 involved some form of information transfer problem that compromised the safety of a flight. Although anecdotal, these types of data illustrated the level of interdependence between components of the aviation system and, therefore, the need to consider resource management from a broader perspective.

The realisation that human performance failures existed beyond the flight crew led to a shift in the emphasis in aviation from *Cockpit Resource Management* to *Crew Resource Management*. This change reflected an increasing understanding that the performance of the flight crew depended on the performance of other members of the industry, including ATC, baggage handlers, refuelers, and flight attendants (Chute & Wiener, 1994, 1996; Helmrcich, 1993).

This change in the understanding of resource management paralleled changes that were occurring in the context of *system safety* where failures in complex, dynamic environments were considered the product of failures that occurred at a number of different levels within the system. Therefore, safety initiatives needed to be directed towards more areas than those directly associated with the operational environment, including management and regulatory bodies.

Together with the role of non-operational personnel, the role of passengers in resource management was highlighted by a number of cases where passengers possessed information that might have prevented an aircraft accident. The most significant of these accidents occurred in 1989, when British Midlands Flight 92, a Boeing 737-400, crashed approximately 900 metres from runway 27 at East Midlands Airport (Air Accidents Investigation Branch, 1990). The aircraft had departed Heathrow en route to Belfast, when the Number One (left) engine suffered a series of compressor stalls, causing airframe shuddering and black smoke to enter the cabin.

Due to a number of factors, including poor coordination between the flight crew, they assumed that the Number Two (right) engine had failed, and subsequently shut down this engine. To reassure the passengers, the captain broadcast on the public address system that they were having trouble with the right engine (No. 2) and that they had shut down this engine as a precaution. This presented some confusion to the passengers who had seen smoke emanating from the left engine (No. 1). However, despite their confusion, the passengers were disinclined to bring the issue to the attention of the crew (Air Accidents Investigation Branch, 1990).

Even when information is brought to the attention of cabin crew, there can be limitations that impede the capacity of the cabin crew to communicate information to the flight crew. For example, British Airways Flight 762 was operating from Oslo, Norway, to London Heathrow on May 24, 2013 when the fan cowl doors from both engines separated during take-off. During the take-off roll at Oslo, a number of passengers observed the flapping doors and alerted the cabin crew. The senior cabin crew member attempted to communicate with the flight crew through the intercom but the communication occurred just as the pilots had retracted the undercarriage with the aircraft still below 1000ft. Therefore, the pilots elected to ignore the communication from the senior cabin crew member as they directed attention to the management of the aircraft (Aircraft Accidents Investigation Branch, 2015).

Given the level of interdependence necessary in managing complex systems, the use of terminology that targets specific operational contexts potentially constrains the management of resources to the operational level. Consequently, a number of researchers, including Johnston and Maurino (1996) and Brown (1997), coined terms such as *Company Resource Management* and *Corporate Resource Management* to reflect the breadth of the responsibility in ensuring that the appropriate resources are readily available and managed where and when necessary.

Resource management has evolved from the aviation context and has since been applied across a wide range of domains, including Formula One motorsport, mining, shipping, and medicine (O'Connor, Papanikolaou, & Keogh, 2010) (see Figure 16.1). As the principles of resource management were applied in new domains, new issues emerged, including the impact of culture (Merritt & Helmreich, 1996), and the development of strategies to evaluate the outcomes of resource management initiatives (Prince et al., 1993). Furthermore, there has been a growing awareness that strategies associated with effective and efficient resource management should be introduced and developed at the earliest stages of technical skill development as a means of inculcating appropriate attitudes, concepts, and skills (Flin, Patey, Glavin, & Maran, 2010; Smith, 1993; Telfer, 1993).

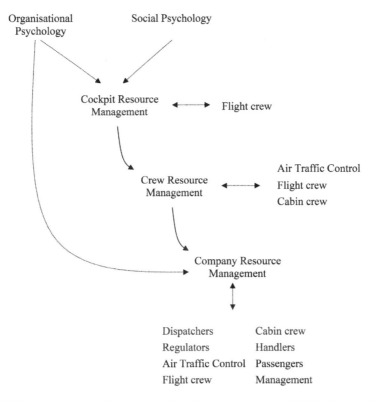

FIGURE 16.1 A diagrammatic representation of the development of CRM training initiatives.

Meta-analyses suggest that, in general, training initiatives that are directed towards improving the effectiveness of resource management have a significant impact on participants' knowledge and behaviour (O'Dea, O'Connor, & Keogh, 2014). However, performance at an organisational or industry level has been difficult to establish given the range of variables that impact performance in the operational environment. Nevertheless, improving the effectiveness and efficiency of resource management in safety-critical environments has broad appeal and has likely served as an important safety initiative in a range of industrial settings.

16.4 RESOURCE MANAGEMENT AND HUMAN FACTORS

Although there is some debate concerning the relationship between *resource management* and *human factors*, there appears to be a general consensus that resource management represents a component of the more encompassing notion of *human factors* (Jensen, 1997). However, it is important to note that the boundaries between resource management and human factors are not particularly clear (see Figure 16.2). System safety, for example, is a construct that is applied both in the context of resource management, and in the broader domain of human factors. Similarly, strategies for Instructional Systems Design (ISD) have application within both domains.

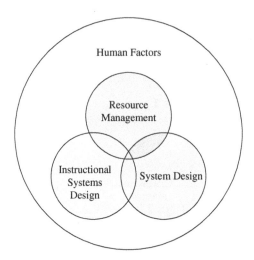

FIGURE 16.2 A conceptual representation of the relationship between human factors, resource management, system design, and instructional systems design.

On the basis of this argument, it is perhaps most appropriate to conceptualise the principles that underpin resource management and human factors as complementary, since both are designed to foster improvements in human performance. However, it also ensures that system design is considered in the development and maintenance of resource management skills and capabilities. For example, the development and application of well-crafted checklists and procedures encourages communication, and provides an opportunity to develop joint situational awareness between operators (Wang, Wan, Lin, Zhou, & Shang, 2018). Similarly, the design of a physical working environment has the potential to enhance or impede resource management by obscuring operators from one another, restricting information, and/or creating overly complex systems that constrain opportunities for communication and collaboration.

Evidence to support the role of design in enabling communication can be drawn from Costley, Johnson, and Lawson (1989) following an evaluation of the level of flight-crew communication in Boeing 737 and Boeing 757 aircraft. They noted that relatively less communication tended to occur in the more advanced aircraft. Further, Wiener (1989) noted a number of cases where the delineation of responsibility between flight crew became confused as they tried to interact with complex and poorly designed automated systems. While Wiener, Chidester, Kanki, Palmer, Curry, and Gregorich (1991) suggest that this is a reflection of the inherent flexibility of automated systems, it also demonstrates the relationship that exists between system design and the management of resources.

16.5 THE REGULATION OF RESOURCE MANAGEMENT INITIATIVES

The focus of regulatory authorities on knowledge-based examinations of human factors reflects the difficulties that are associated with the identification and assessment of human factors skills. However, the result is that human factors may be perceived by operators as simply a knowledge-based topic that has little application within the applied environment. More importantly, it might be argued that an exclusively knowledge-based approach, to the exclusion of skill acquisition, may fail to provide operators with the skills necessary to recognise and apply appropriate resource management strategies and techniques in practice.

The desire for competencies that reflect appropriate human factors-related and resource management behaviours was evident in the development of the Advanced Qualification Program (AQP) in the United States. Proposed as a joint initiative between the United States Government and the US airlines, the AQP was designed to give airlines a greater level of flexibility in the way in which flight training was conducted (Helmreich, Merritt, & Wilhelm, 1999). However, as part of the proposal, airlines were also expected to develop resource management as an integral part of the training program (Seamster, Boehm-Davis, Holt, & Schultz, 1998).

Part of the philosophy that underscored the development of AQP was the contention that the level of performance, rather than the training process itself, was most significant in determining outcomes for the industry. By targeting performance outcomes, it was possible to argue that different individuals required different levels of training, depending on their individual needs. This type of strategy enables a greater level of flexibility in the training regime and is a basic tenet of competency-based initiatives (Franks, Hay, & Mavin, 2014).

An important part of the process of AQP was the development of clear, objective behavioural outcomes against which the performance of operators could be assessed. In the context of resource management, the requirement to define performance outcomes led to the development of the Advanced Crew Resource Management (ACRM) program which was designed to assist organisations in identifying appropriate resource management standards of proficiency (Seamster et al., 1998).

Consistent with the general approach towards AQP, the development of ACRM procedures was oriented towards the individual requirements of an organisation (Bent, 1997). Therefore, it might be expected that the resource management procedures employed within one organisation might be considerably distinct from the procedures employed in another. Provided that the procedures met the minimum requirements of the regulatory authority, both approaches could be considered acceptable.

Seamster et al. (1998) suggest that the most appropriate method of developing ACRM skills within an organisation is to integrate resource management behaviours within existing Standard Operating Procedures (SOP). This ensures that skills are enacted as a part of normal operations. In addition, the integration of resource management behaviours with SOPs inculcates resource management as an integral part of the normal operating context, rather than as a series of strategies that are only applicable during emergency situations.

The implementation of appropriate resource management procedures occurs as part of a complex process involving an assessment of the organisational needs, the identification and assessment of relevant procedures, and the production of appropriate tools and guidelines (Seamster et al., 1998). An important part of this process is the development of training systems that have the capacity to engage participants and encourage the acquisition of the appropriate knowledge and skills. The ultimate aim of the training strategy is to enable the development of a system that provides the learner with the appropriate combination of awareness, knowledge, and, finally, resource management skills.

Ideally, the evaluation of resource management knowledge, attitudes, and skills should occur at various stages of development in conditions that correspond to the operational context (Man, Lam, Cheng, Tang, Tang, 2020). In high-risk environments, this might involve an assessment of individual performance in a simulated, multi-crew context operating under real-time conditions. A number of elements of performance can be examined including the extent to which communication norms are established and practised, the level of assertiveness evident, and the extent to which members of the team engage in self-critique (Turkelson, Aebersold, Redman, & Tschannen, 2017). The aim is to provide a valid and reliable assessment of the extent to which appropriate resources procedures are practised in both normal and non-normal situations.

Validity and reliability are important principles that underpin the development, implementation, and evaluation of resource management initiatives. In the case of evaluation, 'reliability' refers to the extent to which a technique has the capacity to achieve consistent results both across different evaluators, and over repeated evaluations (Malec, Torsher, Dunn, Wiegmann, Arnold, Brown, & Phatak, 2007). Complementing reliability, the validity of evaluative tools needs to be established, preferably engaging both subjective and objective measures and in conditions that present demanding characteristics (Nishisaki, Keren, Nadkarni, 2007). This provides a degree of confidence that resource management initiatives have improved performance. However, it also provides a basis for subsequent assessments, enabling the application of appropriate remedial strategies where a deterioration in performance is identified.

16.6 COMPETENCY-BASED RESOURCE MANAGEMENT

In an effort to develop competency-based criteria for resource management in the aviation context, the European Joint Aviation Authority (JAA) and the United States Federal Aviation Administration (FAA) identified a series of behaviours that were thought to reflect the application of resource management skills in team-based environments. In the case of the Federal Aviation Administration, ACRM was advocated, since it specified performance standards for training and assessment (Seamster et al., 1998). These standards were developed on the assumption that resource management is equally applicable under normal and non-normal conditions. Therefore, the evaluation of these behaviours ultimately required an assessment of performance in the operational environment.

An important principle that underscored the development of resource management competencies at the time was the identification of procedures that were specific to a particular organisation or operating context. For example, Seamster et al. (1998) suggested that organisations should identify issues that had occurred over a given period of time and that related to poor resource management, and should use these problems as the basis for the development of appropriate remedial strategies. This, in effect, is the approach taken in the development and application of resource management skills in the medical context (Sundar, Sundar, Pawlowski, Blum, Feinstein, & Pratt, 2007). While these procedures should provide the basis for improvements in performance, care should be taken to avoid the development of excessive and/ or inappropriate procedures (Helmreich & Davies, 1997; Salas, Wilson, Burke, Wightman, and Howse, 2006).

The types of procedures that might be introduced as part of a resource management initiative might include additional checklists, callouts, or communication strategies. However, it is important to note that resource management behaviours must be observable and measurable by skilled evaluators. This principle is consistent with the philosophy that underlies competency-based training more generally, since it facilitates the implementation of a valid and reliable assessment strategy. An example of resource management behaviours that relate to situational awareness is provided in Table 16.1.

Consistent with the approach taken by the FAA, the JAA also undertook a concerted strategy to identify resource management competencies for the purposes of training and evaluation. However, where ACRM was based upon the assumption that an organisation should develop its own resource management procedures, the JAA approached the problem from a more prescriptive perspective by specifying the categories of CRM skills that ought to be taught and evaluated (see Figure 16.3). Referred to as Non-Technical (NOTECHS) skills, the categories included cooperation, leadership and managerial skills situational awareness, and decision-making (Flin, Martin, Goeters, Hormann, Amalberti, Valot, & Nijhuis, 2003).

Each of the categories of NOTECHS can be further divided into elements that contribute to the category. For example, cooperation can be further divided into 'consideration of others', 'support of others', and 'problem-solving' (Flin et al., 2003). These elements are then associated with behavioural markers that are presumed indicative of good and bad practice within the operational environment.

According to Flin et al. (2003), social skills such as cooperation, leadership, and managerial skills can be observed directly through the communication between crew members. An example might include the 'support of others' element of the category 'cooperation', where good practice is reflected in the provision of personal feedback.

TABLE 16.1
A Sample of ACRM Competencies for Situational Awareness amongst Airline Flight Crew

Situational Awareness Observable Behaviours

(a) Crew discusses weather at destination and alternate prior to approach
(b) Crew discusses terrain prior to approach
(c) Crew manages workload and distractions
(d) Crew manages time and resources effectively

Source: Adapted from Seamster et al. (1998).

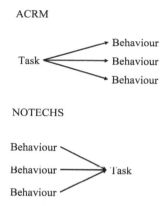

FIGURE 16.3 An example of the differences between ACRM and NOTECHS. Where ACRM used the task as the stimulus for the identification of appropriate behaviours, NOTECHS identified appropriate behaviours and presumes that the successful application of these skills will facilitate successful performance within the operational environment.

Poor practice in the 'support of others' element might be indicated by the failure to provide personal feedback or showing no reaction to other crew members.

In the case of cognitive skills such as situational awareness and decision-making, the appropriate skills must be inferred from specific forms of behaviour. For example, elements of situational awareness must be inferred from behaviours such as monitoring and reporting changes in the system state (Flin, Slaven, & Stewart, 1996). Similarly, elements of decision-making must be inferred from behaviours such as the extent to which information is acquired to identify a problem that might arise (see Figure 16.3).

One of the advantages associated with the NOTECHS approach is that the resource management skills necessary to perform successfully within the operational environment are clearly specified, and are standardised across different environments, tasks, and organisations. Equally, however, the approach might be criticised as being too broad, to the extent that certain competencies may not be appropriately specified for a particular operation or organisation. However, it should be noted that there is no impediment to adapting the NOTECHS skills so that they do reflect the demands of a specific organisation (Edkins, 2002).

16.7 THREAT AND ERROR MANAGEMENT

Threat and Error Management (TEM) emerged as an outcome of the extensive research initiative that accompanied the development of resource management as a construct. It reflects the fact that, in the context of normal operations, practitioners will confront threats to safety, the emergence of which are beyond the control of operators (Helmreich, & Musson, 2000; Thomas, 2004). For example, a surgical patient may have an underlying pathology that only emerges once the surgeon commences an operation. Similarly, miners might only become aware of a geological flaw in a coal seam once mining has commenced.

In the context of TEM, the management of these external threats is a cyclical process based on the application of resource management skills, including the recognition of a threat, communication concerning a threat, cooperative decision-making in determining a response to a threat, and coordination in responding to a threat (Helmreich & Davies, 2004). The process then continues, ensuring that any additional threats are identified and managed efficiently and effectively to maintain the safety of the system (Brennan, De Martino, Ponnusamy, White, De Martino, & Oeppen, 2020).

In addition to external threats, internal threats can also emerge and impact safety, particularly through the commission of errors. TEM is based on the proposition that errors are likely to occur in the context of complex, dynamic operations and that skilled resource management involves recognising and responding effectively to these errors as they emerge (Thorogood & Crichton, 2014). This requires a combination of strategies that are intended, in the first instance, to avoid errors. They might include additional training, checklists, communication protocols, and/or the implementation of technological solutions.

If errors occur, they should be captured to the extent that they are prevented from impacting the system. This might include warnings and/or cross-checking procedures that constitute hurdles or preventing stages of a process being initiated where an error may occur. Finally, where errors are not trapped, their impact on the system should be minimised to the greatest extent possible, either by designing systems that are error-tolerant or by ensuring that sufficient resources are available to recover from the consequences of errors (Brennan et al., 2020).

The capability for threat and error management within an organisation can be assessed through safety audits that are designed to collect data concerning the frequency and nature of threats that are experienced by practitioners, errors that are committed, and the extent to which they are identified and managed (Helmreich, 2000). Changes in performance against a baseline and/or benchmarked assessments provide opportunities for intervention, much as occurs in the context of risk assessment. These types of safety audits also provide an assessment of organisational resilience, since in effect the data reflect the capability of practitioners to respond efficiently and appropriately to threats that are an inevitable aspect of the operational environment (Dawson, Chapman, & Thomas, 2012).

Part V

Human Factors Tools and Techniques

17 Hazard Analysis

17.1 HAZARDS AND INCIDENTS

Hazard analysis is a proactive means of identifying the health or danger associated with a system. It comprises a collection of strategies, the implementation of which is based on an assumption that an increase in the frequency and/or probability of hazards is associated with an increase in the frequency of incidents, which, in turn, is associated with an increase in the frequency and/or severity of accidents.

While a relationship between hazards, incidents, and accidents is reasonably well-accepted, the nature of the relationship is likely to vary, depending on the complexity of the system. Therefore, hazards, like incidents, should be considered symptoms of inherent threats within the system. Hazard analysis complements strategies such as accident and incident analysis to mitigate the impact of broader threats that might include one or more of the following:

(a) Mechanical
(b) Thermal
(c) Fire
(d) Chemical
(e) Electrical
(f) Radiation
(g) Noise
(h) Fatigue
(i) Stress

Since hazard analysis should normally constitute part of a cyclical process that includes a combination of accident investigation and preventative measures (see Figure 17.1), it reflects a continuous process of identification, analysis, and review. The primary aim in identifying 'weaknesses' or 'threats' within the system is the opportunity to allocate resources and implement strategies to improve the overall health of the system in advance of incidents and/or accidents (Kaber & Zahabi, 2017).

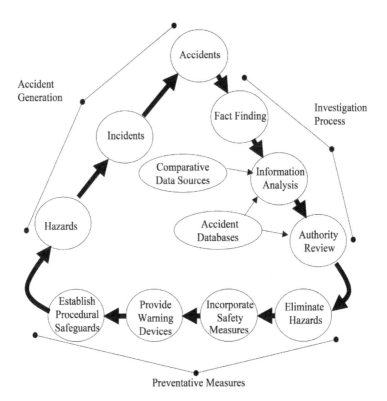

FIGURE 17.1 A diagram of the relationship between accident generation, investigation and prevention. (Adapted from Diehl, 1991, p. 98.)

17.2 THE PROCESS OF HAZARD ANALYSIS

Hazard analysis comprises four broad stages, including a preliminary analysis, a systems analysis, an operating analysis, and the analysis of accidents and incidents. Typically, the process begins with a task analysis of the environment. The task analysis is designed to identify the various functions that comprise successful performance and establish the resilience of a system or task (Cameron, Mannan, Németh, Park, Pasman, Rogers, & Seligmann, 2017; Chandrasekaran, 1990). The result is an analysis of both the skills and knowledge required by operators, and the demands imposed by the task itself (Wang, 2002).

The systems analysis phase of hazard analysis involves a consideration of the validity of the data arising from the task analysis (Stanton, 2006). This may be accomplished through a simulation of the task (either conceptual or actual), or some other means of cross-referencing the data. This process is allied with the operational trials and evaluation component where the system or task is observed and examined as it would function within the operational environment (Stewart, 1993).

The final process of hazard analysis is based on the information derived from incidents and/or accidents. Generally, this involves an examination of the factors involved in an incident/accident and the classification of these features into 'hazard patterns'. Drury and Brill (1983) used this technique successfully in an analysis of the hazards involved in consumer products.

Hazard patterns can be grouped using the nature or extent of injuries, the behaviour involved, and/or the factors that precipitated the failure. This process enables the identification of predominant features that characterise the nature of the environment within which most failures occurred. For Drury and Brill (1983), an examination of the nature of accidents involving glass indicated that victims could be described as predominantly male, dealing with defective glass, with injuries resulting predominantly to the upper limbs. This type of information provides a basis for the elimination and/or mitigation of hazards. The recommendations included a reduction or elimination of incidences of defective glass, thereby removing exposure to the hazard and preventing injuries.

Following the identification of the hazards associated with a system, Salvendy (1997) advocated the application of a systemic model which considers the management of hazards from the initial design process through to the training process (see Figure 17.2). It constitutes a hierarchy where generally the design or removal of a hazard tends to be more effective in preventing errors than providing warnings or training to operators (Norman, 1993)

While the process illustrated in Figure 17.2 is optimal, it is also based on the assumption that a system or product is subject to a hazard analysis prior to implementation within the operational environment. This allows for the development, application, and evaluation of the intervention to ensure that exposure to hazards is minimised to the greatest extent possible. However, in many cases, hazards only become evident once the complexities associated with the operational environment are introduced. Therefore, in many cases, guards and warnings will need to be implemented since it is no longer possible to redesign the system or product nor remove features that might constitute hazards.

Once a system or product becomes operational, the identification of hazards is critical in ensuring that errors and injuries are minimised. While operational audits are useful, they can be costly to implement and interpret. However, the use of advanced technology systems offers a useful, cost-effective strategy enabling the acquisition of data accurately and in real-time. Although the interpretation and use of these data need to be managed carefully to ensure that the outcomes are valid, reliable, and

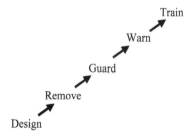

FIGURE 17.2 A diagrammatic representation of the process through which hazards can be managed. (Adapted from Salvendy, 1997.)

ethical, they offer a means of ensuring that potential hazards can be identified and remedied as quickly as possible.

In the airline industry, Quick Access Recorders (QAR) are employed as an efficient means of identifying hazards that occur during the day-to-day operations of an aircraft (Wang, Zhang, Dong, Sun, & Ren, 2019). In United Airlines, QARs were installed in all Boeing 777 aircraft to monitor a variety of aircraft systems (Phillips, 1996). This originally occurred as part of a broad-based Flight Operations Quality Assurance (FOQA) Program initiated by the Federal Aviation Administration in the United States.

The QAR offers a means of identifying and maintaining operating standards, auditing training and route check procedures, and monitoring the constraints imposed by operational demands (Wang, Wu, and Sun, 2014). The major advantage associated with the QAR is that there is no requirement to extract the Flight Data Recorder to monitor the overall performance of the system. Therefore, it provides an early indication of hazards that might impact operational performance.

Where the QAR tends to be used at an organisational level, techniques have also been developed to identify hazards at an industry level. For example, the INDICATE (Identifying Needed Defences in the Civil Aviation Transport Environment) system was based on a proactive approach for the recognition of operational hazards, where potential hazards could be communicated directly with responsible safety authorities and regulators (Edkins, 1998).

The implementation of the INDICATE system comprises four stages. The first is a series of focus groups, the purpose of which is to identify the hazards that are likely to exist in the operational environment (Edkins, 1998). During this stage, participants are asked to rank the significance of the potential hazards that might impact performance.

The second stage involves the identification of defences that are currently in place and that prevent the hazard from impacting the safety of the operation. This provides the basis for the third stage of the process where deficient defences are identified by participants, and additional defences and/or changes to existing defences are proposed. In the aviation context, this process has resulted in the identification of a number of operational hazards including:

- High workload during passenger boarding
- Poor communication from Air Traffic Services
- Violent passengers
- In-flight turbulence
- Poor communication between operational areas
- A lack of Licenced Aircraft Maintenance Engineer retraining
- Poor cross-checking
- Poor passenger control on the tarmac
- Damage to aircraft during towing
- Inclement weather (Edkins, 1998)

Once the key hazards that are likely to impact performance have been identified, their occurrence can be monitored electronically with the data communicated to safety

Hazard Analysis

regulators. Changes in the frequency of interactions and/or the pattern of hazards can be used to trigger an industry-wide intervention in the absence of an incident or accident report.

17.3 MINIMISING THE IMPACT OF HAZARDS

There are a number of strategies that can be employed to mitigate the impact of hazards on performance. While the optimal approach involves either the removal of the human from the hazardous environment, or the removal of the hazard from the operational environment, the latter tends to be more difficult, particularly in complex environments. Therefore, there continues to be a proliferation of automated and robotic systems that are designed to cope with the specific demands associated with hazardous environments and that obviate the requirement for direct human involvement.

Where exposure to hazards or the removal of hazards is not possible, other strategies need to be considered, including the provision of guards that are intended to prevent injuries (see Figure 17.2). For example, construction is a work environment that is inherently hazardous, and guards such as hard hats, barriers, and ear protection are necessary to prevent injuries. In addition to guards, warnings may be provided in the form of labels or signs. These might be used in situations where the hazard is periodic such as at traffic lights or pedestrian crossings.

17.4 HAZARD WARNINGS

In most cases, warnings of hazards are designed to modify personal behaviour to reduce the incidence and, therefore, the risk of injury (Meyer, 2004). Behavioural analysis suggests that there are three major mechanisms to modify human behaviour, including the provision of a reward, a warning, or punishment (Long & Kearns, 1996). The impact of these mechanisms on human behaviour is summarised in Table 17.1.

Warning signs are used in almost every facet of daily life as a means of modifying human behaviour. Typically, they are divided into four distinct categories on the basis of the consequences of violating each sign. Examples of these categories are provided in Figure 17.3.

The precise nature of warnings and warning signs has been the subject of considerable research, and there are a number of recommendations for the construction of warning signs within particular environments. For example, Aurelio & Newman (1997) examined the nature of the cautionary signs that warn against the wide

TABLE 17.1
A Summary of the Impact of Behavioural Mechanisms on Behaviour

(a) Reward	Induces behaviour
(b) Warning	Discourages behaviour
(c) Punishment	Prohibits behaviour

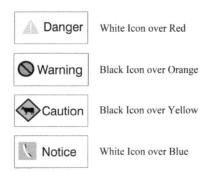

FIGURE 17.3 Examples of the various categories of warning signs and the standard colours with which they are associated. (Adapted from Aurelio and Newman, 1997.)

turns made by trucks. The authors noted that a wide variety of signs had been used, including text-based and icon-based designs, and that there was only limited standardisation. According to the American National Standards Institute (1991), a warning must contain *at least* a signal word, an indication of the hazard, and the consequences of the failure to comply.

Clearly, labels are not the only mechanism for communicating warnings, and Thompson (1995) cites the example where a verbal warning is provided to passengers prior to disembarkation from aircraft. In this case, passengers are typically advised to: 'open the overhead lockers with caution, as the contents of the locker may have shifted during the flight' (Thompson, 1995, p. 25).

In the absence of other information, this verbal warning provided to passengers violates the minimum standards for warning design, since there is no indication as to:

(a) The consequences of opening the storage bins;
(b) The seriousness of the consequences in injuries; and
(c) The behaviours that can be initiated by passengers to prevent the contents from either moving or falling.

Irrespective of the particular warnings provided to safeguard human performance, it might be argued that the necessity for a warning is an indicator of an inadequate design and that attention should be drawn to strategies that might include the modification of the design. Ideally, this obviates the requirement for a warning or other safeguards against hazards.

18 Cognitive Task Analysis

18.1 ANALYSING BEHAVIOUR

Part of the difficulty associated with the development of valid and reliable assessments of human behaviour is the level of interdependence that characterises the operational environment. Therefore, one of the mechanisms employed in exploring human performance in this context involves the deconstruction of a complex activity into a series of components. These components tend to be more manageable and are simpler to assess and develop than more complex human behaviour.

An example of the deconstruction of human behaviour can be drawn from the task of driving a motor vehicle. This is a complex activity involving a number of cognitive, perceptual, and psychomotor skills including the capability to:

- Judge distances
- Judge rates of closure
- Depress an accelerator
- Depress a brake and
- Judge the transition between gears

These skills can be deconstructed further so that, for example, judging distances might require the capability to:

- Perceive images
- Distinguish target information from distractors
- Relate the image to a known distance
- Recall a known distance from memory

By isolating the constituents of a more complex task, assessments of the relationships between stimuli and behaviour are presumed more reliable and accurate. Once the constituents have been isolated, they can be targeted for intervention and the impact assessed on overall performance (Hignett, 1996; Wei & Salvendy, 2004).

18.2 TASK ANALYSIS

Task analysis is a term that describes a collection of tools that can be used to identify the components of human performance within a complex system. It enables the acquisition of information pertaining to cognition and/or behaviour that contribute to the performance of a task. Data arising from a task analysis can be used for a number of purposes, including the identification of problematic elements of human performance, the development of guidelines for Instructional Systems Design (ISD), proposals for efficiencies, and/or the generation of design specifications (Kaempf, Klein, & Thordsen, 1991; Means, 1993; O'Hare, Wiggins, Williams, & Wong, 1998).

Task analysis is not necessarily initiated by the potential of a system fault. Rather, it is based on an assumption that certain activities must take place within a system, and that every activity needs to be performed successfully to ensure that the task is performed accurately and efficiently (Russ, Militello, Glassman, Arthur, Zillich, & Weiner, 2019). Therefore, one of the aims of task analysis is to identify the components of human performance, as a means of establishing those elements that are most susceptible to failure (Marti, 1998).

Task analysis is a top-down method of investigation where a complex outcome is deconstructed to identify its constituents. It is designed to target the relationship between human performance and the contribution to the overall performance of the system (White, Braund, Howes, Egan, Gegenfurtner, van Merrienboer, & Szulewski, 2018). The focus on this relationship is encapsulated within an approach referred to as the Technique for Error Rate Prediction (Nuclear Regulatory Commission, 1975).

18.3 TECHNIQUE FOR ERROR RATE PREDICTION

The Technique for Human Error Rate Prediction (THERP) was developed as a top-down strategy to estimate the probability of human error associated with the performance of a specific task (Dougherty & Fragola, 1988). It is an engineering technique where a task is deconstructed into discrete, observable actions (Ribeiro, Sousa, Duarte, & Frutuoso e Melo, 2016). These actions are listed in series that, combined, reflect a linear progression towards the achievement of a goal (see Table 18.1).

Where the performance of an activity does not lead directly to the achievement of the goal, an error is said to occur. According to the THERP framework, errors can be identified as errors of omission, errors of misdiagnosis, or errors of commission (Reason, 1990; Strater & Bubb, 1999; Swain & Guttman, 1983). The latter can be further subdivided into the performance of a task in the incorrect sequence (sequence error) or the performance of a task at an inappropriate time (timing error) (Dougherty, 1993).

Each step in the performance of a task can be examined based on the probability of an error of omission or commission. In the case of recharging a hybrid electric vehicle, an analyst might be interested in determining the probability that a low electrical charge warning will be overlooked (error of omission), or that the warning will be misinterpreted as a low fuel warning (error of commission). The cumulative

TABLE 18.1
A Task Analysis for Switching Fuel Tanks in a Piston-Engine Light Aircraft Using a THERP-Based Approach

Step	Means
Diagnose Event	
Detect the need to switch tanks	Time elapsed
Observe fuel flow	Fuel flow indicator
Observe fuel tank	Fuel tank gauge
Observe oil pressure	Oil pressure indicator
Observe oil temperature	Oil temperature gauge
Switch Tanks	
Engage fuel pump	Fuel pump switch
Change tanks	Fuel selector
Identify Fuel Flow	
Identify tank	Fuel selector
Observe fuel flow	Fuel flow indicator
Disengage fuel pump	Fuel pump switch

Source: Adapted from Dougherty and Fragola (1988, p. 60).

probability of the failure to adhere to the sequence of steps identified in the task analysis can be summarised as:

$$P[Task\ Failure] = \sum_i P[Step\ i\ Failure]$$

where P refers to the probability of an event [x] and Σ_i is the sum over i (Dougherty & Fragola, 1988). Therefore, the probability of a failure to successfully respond to the low electrical charge warning will comprise the sum of the probabilities for each step in the process.

It should be noted that this cumulative approach to probability assessment is based on a model of system performance where there is a limited degree of redundancy (such as a passenger noticing and interpreting the low-electrical charge warning correctly). However, most systems, irrespective of their complexity, will incorporate a level of redundancy such as self-monitoring of performance, the application of a checklist or mnemonic, and/or the observation of performance by another operator. As a result, the cumulative probability of a failure is more accurately reflected as:

$$P[Task\ Failure] = \sum_i (P[Initial\ Step\ i\ Failure] \times P[\text{Re}dundant\ Step\ i\ Failure\ |\ Initial\ Step\ i\ Failure])$$

where P refers to the probability of an event [x] and $P[x\ |\ y]$ refers to the probability of [x] given that [y] occurs. Therefore, the probability of a failure to successfully respond to a low-charge condition in a hybrid motor vehicle now comprises, for each step, the product of the probability that an initial failure will occur and the probability that a redundant failure will occur given an initial failure.

Complicating the matter further is the introduction of Performance Shaping Factors (PSF) to the model. PSFs such as anxiety, workload, and/or limited experience will increase the probability of a step failure, while a moderate workload and significant experience will tend to reduce the probability of a failure.

Swain and Guttman (1983) suggest that the PSFs associated with each step in the performance of a task need to be identified and that their impact needs to be incorporated as part of an estimate of the overall probability of a task failure. Therefore, the probability that a task will fail can be depicted as:

$$P[Task\ Failure] = \sum_i (P[Initial\ Step\ i\ Failure] \times \alpha \times P[Re\ dundant\ Step\ i\ Failure\ |\ Inital\ Step\ i\ Failure])$$

where a refers to the product of values of the PSFs that are associated with each step in the task. These values, and the probabilities associated with the initial and redundant failures, are normally obtained from a standard dataset that establishes the probability of a failure in highly specific situations. For example, the Human Error Probability (HEP) for an error of omission involving the use of instructions via written procedures, incorporating fewer than ten items using a 'check-off' strategy, has been estimated at 0.001 (Swain & Guttman, 1983). The HEP for the same checklist with more than ten items is estimated at 0.003.

Although revisions of the THERP model of probability estimation have attempted to incorporate cognition as part of the analytical process (Pan, Lin, & He, 2017), the strategies employed in THERP are more consistent with a behavioural approach to the analysis of a task. This is most evident in the quantification of human behaviour as a unitary value, in isolation from the performance of the system.

18.4 COGNITIVE TASK ANALYSIS

Unlike a behavioural approach to the analysis of a task, a cognitive approach is designed to integrate both the cognitive and behavioural elements of human performance. These data are considered in the context of the relationship between the operator and the system. Therefore, the aim of Cognitive Task Analysis (CTA) is to develop an accurate and reliable representation of the process through which an operator functions at all levels within a complex system (Crandall, Klein, & Hoffman, 2006; Seamster, Redding & Kaempf, 1997).

From a human reliability perspective, the difficulty associated with the application of CTA is that it will not necessarily yield a unitary, quantitative value of the HEP associated with a task. It is more likely to provide a combination of qualitative and quantitative data that can be employed to estimate the overall reliability of a system. This information is generally derived and interpreted on the basis of a theoretical model of the relationship between human performance and the operational environment.

The Cognitive Event Tree (COGNET) is one approach to task analysis that incorporates a cognitive model of human performance as a fundamental feature of the process of data acquisition and interpretation (Gertman & Blackman, 1994; Pan et al., 2017). The model posits causal relationships between the system and the human operator using the Skill-Rule-Knowledge (SRK) cognitive framework (Rasmussen, 1983). This gives rise to a taxonomy of the types of errors that are likely to be evident during the performance of a task, including slips, lapses, small mistakes, and mistakes.

The SRK model of cognitive performance is consistent with both symbolic and case-based models of cognition (McClelland, 1988). In the case of the former, skill-based performance might equate with the activation of procedural memory (productions), where performance is relatively rapid and accurate (Anderson, 1987). In contrast, knowledge-based behaviour is consistent with the activation of declarative memory, whereby task-related information is retrieved and processed in short-term memory. At this stage of information processing, human behaviour tends to be both time-consuming and error-prone.

Contrasting skill-based and knowledge-based behaviour, rule-based performance appears to be more consistent with a case-based approach to cognition, since rule-based performance involves a level of control greater than is expected to be exercised during skill-based performance. Similarly, case-based reasoning appears to involve a level of cognitive processing that is greater than simply a nonconscious strategy of matching cues (Burke & Kass, 1996).

The integration of symbolic and case-based models of cognition represents a hybrid approach to the explanation of human performance. Therefore, it embodies an assumption that both cases and production rules can be used to direct subsequent human behaviour. One of the aims associated with CTA is to elucidate such information as a means of explaining the process through which specific tasks are performed within an applied environment (Militello & Hutton, 1998).

There are a number of methods available to access operator knowledge for the purposes of CTA, including the use of structured and semi-structured interviews, verbal protocols, and by observing operators as they use their knowledge in the performance of the task (Russ et al., 2019; White, Braund, Howes, Egan, Gegenfurtner, et al., 2018). In many cases, these methodologies will be applied in combination to enable the acquisition of a broad-based overview of the cognitive and behavioural features involved in the performance of a task. This multifaceted approach to information acquisition enables the data arising from one form of data collection to be used to validate or triangulate the data arising from the application of another methodology (Rowe, Cook, Hall, & Halgren, 1996).

Since the goal of CTA is to clearly describe the mechanisms that underpin human performance, it is assumed that human responses are contingent on the behaviour of the system. Where there is change in the system state, a shift can also be expected in the response of the operator. This assumption provides the basis for the Critical Decision Method (CDM), where a non-routine event is used to elicit a response amongst operators that can be examined subsequently (Hoffman, Crandall, & Shadbolt, 1998; Klein, Calderwood, & Macgregor, 1989).

Typically, the application of CDM occurs in four stages. In the first stage, a non-routine event is used as a trigger to elicit a response, usually amongst Subject-Matter

Experts (SME). The SME is encouraged to describe the event in as much detail as possible to enable the development of a timeline and various decision points. The development of the timeline constitutes the second stage of the process of CDM (Kaempf et al., 1991; O'Hare et al., 1998).

From a cognitive perspective, the third stage of CDM is most significant, since it involves the introduction of cognitive probes as part of a broader cognitive interview. Cognitive probes are designed to elicit information such as the level of situation assessment at various stages of the process, any goals that may have been specified, any cues that were particularly significant, and any rules that may have been applied. This information provides the basis for the final stage of CDM in which any errors are identified in the knowledge elicitation process.

The CDM approach to the analysis of human performance has been employed successfully by Kaempf et al. (1991) to investigate the responses, amongst flight crews, to a fuel leak during a simulated flight in a Boeing 727. The performance of crews was recorded, as were any verbal utterances that were made during the scenario. In combination with post-task interviews, storyboards were developed that integrated the data and described the process of problem-resolution.

Although the quantitative performance of the flight crews indicated that there were difficulties involved in the management of the fuel leak, the additional data afforded by CDM revealed the precise nature of these difficulties. For example, it appeared that flight crews experienced some difficulty in both diagnosing the basis of the fuel leak, and minimising the impact of the leak (Kaempf et al., 1991). The cognitive interview revealed that this was due, in part, to an inadequate or inappropriate mental model of the function of the fuel system.

18.5 THE PRECURSOR, ACTION, RESULT, INTERPRETATION METHOD

Unlike CDM, the Precursor, Action, Result, Interpretation (PARI) method of CTA does not rely on a non-routine event to initiate the process of knowledge elicitation. Rather, two experts are used, one of whom poses a problem, while other provides a solution (Riggle, Wadman, McCrory, Lowndes, Heald, Carstens, & Hallbeck, 2014). Part of the PARI strategy involves the identification of an appropriate problem based on the experts' understanding of the nature of the operational environment. The problem is subsequently deconstructed into procedures that specify actions, and precursors and interpretation data that specify the context in which the actions occur (Clark & Estes, 1996).

18.6 CTA AND SYMBOLIC ARCHITECTURES

The PARI method, like other CTA strategies, is designed to elucidate the components of cognitive performance in a form that is usable, either for the purposes of design, or to direct subsequent performance. Therefore, symbolic architectures represent an attractive basis for CTA, since the features of cognitive performance can be described in the form of declarative or procedural (productions) knowledge. This type of symbolic architecture appears to provide an underlying cognitive structure on which the process of CTA is based (Roth, Woods, & Pope, 1992; Tofel-Grehl & Feldon, 2013).

Cognitive Task Analysis

The main advantage associated with the identification of productions is that it becomes possible to establish the probability with which an action (THEN) will necessarily follow a condition (IF), given various performance shaping factors. These productions may be highly developed, as in skill-based performance, moderately developed, as in rule-based performance, or relatively ill-defined, as in knowledge-based performance. In the case of skill-based performance, slips, lapses, and small mistakes are the types of errors that are most likely to be evident. At the rule-based level, slips, lapses, and mistakes tend to be prevalent, while at the knowledge-based level of performance, it appears that lapses and mistakes are most apparent.

Using an event or fault tree, it is possible to integrate productions with the types of errors that are most likely to occur at the various stages of an operation (see Figure 18.1 for an example). For example, the misinterpretation of a rule that is correct is regarded as a knowledge-based mistake denoted as K_m. In contrast, the failure to recognise a change on an instrument is a skill-based, small mistake denoted as S_{sm} (Gertman & Blackman, 1994).

For each error that can potentially occur within a system, a reliability estimate can be obtained using standard values in combination with a weighted value that accounts

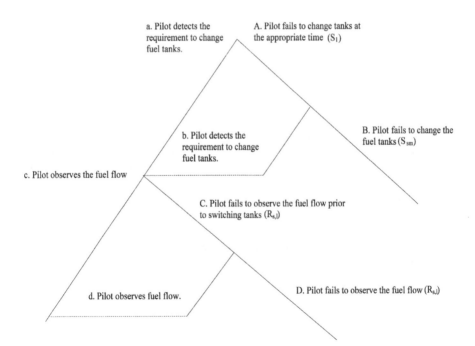

FIGURE 18.1 An event tree for changing fuel tanks in a piston-engine aircraft using the COGENT system. Note the nomenclature in the figure with appropriate performance identified by lower case letters, and inappropriate performance identified by upper case letters. The errors identified are denoted as Skill (S), Rule (R), or Knowledge-based (K) errors, each of which is associated with an error denoted as Slip (s), Lapse (l), Small Mistake (sm) or Mistake (m). (Adapted from Gertman and Blackman, 1994.)

for the impact of performance shaping factors. However, the difficulty associated with a purely quantitative approach to CTA is that it does not necessarily fully integrate the cognitive features that can impact human performance. While quantitative estimates can be calculated as to the impact of cognitive factors, it might be argued that these estimates remain subjective and, therefore, the model might be served equally by a more qualitative description.

18.7 COGNITIVE TASK ANALYSIS AND RELIABILITY

The use of qualitative forms of data analysis and representation has been a relatively common practice in CTA as a means of representing the complexity of the data (Klein et al., 1989; Russ et al., 2019). However, this contrasts with the more quantitative approach that has generally characterised approaches to assessments of reliability. One of the main advantages associated with CTA is that it has the capacity to highlight the performance shaping factors (PSF) that might impact human performance. It also provides a methodology that can contextualise the influence of PSFs. Therefore, CTA may have the potential to enable the development of accurate and reliable numeric weightings.

An example of the utility of CTA for reliability analysis can be drawn from an investigation of the response of two crews to simulated accident scenarios involving a Nuclear Power Plant (NPP). The NPP incidents involved the loss of coolant as a result of a valve failure between a high-pressure system and a low-pressure system (Roth et al., 1992). The valves failed in the open position, the result of which was the potential release of high-pressure radioactive fluid into the environment.

Although the crews eventually solved the problem, Roth et al. (1992) noted that they both experienced difficulties in diagnosing the source of the failure. These difficulties were examined subsequently by tracing the sequence of events using a qualitative, storyboard framework. The results indicated that the crews experienced problems with the integration of the information from what appeared to be a number of disparate sources. Roth et al. (1992) concluded that this was due to a failure to attend to all of the information pertaining to the incident.

From a reliability perspective, the interest in this incident might simply involve the development of an estimate of the probability that NPP crews would fail to attend to all of the information pertaining to an event. However, human performance tends to be a product of factors in addition to those that are directly associated with the operating environment. For example, a more detailed analysis of the results obtained by Roth et al. (1992) indicated that the operating procedures and the organisational culture created an environment that served to focus the crews' attention on a particular problem to the exclusion of other information. Therefore, the performance observed during the scenarios was likely a product of both the limitations associated with human information processing and prior experience in the form of operational procedures and training.

One of main advantages associated with a qualitative approach to CTA is that it facilitates the identification of cognitive PSFs that might otherwise have been overlooked (Joice, Hanna, & Cuschieri, 1998). Although information such as prior experience is difficult to quantify, it nevertheless has significant implications for predictions of human performance. It also provides an opportunity to intervene in an operational environment and change inappropriate organisational factors, thereby improving human performance.

19 Accident and Incident Analysis

19.1 ACCIDENT INVESTIGATION

Approaches to the investigation of accidents and incidents typically reflect an inherent philosophy, either to seek to explain a series of events, or to infer blame and responsibility for failures. From the perspective of the International Civil Aviation Organisation (ICAO), the purpose of accident investigation is fundamentally the explanation of the factors involved in the event, rather than the attribution of blame or liability (ICAO, 1986). The intent is to enable an objective analysis of the factors that may have contributed to an occurrence, unincumbered by personal or political interests, so that steps may be taken to prevent accidents and incidents of a similar nature in the future.

The focus on accident causation, rather than personal liability and responsibility, is significant, since it enables the acquisition of information from individuals who may fear the punitive consequences of their actions (Kee, Jun, Waterson, & Haslam, 2017). To this extent, accident investigation authorities are usually structured as organisations that operate independently of the regulatory authority. In the United States, for example, the National Transportation Safety Board (NTSB) is the organisation that is charged with investigating transport-related accidents and incidents. This organisation is directly responsible to the United States Secretary of Transport and operates independently of regulatory organisations such as the Federal Aviation Administration and the Federal Railroad Administration. A similar structural distinction occurs in United Kingdom, Europe, Canada, Australia, and New Zealand, where investigative agencies operate separately from regulatory authorities.

While the independence of accident investigation authorities facilitates the acquisition of information, there are a number of disadvantages associated with this type of organisational structure including a lack of communication, a lack of direct accountability, and the possibility that information may be subpoenaed in any subsequent court action. The lack of communication and accountability between regulatory and safety organisations is most evident following the recommendations that inevitably arise from accident investigations. Recommendations are the main impetus for change following accidents, although there is often no legal requirement for compliance by the regulatory authority (Newnam, Goode, Salmon, & Stevenson, 2017; Wood & Sweginnis, 1995). Further, the differing nature of the roles of accident investigation and regulatory authorities can lead to differences of opinion as to the optimal approach to resolve an issue (Rollenhagen, Alm, & Karlsson, 2017).

An example of differences of opinion between investigative and regulatory authorities can be drawn from the proposed introduction of oxygen masks for aircraft cabin crew as a mandatory requirement. The NTSB formulated the initial recommendation following the crash of a Pan American Airways aircraft in 1973. This was part of an overall recommendation for the provision of Crew Protective Breathing Equipment (CPBE) in the event of a cabin fire (Barlay, 1990). This recommendation was given further credence following an in-flight fire aboard a Varig Boeing 707-345C near Saulx-les-Chartreau, France in July 1973. The flight crew decided to crash-land the aircraft, since the cabin crew were forced to fight the fire without the aid of oxygen. It was anticipated that deploying the cabin oxygen masks may have further fuelled the fire. While these events forced the FAA to consider the introduction of CPBE as mandatory equipment for both cabin and flight crew, it was only in 1990 that the regulation finally came into effect.

19.2 ACCIDENT INVESTIGATION AND AVIATION

Accident investigation has a long history in aviation, and has provided one of the most significant sources of information leading to improvements in safety. The first accident report was submitted in 1909, and involved the crash of the Wright Flyer. Orville Wright was conducting a flight at Fort Myer, Virginia, when the propeller blade separated from the aircraft. The loss of thrust resulted in a series of aerodynamic stalls that culminated in the crash of the aircraft and the death of the only passenger on board (Kargl, 1994).

Even by contemporary standards, the accident report submitted in respect of the crash of the Wright Flyer was extremely detailed, and provided a guide towards structural improvements in aircraft design. However, while there were a number of witnesses to the crash of the Wright Flyer and the significant factors were readily apparent, the vast majority of accidents require a retrospective analysis and the reconstruction of events based on probability. An example of this process of accident investigation involved the loss of a series of de Havilland Comet aircraft following World War II.

The Comet had a relatively chequered history, and was introduced in 1949 as the first of the jet airliners to undertake regular passenger services. The Comet Series 1 carried 36 passengers and was powered by four turbojet engines positioned within the wing roots. In addition, cabin pressurisation enabled the aircraft to cruise at high altitude, thereby ensuring a level of passenger comfort, unrivalled at the time.

The first of the accidents involving Comet aircraft occurred during a British Overseas Airways Corporation (BOAC) flight from Calcutta (Kolkata) to Delhi on the anniversary of the introduction of the aircraft. Wreckage from the aircraft was found scattered over a 5-kilometre area, suggesting that the aircraft had disintegrated at approximately 7,000ft during the initial climb. The subsequent investigation indicated that the aircraft had suffered structural damage as a result of excessive loading of the elevator spars. This was consistent with a 'pull-up' manoeuvre which may have been exercised by the pilots when they encountered a severe downdraft caused by thunderstorms in the area (Brady, 2017).

In January 1954, a second BOAC Comet, registered G-ALYP, was lost 44 miles off the Island of Elba, during a flight from London to the Far East. On April 8, another BOAC Comet disintegrated, prompting the suspension of services pending the outcome of an investigation. What followed was an exhaustive recovery and reconstruction of the disintegrated fuselage of G-ALYP in an effort to determine the cause of the failure (Brady, 2017). Sabotage was quickly discounted and attention centred around the process of pressurisation and depressurisation.

The violent nature of the disruptions suggested that rapid decompression may have been a factor which precipitated the failure. As a result, investigation authorities immersed the fuselage of a Comet aircraft in a water tank at the Royal Aircraft Establishment at Farnborough which was designed to simulate the external pressures on the fuselage that would normally exist at altitude. The aircraft was depressurised and pressurised repeatedly as if it were operated under normal conditions.

Following an exposure to conditions equivalent to 9,000 flying hours, the aircraft testbed at Farnborough finally revealed the cause of the failures. Combined with evidence that had emerged from the recovery of G-ALYP, it appeared that an explosive decompression had occurred due to fatigue cracks emanating from rivet holes around the forward cabin window, the escape hatch, and around the Automatic Direction Finder (ADF) antenna.

While the aircraft testing facility was primarily designed to determine the cause of the failures associated with the Comet, the data also revealed a number of significant features associated with cabin pressurisation. For example, the Comet was designed so that the windows and antennae were 'cut-out' of the fuselage. It emerged that this manufacturing technique had the potential to create hairline fractures within the fuselage. Subsequent aircraft were designed with smaller windows and strengthened hulls to ensure that the impact of fractures was minimised.

The circumstances involved in the Comet crashes illustrate the significance of accident investigation in initiating improvements within complex systems. However, they also reveal the difficulties faced in isolating the myriad factors that might contribute to an accident.

19.3 THE PROCESS OF ACCIDENT INVESTIGATION

Due to the role of accident investigation as an integral part of performance improvement, together with increasing demands for safety, the aviation industry has developed arguably the most sophisticated models and approaches to accident investigation, influencing policies and procedures in other domains, including in medicine and the maritime industry. For example, international guidelines are published by the International Civil Aviation Organisation (ICAO) as part of *The Manual of Aircraft Accident Investigation* (ICAO, 1986). While the process of accident investigation is subject to the national laws of the state within which an accident occurs, the principles are generally applied internationally. For example, in the majority of jurisdictions, the transcripts from both the cockpit voice recorder (CVR) and flight data recorder (FDR) remain the property of the investigating authority at the moment of a crash. The intention is that the information arising from these recordings will be used for the purpose of investigation, rather than litigation. Nevertheless, in some countries, legal authorities have the power to subpoena both the CVR and FDR transcripts, if required (van Delan, Legemaate, Schlack, Legemate, & Schijven, 2019).

According to ICAO (1986), an aircraft accident is defined as:

> an occurrence associated with the operation of an aircraft with the intention of flight until such time as all such persons have disembarked, in which a person is fatally or seriously injured, the aircraft sustains damage or structural failure, or the aircraft is missing or completely inaccessible.
>
> (p. 4)

A 'serious injury' involves hospitalisation for at least 48 hours, a fracture, severe haemorrhage, injury to an internal organ, second or third degree burns over more than 5 per cent of the body, or verified exposure to infectious substances or radiation. From a technical perspective, these definitions include a wide range of events, from the complete destruction of an aircraft to relatively minor ground collisions.

An aircraft incident does not necessarily involve aircraft damage, and is defined as 'an occurrence, other than an accident, associated with the operation of an aircraft which affects or could affect the safety of the operation' (p. 4). This might include a near miss, a failure to comply with air traffic control instructions, or an emergency landing.

Beyond aviation, an occupational accident is defined by the International Labour Organisation (ILO) as 'an occurrence arising out of, or in the course of, work which results in a fatal or non-fatal injury, e.g. a fall from a height or contact with moving machinery' (ILO, 2015, p. iv). It is distinguished from a dangerous occurrence, where an event occurs that has the potential to cause injury. Like an incident, a dangerous occurrence presents an opportunity to prevent an accident, provided that the process of investigation is sufficiently detailed and robust to identify those features that may have contributed to the occurrence.

19.4 PRINCIPLES OF INVESTIGATION

From an applied perspective, the most important aspect in the initial response to an accident is the security of the site and the removal of any victims. In most industries, the security of the accident site becomes the responsibility of the investigating authority (Choudhry & Fang, 2008). However, there are occasions where accident sites have been disturbed, particularly when access to the site is delayed. This can result in the loss of information critical in establishing the factors that may have contributed to an occurrence.

Having secured the accident site, the complexity of the situation and the need to acquire relevant information quickly and accurately can be challenging and requires a degree of organisation and planning. In part, this involves the establishment of standard practices and procedures to ensure that all of the information necessary for the investigation is coordinated and processed systematically. However, it

Accident and Incident Analysis

also requires a capability to recognise those conditions when specialist assistance is required, either in the recovery of material and/or in the interpretation of information arising from the material collected. For example, where metal fatigue might be suspected, the services of a metallurgist might be sought during the collection of material to ensure that potential evidence is not lost in the transfer of material from the accident site.

Some of the other activities associated with the process of accident investigation include:

- The establishment of safety rules;
- Photographing the accident site;
- The collection of perishable items;
- The creation of a wreckage inventory;
- The development of a wreckage diagram:
- The recovery of the wreckage; and the
- The reconstruction of the wreckage (Wood & Sweginnis, 1995).

From an organisational perspective, accident and incident investigation requires the prioritisation of tasks, together with the cautious collection of information. It also requires a detailed process of information management that enables both access and a capability to cross-reference information. Finally, there is a requirement for a periodic assessment of the information collected and an analysis of the significance of this information in the context of the investigation.

19.5 INFORMATION SOURCES

The information available for accident investigation is normally divided into two discrete sections: Primary Sources and Secondary Sources. The former includes the information derived from witnesses, the accident site, and any pathological results. Primary information may also be derived from the equipment used by other personnel engaged in similar or related roles. Finally, the investigator may have access to video recordings and/or communication transcripts, including telephony devices.

Evidence from the accident site provides a number of significant clues as to the sequence of events that might have preceded the occurrence. For example, the nature of the injuries experienced by motor vehicle drivers and passengers will indicate whether or not restraints were worn prior to the collision (Staff, Eken, Hansen, Steen, & Søvik, 2012). Similarly, information can also be derived from the distribution of wreckage. In the aviation context, a greater dispersion of wreckage is typically associated with a shallower angle of impact. Figure 19.1 provides an illustration of an aircraft where the wreckage is contained suggesting that the aircraft impacted terrain at a steep angle. The nature of the damage to the aircraft and, in particular, the twisting of the fuselage, suggests that the aircraft impacted the ground while in a spin. This concurs with the high impact angle, since an aircraft in a spin would normally be travelling near the vertical.

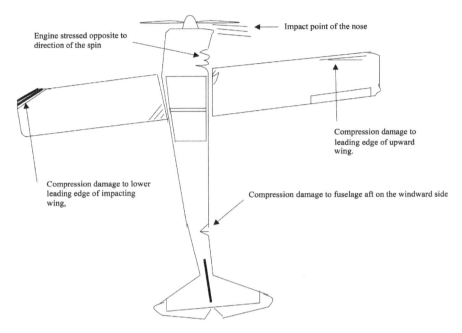

FIGURE 19.1 A diagram of the features that might be evident following the crash of a spinning aircraft. (Adapted from Wood and Sweginnis, 1995.)

Where the remains of a vehicle are dispersed, it is important to develop an overall perspective of the sequence of events that may have occurred during the impact. Normally, a diagram is appropriate in this type of situation, since a visual depiction of the sequence of points of impact should emerge. In the case of motor vehicles, evidence of tyre marks would also indicate the point at which the driver reacted to the situation.

In accident investigation, the most common diagram is referred to as the Straight Line Method, and it simply depicts the position of the various features of the accident site along a straight line (see Figure 19.2). In the case of Figure 19.2, the occupants appeared to be thrown clear of the vehicle shortly following the initial impact. Wreckage from the vehicle comes to rest beyond the remains of the victims, suggesting that either the occupants were unrestrained or that the impact was of such force that the vehicle disintegrated and dislodged secure points that would have otherwise restrained the occupants.

Secondary sources of information employed in explanations of accident causation include statistical analyses, accident databases, simulations, and empirical research. This information is intended to provide additional information that might assist in explaining the events that preceded an occurrence. However, as Harle (1994) notes, some caution should be exercised when applying secondary sources of information, since there are limitations concerning the extent to which information from secondary sources can be generalised. For example, interpretations based on confidential reporting systems rely on subjective reports that may not capture the broad range of issues that might be associated with an event. Similarly, the outcomes of empirical

Accident and Incident Analysis

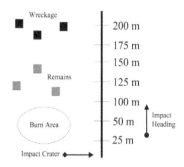

FIGURE 19.2 An example of the straight-line method of the depiction of an aircraft accident site. (Adapted from Wood and Sweginnis, 1995, p. 52.)

research may be of limited application where it fails to account for the multiple interactions that precede occurrences in practice.

In comparison to highly controlled experimental designs, simulation has the potential to provide a greater degree of ecological validity in recreating and thereby explaining the factors that may have contributed to an accident or incident. However, it also enables the evaluation of propositions that may otherwise remain speculative. For example, the use of simulation following the crash of Boeing 707 aircraft at Los Angeles, Pago Pago, and Tokyo revealed that, in each case, there was a tendency amongst pilots to land short of the runway, apparently under normal flight conditions (O'Hare & Roscoe, 1990). By testing systematically the nature of the various conditions under which this behaviour occurred, it was possible to deduce that the tendency to land short of the runway occurred under perceptual conditions where there was lower luminance in the foreground.

For pilots, approaching a runway with lower luminance in the foreground is associated with a visual illusion whereby the approach angle is perceived as greater than is actually the case. As a result, there is a tendency to correct for the apparent high angle of approach by reducing the altitude of the aircraft, eventually reaching a situation where the aircraft is at too low an altitude to reach the runway safely. Referred to as the 'black-hole effect', it is a particularly powerful visual illusion that remained poorly understood in the absence of systematic analyses using high-fidelity simulation (Roscoe, 1982).

19.6 THE WRITTEN REPORT

Accident reports should normally comprise two main sections: A factual information section and an analysis section. A third section may be included for concluding statements. The factual information section of the report generally includes the chronological history leading to an occurrence, in addition to a summary of the injuries and damage sustained. Where appropriate, other subsections may include:

- Personnel information
- Wreckage and impact information

- Meteorological information
- Communications
- Medical and pathological information
- Fire reports
- Survival aspects
- Tests and research (ICAO, 1986)

The analysis section of an accident report contains information pertaining to the interpretation of events that may have preceded the occurrence. This might include the behaviour of the operators, their interaction with equipment or tools, and/or the involvement of other parties. In effect, the information is deduced from the factual information, the intention of which is to establish a reasoned explanation of the occurrence.

An illustration of the sequence from factual information through to a concluding statement can be drawn the following example relating to the use of Personal Protective Equipment (PPE):

FACTUAL INFORMATION
On inspection, the harness that the victim was wearing for working at height was not torn nor frayed and appeared to be in normal working order.

ANALYSIS
Inspection of the harness following the fall suggests that the victim may not have attached the harness prior to beginning his work at height.

CONCLUSION
The fall is likely to have occurred following the failure of the victim to attach his harness, prior to working at height.

The sequence in which the accident report is presented is intended to reflect a systematic approach to the analysis of an occurrence. However, this systematic approach also enables comparative analyses across different reports to ensure that the similarities and differences between occurrences can be discerned.

19.6.1 TEST FOR EXISTENCE

Inevitably, the post-hoc nature of accident investigation results in a level of subjectivity in the interpretation of the information available. Amongst teams, this can be especially problematic, particularly where investigators are involved from different backgrounds. The solution involves subjecting the information available to a series of tests to determine the accuracy and reliability of the conclusions drawn. The most important of these tests is the *test for existence* which is intended to establish the extent to which a particular factor was evident in an occurrence.

The test for existence begins by ensuring that all of the possible factors involved in the occurrence have been considered. This typically involves a meeting of the various members of the investigative team and a systematic analysis of the sequence

Accident and Incident Analysis

of events. At each stage of the sequence of events, the factors that are presumed to be involved are identified and considered.

Each of these factors is subsequently weighted in the context of the sequence of events. Empirical evidence might be used at this stage to support a particular assertion as to the likelihood that one or more factors may have been evident in an occurrence. Further, the circumstances might be compared to other accidents or incidents of a similar nature. The outcome of this process is a consideration of the likelihood that a particular factor was, or was not, involved in an occurrence (ICAO, 1993).

19.6.2 Test for Influence

Having established that a particular factor was involved in an accident or incident, the next stage requires a test of the extent to which the factor was influential in precipitating the occurrence. This process involves a more detailed consideration of the empirical research in the area against the characteristics that are likely to occur. This approach is particularly applicable in the context of human performance where information can be drawn from research across a broad range of topics to support an explanation for a sequence of events.

An illustration of the importance of empirical research in tests of influence can be drawn from the collision of a passenger train at Norwich in the United Kingdom on July 21, 2013. Operated by Greater Anglia, the train was a two-car diesel unit operating as Train 2C45, a scheduled passenger service from Great Yarmouth to Norwich. Since Norwich is a terminating station, more than one train can be stabled on a platform at any one time.

The service was approaching Platform 6 at Norwich which was already occupied by stabled trains. The driver of 2C45 indicated that he was aware that trains were stabled at Norwich and that, in his experience, this was not an unusual occurrence. However, at 0011 hrs, 2C45 collided with the stationary train at approximately 13 km/h. Eight passengers required hospitalisation.

The Rail Accident Investigation Branch (2014) concluded that one of the factors that likely influenced the driver's failure to stop the train prior to the collision was a lack of concentration. In explaining the reasoning associated with this conclusion, the RAIB (2014) noted that maintaining concentration constitutes a non-technical skill required by drivers, and that the Rail Safety and Standards Board had published recommendations as to how non-technical skills, including maintaining concentration, could be included in assessments of driver competency. These recommendations emerged from the outcomes of an analysis of the integration on non-technical skills as part of the safety management systems employed within selected rail companies. However, at the time of collision, Greater Anglia had yet to incorporate non-technical skills in its competency management system.

19.6.3 Test for Validity

The final test involves an examination of the validity of the conclusions drawn following the investigation of an occurrence. Since the aim of accident and incident investigation is the prevention of future occurrences, an assurance is necessary that any learning outcomes or initiatives that emerge from the process of investigation are based on a comprehensive, reasoned argument. Therefore, the main criteria for this test are: (a) the extent to which an analysis is systematic; and (b) the extent to which empirical evidence was applied that tests hypotheses that may have emerged during the investigation. The main aim is to ensure that causal relationships that are inferred as explanations of an occurrence are based, as much as is possible, on the evidence available (ICAO, 1993).

19.7 ACCIDENT INVESTIGATION PROTOCOLS

Irrespective of the domain, investigators often note that there are no new accidents; simply different versions of accidents that have occurred previously. These similarities between occurrences enable comparisons and the examination of trends in the frequency of different types of occurrences. This, in turn, enables investigations to identify, at a systemic level, potential latent conditions that may have contributed to a series of incidents and/or accidents.

Accurate and reliable comparisons between occurrences requires that features are categorised using a consistent classification or protocol. From a human factors perspective, protocols are typically based on a conceptualisation of human performance that locates features within categories that are based on functional constructs. For example, Nagel (1988) used a broad classification based on a three-stage model of information processing. In this case, features of human performance associated with an occurrence can be related to functional constructs incorporating a stimulus component, a response selection component, and a response execution component. These are labelled information, decision, and action categories respectively.

A more detailed taxonomy has been developed by Wiegmann & Shappell (2003) in an analysis of human error-related accidents involving United States Navy and Marine Corps aviators. Referred to as HFACS, it was developed by integrating a four-stage conceptualisation of information processing (Wickens & Flach, 1988), Rasmussen's (1982) model of human internal malfunction, and Reason's (1990) model of unsafe acts.

In developing HFACS, 1,970 aircraft accidents were examined and classified according to 289 standardised pilot-causal factors. A maximum of three causal factors could be assigned to any single accident. The process resulted in 4,279 pilot-causal factors that were analysed subsequently using each of the three classification protocols. The results indicated that 80 per cent of the pilot-causal factors could be classified according to the four-stage information-processing model, where 84 and 88 per cent of these factors could be classified according to the unsafe acts and internal malfunction models respectively (Wiegmann & Shappell, 2003). However, the distribution of results was relatively consistent across the three models, with procedural and response-execution errors accounting for the majority of pilot-causal factors.

Consistent with previous research, the results obtained by Wiegmann and Shappell (2003) revealed a strong delineation between the factors involved in minor and major aircraft accidents (O'Hare, Wiggins, Batt, & Morrison, 1994). Specifically, procedural errors tended to be more apparent in minor accidents, while judgement and decision

Accident and Incident Analysis

errors tended to be more apparent in major accidents. From an accident investigation perspective, this suggests that the more severe the consequences of the accident, the greater likelihood that judgement factors were involved.

While the application of these protocols is normally undertaken retrospectively, it should be noted that protocols are also applied for the purposes of accident prevention. Rasmussen (1982), for example, developed a model of information-processing failures that was subsequently developed into a proactive protocol by O'Hare, Wiggins, Batt, & Morrison (1994).

19.8 PROBABLE CAUSE VERSUS SIGNIFICANT FACTORS

One of the most important issues facing contemporary approaches to accident investigation is the notion of 'probable or root cause' versus 'significant factors'. The term 'cause' is defined by ICAO (1986) as 'actions, omissions, events, conditions, or a combination thereof, which led to the accident or incident' (p. 6). Clearly, there are a number of issues arising from this definition, not the least of which is the assumption that accidents and incidents are the product of a single cause. According to a number of researchers including Reason (1990), human error-related occurrences are a product of a series of failures, each of which contributes to the final outcome.

A reference to 'cause' in the context of an accident or incident is a statement of fact, rather than an explanation of the factors involved. While there has been a considerable level of criticism associated with the use of the term 'probable or root cause' (e.g. Dekker, Cilliers, & Hofmeyr, 2011), a number of investigative authorities, including the National Transportation Safety Board in the United States continue to include a statement to this effect at the conclusion of accident reports. In other jurisdictions, the term is no longer applied and has been replaced by the term 'significant factors'. This is intended to take into account the diversity of factors that may have contributed to an accident and incident, rather than highlighting a singular, principal causal factor that might subsequently impede opportunities to prevent future occurrences.

Drawing on significant factors as part of an accident investigation is an approach advocated by the International Civil Aviation Organisation (1993) and should be based on the following principles:

- All causes should be listed, usually in chronological order;
- Causes should be formulated with corrective and preventative measures in mind;
- Causes should be linked and related to appropriate safety recommendations; and
- Causes should not apportion blame or liability (p. 32).

19.9 HUMAN FACTORS AND ACCIDENT INVESTIGATION

Typically, the role of the human performance accident investigator is one of adviser to a larger, investigative team. This is consistent with the multidisciplinary approach typical of contemporary accident investigation. It also reflects an assumption that accidents and incidents are often a product of human performance and latent conditions that breach the defences that exist within a system. This systemic approach to accident investigation has shifted the focus from the unsafe acts that might occur at the operational level towards the latent conditions that occur at an organisational level and that might have contributed or provided the opportunity for unsafe acts.

This change in the philosophy of accident investigation was particularly evident in the approach to aircraft accident investigation undertaken by the then Bureau of Air Safety Investigation (BASI) in Australia. For example, as early as 1990, BASI began to use the term 'significant factors' in preference to 'probable cause' when referring to the concluding statements of an accident report. This signified the recognition that aircraft accidents were generally the product of a number of causal factors, each of which contributed to the sequence of events.

Furthermore, BASI and its replacement, the Australian Transport Safety Bureau (ATSB) placed considerable emphasis on the identification of factors, in addition to operator errors, since these are presumed to promote the environment within which unsafe acts occur at the operational level. This emphasis is highlighted through a special investigation report issued by BASI, concerning the near collision between two Boeing 747 aircraft in September 1990.

The incident occurred when a Qantas Boeing 747-300 was towed across the path of a departing Cathay Pacific Airways Boeing 747-300 at Sydney (Kingsford Smith) Airport (BASI, 1990). The Qantas aircraft was cleared by the Surface Movement Controller (SMC) to cross Runway 34 under tow. The section of the runway north of the intersection between two crossing runways was normally under the control of the SMC. All other areas of the runway were under the control of the Aerodrome Controller (ADC). The ADC took control of the full length of Runway 34 when the entire length of the runway was required. Such was the case for the Cathay Pacific Boeing 747 (BASI, 1990).

Under normal operating conditions, the ADC would select the 'runway 16 in use' switch to indicate that:

- There was a change in the system status;
- An aircraft required the full length of Runway 34;
- The northern end of Runway 34 was not available to the SMC; and
- The ADC required a verbal acknowledgement of the status change from the SMC (BASI, 1990, p. 15).

In this case, the ADC had failed to select the 'Runway 16 in use' switch, and had assumed that the Qantas aircraft would remain stationary at the holding point. The Cathay aircraft was subsequently cleared for take-off as the Qantas aircraft was being towed across the runway. As a result, the aircraft came within 36 metres of a collision (BASI, 1990).

While the lack of communication between the ADC and SMC was clearly a significant factor in precipitating the incident, the investigators noted that there were a number of other significant factors that either contributed to the incident, or failed to

Accident and Incident Analysis

prevent the incident from occurring. For example, reference was made to the lack of documentation regarding the coordination of procedures between the ADC and the SMC. In addition, these procedures were not failsafe and were considered susceptible to misinterpretation.

From an accident investigation perspective, the report is illustrative of the shift towards factors beyond the direct performance of the operators involved. Nevertheless, this change has required the development of additional knowledge cues and a broader understanding of the factors that may impact on human performance within the operational environment.

To assist in identifying human performance factors in accident investigation, a number of checklists have been developed, the aim of which is to guide decisions concerning the relative impact of various factors and thereby prioritise features that may require additional analysis. Typically, the checklists of factors are derived from broad categories including behavioural factors, medical factors, operational factors, equipment design factors and environmental factors. An example of part of a checklist based on the SHELL model is provided in Table 19.1.

As a taxonomy, the application of human performance checklists provides a level of standardisation between both the investigators and investigation authorities, thereby enabling analyses at a systemic level.

TABLE 19.1
An Example of Part of an Aircraft Accident Checklist Based on the SHELL Model

Liveware-Hardware*	Liveware-Environment
Equipment	**Internal**
Switches, Controls Displays	*heat, cold, humidity*
instrument/controls design	*ambient pressure*
instrument/ controls location	*illumination glare*
instrument/ controls movement	*acceleration*
colours, markings, illumination	*noise interference*
Workspace	*vibrations*
workspace layout	*air quality*
workspace standardisation	**External**
communication equipment	Weather
eye reference position	*weather briefing*
seat design	*weather: actual & forecasts*
restrictions to movement	*weather visibility*
illumination level	*turbulence*
motor workload	*whiteout*
information displays	

Source: ICAO (1993, p. 44).
* Human–machine interface.

19.10 CRASH 'RECORDERS'

Within the commercial aviation environment, the cockpit voice recorder (CVR) and the flight data recorder (FDR) are two of the most significant resources available in the event of an aircraft accident. Known colloquially as 'black boxes', the CVR and FDR are designed to record a series of aircraft and flight-crew parameters on a continuous basis (Hughes, 1996). Older recorders used magnetic tape for storage purposes, while the newer models use digital recordings.

TABLE 19.2
Transcript of a Sundstrand V-577 Cockpit Voice Recorder S/N 2282 Removed from an Air Florida B737 Which Was Involved in an Accident at Washington, DC, on January 13, 1982

Intracockpit		Air-Ground Communications	
Time and Source	Content	Time and Source	Content
		1530:48	
CAM-2	* Figure it out	TUG	You have your brakes on right?
CAM-2	We're too heavy for the ice	INC-1	Yeah, brakes are on
CAM-2	They get a tractor with chains on it? They got one right over here ((PA Announcement relative to pushback))		
		1531:33 AOPS	Palm ninety from American Operations
		1531:36 RDO-02	Palm ninety, go ahead
		1531:38 AOPS	Okay, your agent just called to tell me to tell you to amend your release showing nineteen twenty-five zulu per initial RH
		1531:51 RDO02	Okay, nineteen twenty-five Romeo hotel thanks

Source: Adapted from NTSB (1982, p. 101).

Notes: TUG = tractor; INC = intercom; AOPS = American operations; PA = public address system; * = unintelligible word; (()) = editorial insertion. All times are expressed in local time based on the 24-hour. clock.

Each recorder is enclosed in a crashworthy casing, usually coloured orange to enable recovery following impact. In addition, recorders are typically situated in the rear of the aircraft to ensure that they are subject to the least disruption during the accident sequence (Reynolds & Bayan, 1999).

Outside aviation, recorders are also provided in trains, commercial motor vehicles, and marine vessels. Referred to as event data recorders, they operate similarly to flight data recorders and are intended to capture activity pertaining to the vehicle or vessel.

19.10.1 THE COCKPIT VOICE RECORDER

In the aviation context, the CVR operates on a continuous 30-minute loop, and is designed to record the auditory information within the cockpit area, including the communication between flight crew, engine noise, warning devices, and communication with services including air traffic control (Reynolds & Bayan, 1999). In the event of an aircraft accident, a written transcript of the recording is normally compiled and is listed in chronological order (see Table 19.2 for an example).

In most cases, the CVR transcript is treated more sensitively than other factual information pertaining to an accident or incident. This reflects the fact that personal information may be discussed that may not be directly pertinent to the investigation. Consistent with this perspective, many aviation companies only allow access to the CVR in the event of an aircraft incident or accident.

19.10.2 FLIGHT DATA RECORDER

The flight data recorder (FDR) and the digital flight data recorder (DFDR) are designed to record parameters pertaining to the state of the aircraft (Reynolds & Bayan, 1999). DFDRs are capable of monitoring a considerable number of parameters continuously over a 25-hour period. The output from a DFDR is a digital readout that can be interpreted subsequently to determine the precise position of an aircraft within a three-dimensional environment (Hughes, 1996).

While the FDR provides an extremely useful aid for accident investigators, the significant amount of information that is generated often results in a time-consuming process of data analysis. This process, however, must also be tempered by the possibility of erroneous interpretations, particularly arising from digital recorders. Consequently, Reeves (1995) has advocated the application of FDR data in conjunction with alternative sources of data to avoid misleading conclusions.

20 System Evaluation, Usability, and User Experience

20.1 SYSTEM ASSESSMENT

The assessment of a system can be a very complex and time-consuming process that itself is subject to a range of human performance issues that increases the potential for errors. However, it also represents one of the most important tools necessary to anticipate and respond appropriately to deficiencies that, unchecked, may lead to system failures.

The significance of systems assessment was illustrated in an incident involving the loss of an engine on a Nigerian 707-320C freighter over Southern Belgium in 1998 (De Wulf, 1998). Prior to the incident, the aircraft had conducted an emergency landing at Ostend Airport where maintenance engineers and the airport authority had expressed reservations concerning the condition of the aircraft. However, a subsequent inspection by the Belgian Civil Aviation Authority indicated that the aircraft complied with international airworthiness regulations. This occurred despite the fact that the aircraft was overdue for a maintenance overhaul.

20.2 INDICES

There are a number of mechanisms available to assess the accuracy and efficiency of a system, each of which yields an index of system performance. In isolation, these indices provide a relatively narrow representation since they typically constitute elements of a complex system. For example, while profit-loss statements may attest to the relative success of a construction company, an analysis of the rate at which accidents or incidents occur may paint a very different picture. Similarly, an assessment based on a limited range of events or over a limited time frame may not necessarily reflect performance given the complexity of the operational environment. Therefore, it is important to ensure that:

(a) Performance indices are appropriate for the particular context examined; and that
(b) Information is sought from a range of indices, and over an appropriate time period.

Another characteristic of indices that warrants consideration involves the extent to which an index is subject to stochastic variability. For example, the frequency with which marine accidents occurred in Aegean Sea between 1999 and 2009 ranged from 9 in 2004 to 106 in 2009 (Ventikos, Savrou, & Andritsopoulos, 2017). Over this period, the average annual frequency of marine accidents in the region was approximately 38. The extent to which the frequency of marine accidents revolves around the statistical mean reflects stochastic variability. There are a number of factors that contribute to a marine accident, and it would be unreasonable to suggest that the frequency of accidents within a limited period or across a limited selection of companies represents an accurate reflection of the safety of the industry more broadly. The nature of stochastic variability is such that it is impossible to draw valid conclusions from indices that reflect performance over a narrow period and/or where there are organisational or operational constraints.

20.3 PRODUCTIVITY

Where most assessments of a system are based on mitigating the potential losses that may arise through system failure, a productivity-based approach to system assessment is focused on the potential gains that can accrue from a system that meets the needs of users (Simpson & Mason, 1995). For example, if it is agreed that there is some level of cost associated with an inability to use a system accurately and efficiently, then it also must be agreed that there are potential gains that might be afforded by improving the relationship between the system and the operator.

On the basis of this assertion, it might be argued that whether or not an accident or incident actually occurs is largely irrelevant. For instance, if the normal operation of a system requires that the average user employ cognitive and/or perceptual resources beyond an optimal level, then these resources are being drawn from other, perhaps equally important, areas of the system. As a consequence, the user may be less productive than might be the case in a restructured operating environment.

An allied issue involves the rate at which gains can occur within an environment that is restructured to meet the needs of operators. For example, there is evidence to suggest that relatively minor changes within the operating environment can result in significant improvements in system productivity. Hendrick (1998) recounts a situation in which a $300 per unit investment in the ergonomic layout of tractor-trailers resulted in a $2000 per unit saving in downtime due to a reduction in accident damage caused by poor visibility from the operator's compartment.

The interdependent nature of socio-technical systems is such that the implications of changes within an operating system often have an impact within other areas of the operating environment. For instance, the redesign of a display terminal for directory assistance operators at Ameritech resulted in a 600 millisecond reduction in the duration of an average call. Although this constitutes a relatively small figure in isolation, extrapolating over an annual basis, the saving in time is significant for both the company and the caller. As a result, changes at one level of the system enabled improvements at another level, in this case, for the customer.

Productivity within an operating environment can be established using a number of mechanisms, depending on the types of issues under investigation. Where the aim is to determine the costs associated with a system, the life-cost can be calculated as:

$$\text{Total Life Cost} = n(x + Lt)$$

where n = the number of machines in use, x = the capital cost per unit, L = the life expectancy of the machine, and t = the operating costs of the machine per annum (Simpson & Mason, 1995). This provides an estimate of the total costs that are expected to accrue over the lifetime of a system.

The main advantage associated with the life-cost approach to productivity is that comparisons can be made between different systems on the basis of a relatively objective assessment of performance outcomes. For example, if the design of a system is such that the operating costs exceed the costs associated with an alternative system, despite comparable numbers of machines, capital cost, and life expectancy, then it is possible to establish the superiority of one system over another from a human factors perspective.

Despite the apparent simplicity of the life-cost approach to productivity, in many cases, the comparison between systems is based on a trade-off between a number of factors, many of which may be subjective. Therefore, users will not necessarily select a system on the basis of the extent to which it is consistent with appropriate human factors principles. For example, Nielson and Levy (1994) report a correlation of 0.46 between user preferences and performance. This indicates that, in many cases, users prefer systems that do not necessarily perform at an optimal level.

Nielson and Levy (1994) explain this apparent anomaly as a function, in part, of a lack of knowledge on the part of operators concerning the relationship between design and human performance. It appears that functionality may be important up to a point, beyond which other factors, such as cost and aesthetics, tend to take precedence in the decision-making process. Norman (1988) provides a similar explanation for the apparent discrepancy and suggests that the difficulty lies in determining the appropriateness of a design in the absence of personal experience. For example, simply examining a system visually, in the absence of a detailed test within the operational environment, is unlikely to reveal deficiencies in human performance.

Tognazinni (1992) provided further support for the dissociation between user preference and performance following an analysis of two different strategies for replacing text in a computer program. In one condition, a mouse was used to replace the text, and in the other, a set of keystrokes was used. Although the results revealed a more rapid mean response when using the mouse, users tended to prefer the use of keystrokes for this type of function. This is consistent with other evidence arising from analyses of users' preferences for particular displays and the use of colour (Andre & Wickens, 1995). Typically, users prefer the use of colour when displaying information despite the fact that there is little evidence to suggest that it improves performance. Moreover, Christ (1972) noted that, in some cases, the use of colour may, in fact, degrade performance.

In addition to the problem of dissociation, another difficulty associated with the life-cost approach to productivity assessment is the reliance on historical data to assist future decision-making. Alternative strategies include the development and implementation of pilot studies, the application of task analytic procedures, and/or the use of analogous systems. In each case, a level of caution needs to be exercised to ensure that the data arising from these analyses are both valid and reliable.

Where the aim of life-cost productivity assessment is to estimate the potential losses that might accrue from a failure to adhere to appropriate principles of design and usability, the increased revenue calculation provides an estimate of the potential gains that can accrue through an intervention. The application of the increased revenue calculation might occur where a customer is interested in determining the extent to which changes within a system will have a financial impact. For example, if the ergonomic characteristics of a machine are such that the additional time required to perform a particular function is calculated as six minutes (0.1/hour), then the total additional time for 100 employees on an annual basis can be calculated as:

$$0.1 \times 100 \times 1920 \text{ hrs (40 hrs/week for 48 weeks)} = 19,200 \text{ hrs}$$

If the total revenue per hour is calculated as $500.00, then the total costs attributed to the ergonomic deficiencies associated with the machine will amount to $9.6 million per year. Given that there are 50 machines in place, the cost per machine can be calculated as $192,000 per year. If it is assumed that the additional time to perform the task can be converted directly to income generation, then a simple cost-benefit analysis can be undertaken to determine whether the costs associated with improving the machine can be offset against the estimated gains. In doing so, a number of issues need to be taken into account, including the downtime to improve the machine, and whether or not the improvements are likely to ameliorate the problem. Notwithstanding these issues, the example demonstrates the potential losses in productivity that can occur as a result of a failure to optimise the relationship between the user and a system.

20.4 ORGANISATIONAL FACTORS AND SYSTEMS ASSESSMENT

As a system becomes more complex and decentralised, there is often a resulting loss in the capability of an organisation to monitor the system effectively. The consequence is an increase in the probability of a systemic failure that has an impact at a number of levels. An example of this type of situation can be drawn from a case described by Edmonson (1996), in which a hospitalised patient had been administered an inappropriate drug following heart surgery. Although the error was eventually noted and rectified, it occurred more than 24 hours after the patient had been admitted. During this time, the responsibility for the care of the patient had involved up to six healthcare professionals.

The number of individuals involved in this is incident is also compounded by the fact that the system of drug administration itself is a particularly complex process and is open to error. Bates, Leape, and Petrycki (1993) have identified ten points in the process of drug administration during which errors can occur, including the:

- Prescription
- Delivery of the prescription to the unit secretary
- Transcription of the order
- Collection of the order
- Verification of the order
- Transfer of the order to the pharmacist
- Dispensing of the medication
- Collection of the medication
- Administration of the medication
- Receipt of the medication by the patient

The complexity associated with this type of system means that it becomes extremely difficult to monitor effectively at the operational level. According to Sutherland, Canobbio, Clarke, Randall, Skelland, & Weston (2020), the problem lies in the variability and frequency of the processes that occur. Further, the administration of medication is not necessarily a reversible process, and the feedback arising from an error is neither direct nor immediate, inevitably resulting in the repetition of errors.

This type of nonlinear feedback is also evident in other complex environments such as aircraft manufacture. For example, in 1998, the Boeing Aircraft Corporation was forced to issue a service bulletin alert when an error in the production of the tailplane section of Boeing 777-200 resulted in an increase in the potential for corrosion. The error was only identified following an internal investigation of the production process and resulted in a requirement for repairs to at least 18 aircraft. The costs associated with the error amounted to $200,000 per aircraft (Jeziorski, 1999).

In addition to a level of interdependence within complex systems, the performance of users can also be impacted by the culture that pervades an organisation. This was evident in a comparison between eight hospital unit teams where positive relationships were noted between the rates at which errors occurred and characteristics of the organisational environment. It appeared that a greater frequency of errors was associated with a greater level of nurse-manager direction setting, the perceived unit performance, and the quality of the relationships within the unit. These apparently anomalous results were explained as indicative of an operational environment where there was a willingness both to recognise that errors occur, and to report these errors (Edmonson, 1996).

Within an organisation, the active reporting of errors likely reflects a 'just culture' which encourages the reporting of errors in the absence of retribution. However, with the introduction of a just culture, it also follows that the rate at which errors are reported is likely to increase. The consequence is the somewhat counterintuitive situation where an organisation that adopts strategies in an effort to learn from mistakes is also likely to report a relatively greater frequency of errors. Therefore, in assessing accurately the performance of an organisation using indices, it is important to ensure that the context is established within which the data are acquired.

20.5 USABILITY ENGINEERING

The success of a system or product will ultimately depend on the extent to which it meets users' needs. Where a system fails to conform to users' expectations, or the system is difficult to use as intended, it increases the likelihood of error (Norman, 1988). Therefore, the aim of effective systems design is to ensure that there is a level of congruence between the user and the designer in their perceptions and use of a system or product.

Usability engineering incorporates a set of guidelines designed to ensure that a system is usable at an operational level. The first, and perhaps most significant of these guidelines, is the requirement that systems are designed so that they meet clearly prescribed objectives. For example, a Global Positioning System (GPS) interface might include the usability specification that 80 per cent of users will be able to enter six pairs of co-ordinates in less than two minutes with fewer than two errors following a training period of 15 minutes. This type of specificity provides a goal for the design process and establishes clearly the point at which the process of design has been completed.

Despite the apparent simplicity in specifying criteria against which a device or system can be assessed, the development of usability specifications can be a complex process that relies on the identification of indices that accurately reflect performance and for which data can be acquired efficiently. This requires considerable discussion amongst stakeholders concerning the most appropriate levels of usability that might be required for a particular system or device. Typically, the difficulty involves reaching some level of agreement that is considered appropriate for users, but that is also achievable within the constraints of the design requirements. This balance between the 'optimal' and the 'possible' is a theme that recurs throughout the process of design.

A second guideline associated with usability engineering requires that potential users are incorporated as a fundamental element of the design process (Farao, Malila, Conrad, Mutsvangwa, Rangaka, & Douglas, 2020). The aim is to enable a cooperative approach to systems development and ensure that the outcomes of the design process are compatible with users' expectations and perceptions. This process is also consistent with the philosophy of 'user-centred' and 'participatory' design where users are involved from operational need determination through to systems evaluation (Stanney, Maxey, & Salvendy, 1997).

The final guideline advocated as part of usability engineering requires a cost-effective approach to the design process (Carroll, 1997; Gherardini, Renzi, & Leali, 2017). For example, where initial refinements of a system design may result in significant improvements in performance, subsequent changes may yield lesser returns, despite the continued costs associated with usability testing. Therefore, there is a point during the design process where the costs associated with refinement outweigh the impact of improvements.

The cost-effectiveness of usability engineering needs to be considered from a number of different perspectives, including the time involved for prospective users, the costs associated with the redevelopment of a system, and the opportunity cost in extending the duration of the design process. As a means of improving the cost-effectiveness of usability, testing should occur throughout the design process. The intention is to isolate those elements of the system that provide the greatest impact on the usability specifications. For example, during the design of a smart phone, it might be determined that the element that impacts the frequency of errors most significantly is the size of the icons and the distance between icons on the interface. Consequently, if the final mock-up of the design does not meet the usability specification pertaining to the frequency of errors, the designers can target the size of icons or the distance between icons as the elements that are most likely to improve performance. The advantage associated with this type of approach is that the time involved in the redesign of the system is used efficiently to maximise improvements within the system.

20.6 USABILITY AND GENERALISATION

In developing a usable system or tool, the most significant challenge involves ensuring that the design is equally usable amongst the range of intended users. Different users will draw on different skills, different experiences, and will use devices in different ways to achieve their goals. Therefore, what appears usable for one user in a specific context may be less usable for a different user and/or in a different context.

Dempsey and Leamon (1995) illustrated the challenges associated with assessments of usability in an examination of the extent to which bent-handled tools could be employed across settings in an industrial context. Bent-handled tools are designed to minimise the physical stress imposed on the upper body by hand-held tools such as hammers and pliers. In the case of pliers, the extent to which users could twist wire differed for different heights at which the task was to be performed. Therefore, the efficiency and the risks associated with the use of the piers differed across types of tasks, a characteristic about which users needed to remain aware.

A similar challenge was evident in the case of a bent-handled knife for use in a chicken factory. According to Armstrong, Foulke, Joseph, and Goldstein (1993), the angle between the knife handle and the blade was set at an optimal level. Consequently, the efficient use of the knife depended on the characteristics of the operating environment, including the height of the benches, and the type of meat-processing task being undertaken. Where the characteristics of the operating environment differed from the design criteria, the use of the knife tended to be less efficient and, in some cases, could exacerbate the risk of strain and injury.

At a general level, Nielsen (1994) provides a list of principles that can be used to assess the usability of an interface (see Table 20.1). The principles provide the basis for initial assessments of interfaces, referred to as a 'heuristic analysis'. An advantage associated with a heuristic analysis is the capability to identify opportunities for improvement at relatively minimal cost, since it constitutes an audit. This can be especially useful at the early stages of the design cycle. However, as an auditing process, it does not necessarily generate solutions. In this case, solutions are derived from the expertise of the designers and/or interactions with prospective users.

TABLE 20.1
Nielsen's Usability Principles

1. *Visibility*: It should be clear to the user what is occurring through timely, meaningful feedback.
2. *Transparency*: The information presented should correspond to information in the actual environment.
3. *User Control*: Users need to be able to exercise control over an interface to satisfy their goals.
4. *Consistency*: Information should be presented using standardised terminology, using a consistent format.
5. *Error Management*: Defences should be provided that prevent unintentional errors on the part of users.
6. *Recognition*: Information should be salient and meaningful to users, drawing on recognition memory.
7. *Flexibility and Efficiency*: Interfaces should account for individual differences in goals and pathways to achieve goals.
8. *Minimalist Design*: Features of an interface that might distract users should be minimised,
9. *Error Recognition and Recovery*: Error-related information should be presented clearly and succinctly, and should direct the user towards the recovery from an error.
10. *Documentation*: Documentation should be available and direct users towards the achievement of goals.

20.6.1 Usability and User Experience

Where usability targets the capability of a user to operate a system or device as intended by the designer, user experience, or UX, refers to psychological features beyond usability, including enjoyment, engagement, aesthetics, and satisfaction (Tan, Liu, & Bishu, 2009). At a social level, UX may also include perceptions of inclusiveness, acceptance, ownership, and/or support. It is a construct that is employed heavily in interface design, and is particularly important in commercial and information sites, where the interface constitutes a critical conduit for behavioural change, whether it involves the purchase of a product or the implementation of new procedures that are intended to prevent injury.

User experience is typically evaluated through the administration of surveys where prospective users are asked to rate various dimensions using a Likert scale (Peruzzini, Grandi, & Pellicciari, 2017). These data can be augmented with qualitative information arising from interviews, the goal of which is both to triangulate the data arising from surveys, and proffer suggestions that might be associated with improvements in UX. Subsequent interview and survey data can also be employed to determine whether changes in designs result in improvements in the outcomes.

There is a strong association between user experience and usability, with difficulties associated with usability impacting users' experience. A poorly designed interface will likely result in feelings of frustration amongst users. These are responses that may reduce the likelihood that the user will engage with the interface in the future with implications for productivity, safety, and/or sales.

While there is general agreement to the effect that user experience constitutes an important construct in the context of interface design, there is less consistency concerning the means by which user experience should be evaluated. Further, there are differences between professions in the perceived utility of the outcomes of evaluations of user experience (Law, van Schaik, & Roto, 2014). This probably reflects the variability in approaches to the assessment of user experience, and suggests that it is a complex construct, the evaluation of which is likely to be impacted by individual differences amongst users, the nature of the system under investigation, and the context within which the assessment is being undertaken.

20.6.2 DOCUMENTATION AND TRAINING

In addition to an evaluation of the utility of a device or system, usability testing also offers the opportunity to develop user documentation, include maintenance and user manuals. Since there may be limits to the usability that can be incorporated into the design of a system or device, the transparency of the design may be less evident for some users than it is for others. Consequently, the development of user documentation needs to involve prospective users, with the quality of the documentation potentially impacting perceptions of user satisfaction (Gemoets & Mahmood, 1990)

In developing user documentation as part of an assessment of usability and user experience, gaps are likely to become evident between the intended use of a device and the expectations of the minimum capabilities of users. Where changes to the usability of the device are unwarranted or unfeasible, training may be required to 'bridge the gap' between capabilities and expectations. For example, there are inherent dangers associated with firearms, the successful management of which only occurs through a combination of design (e.g. a safety switch that prevents an inadvertent discharge) and training (e.g. the process and necessity to apply the safety lock to prevent an inadvertent discharge).

The effectiveness of training strategies and documentation can be examined as part of a broader analysis of the usability requirements of a system or device. This approach considers the broader context within which a system or device is to be introduced. For example, in large organisations, the introduction of new technologies may have implications for upstream and downstream processes. It may involve changes in how information is managed, and the processes and procedures by which it is managed, which may impact other systems or capabilities. Evaluations of training and documentation can ensure that the transition to new systems is optimised and can provide an opportunity to identify potential latent conditions or barriers to optimal practice.

A secondary, although no less important part of the documentation process is the development and maintenance of accurate records that detail the nature of the system, together with a justification for the various design-related decisions that might have been made, and the basis on which they were made. The intent is to ensure that users and, in particular, maintenance personnel, have a clear understanding of the overall function and operation of the system, thereby enabling accurate and efficient diagnoses in the case of technical failures.

20.7 ARCHETYPES AND PERSONAS

Since users are likely to approach a design from different perspectives, it is important that these differences are taken into account in the context of user testing. However, testing every possible category of user can be both costly and time-consuming. The solution is to develop broad categories that represent a population of users from which a sample can be recruited and examined.

In identifying populations of potential users, characteristics can be defined demographically or from a behavioural perspective (Marshall, Cook, Mitchell, Summerskill, Haines, Maguire, Sims, Gyi, & Case, 2015). A demographic approach involves characterising users on the basis of features including age, sex, marital status, and/or ethnicity. Referred to as a *persona*, groups can be collapsed or subdivided, depending on the level of specificity required (Vincent & Blandford, 2014). For example, while marital status may be largely irrelevant in the context of an online shopping site, age and income may be retained as key features that differentiate the interests of specific user groups.

Complementing personas are archetypes that relate to the behaviour or interests of users. Archetypes can account for differences within groups that might otherwise be considered homogeneous (Marshall et al., 2015). For example, a younger female user might be engaging a banking website in one instance to check the balance in her working account and, in another instance, to invest in commodities. Consequently, identifying populations of users simply on the basis of their demographic characteristics is unlikely to account for differences within these groups that may emerge from time to time.

20.8 USER TESTING

There are a number of purposes associated with the application of user trials in the assessment of products and systems, including:

- Comparisons between competing systems;
- Establishing the optimal length and detail of training programs;
- The appropriateness of a particular style of system interface; and/or
- Building comparative data for marketing purposes.

Given the range of objectives that can be achieved through user trials, the most important requirement is to establish initially the goals and the specific questions that the process is intended to answer. This is consistent with a typical approach to research design and methodology, and acts as a guide for the subsequent investigative process. For example, it enables the identification of the appropriate personas from which a representative pool of users can be recruited. In addition, there is a general indication of the types of trials that need to be employed to provide adequate data.

System Evaluation, Usability, and User Experience

Generally, the number of participants required to test usability issues will vary, depending on the nature of the investigation and the limitations imposed by financial and/or time constraints. However, Virzi (1992) notes that approximately 80 per cent of problems can generally be identified by up to five participants. Additional participants will typically reveal fewer problems such that, eventually, the returns may not be considered sufficiently cost-effective.

Although relatively costly, user trials complement the information derived from less costly methods of evaluation such as heuristic evaluations (Tan et al., 2009). Where heuristic evaluations comprise audits against prescribed standards, user trials allow for greater flexibility, since they permit users to engage different pathways towards goal satisfaction, providing an opportunity for the emergence of problematic issues, than might be available through a more prescribed evaluation.

20.9 A/B TESTING

A/B testing is an approach to user testing that enables comparative assessments of different design solutions. In statistical terminology, the technique is consistent with an after-only or between-groups design. It is particularly useful where randomised groups can interact with different designs simultaneously and under similar conditions. For example, two versions of a sales interface might vary slightly, with users allocated randomly to different interfaces. Assessments of the time taken to complete a sale might reveal that the location or colour of a feature translates to differences in the frequency with which an initial selection translates into a purchase. Over large numbers of transactions, the outcomes could result in significant increases in sales.

While A/B testing offers considerable opportunities for designers in evaluating different designs efficiently and cost-effectively, there are ethical issues associated with the involvement of participants in testing, in the absence of explicit consent (Benbunan-Fich, 2017). A solution involves the development of a pool of participants who consent to their involvement in user research, with debriefing offered following testing. A pool of potential participants would also enable the integration of moderating or mediating factors, thereby enabling more detailed explanations of human behaviour. Inferring causality on the basis of the outcomes of A/B testing can be problematic in the absence of experimental control and where the characteristics of users and the conditions under which behaviour is occurring are unclear.

20.10 HUMAN–COMPUTER INTERACTION

One of the areas where the application of user and use experience methodologies has been most prolific is human-computer interaction (HCI). In part, this has been a response to the recognition that the interface between the operator and the system is the point at which a failure is most likely to occur (Ramkumar, Stappers, Niessen, Adebahr, Schimek-Jasch, Nestle, & Song, 2017). Therefore, improving the user interface is presumed to result in an improvement in the overall reliability of the system.

Wilson and Rajan (1995) suggest that the simplistic view of a human–machine interface is one where the user is considered a simple information processor,

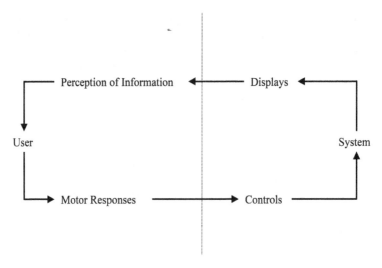

FIGURE 20.1 A simplistic view of the human-machine interface. (Adapted from Wilson and Rajan, 1995.)

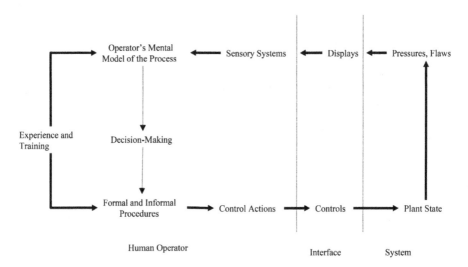

FIGURE 20.2 The more contemporary view of the human–machine interface. (Adapted from Wilson and Rajan, 1995.)

responding to changes in the system (see Figure 20.1). However, a more comprehensive view is one where the process of information acquisition and response is more complex, with the user engaging with the interface as a responsive participant (see Figure 20.2).

The distinction between simplistic and more comprehensive approaches to human–machine interfaces has important implications for user testing, especially in the context of the type and extent of the information sought. For instance, a simplistic approach to user experience might involve a methodology where users are observed by a series of raters, with the frequency and types of errors constituting the basis on which subsequent judgements are made.

Although this type of methodology provides useful information concerning the nature or errors, it is does not necessarily provide an assessment of the type of information processing that led to the errors. Nor does it reflect the extent to which users may have developed an accurate assessment of the system structure. In contrast, the comprehensive approach to HCI is intended to enable the acquisition of data based on users' understanding of the system, rather than simply their response to the system. In addition to more detailed information, this type of approach invariably provides a guide towards improvements in the usability and user experience.

20.10.1 GOMS and Other Models

The Goals, Operators, Methods and Selection (GOMS) model of human–computer interaction has been one of the most influential approaches to human–computer interaction, where the user comprises an interactive element of the overall function of the system (Card, Moran, & Newell, 1983). GOMS is based on the premise that HCI incorporates:

- Goals that the user is attempting to achieve;
- Operators that are available to interact with the system;
- Methods through which the interaction can occur; and
- Selection rules to determine how the interaction can occur.

The model was developed as a means of systematically assessing the extent to which HCI conformed to the principles of human factors design. The main advantage of the GOMS approach is the simplicity and the ease with which data can be acquired (Ramkumar et al., 2017). Each item is assessed based on the time to complete a task, ensuring a degree of objectivity, sensitivity, and reliability. However, despite the advantages afforded by GOMS, the focus on quantitative data has been criticised, since the detail acquired may be more limited than might be desired (Christie, Scane, & Collyer, 1995; John & Kieras, 1996).

An alternative model of HCI was proposed by Marshall, Nelson, and Gardiner (1987), the aim of which was to retain the detail and the structure of interactions between the user and the interface by focusing on psychological constructs such as perception, metacognition, and decision-making. Where Card et al. (1983) had opted for an approach to usability that simplified some of the psychological constructs under consideration, Marshall et al. (1987) took the view that these constructs should be better clarified, so that they could be examined more accurately and reliably.

These two approaches to the modelling of HCI illustrate an important issue concerning the nature of usability and user experience. Specifically, the nature of the framework that underpins an evaluation will determine the nature of the data that are acquired and the types of questions that can be answered. HCI is a cognitive process that will always require some level of information processing. Therefore, the

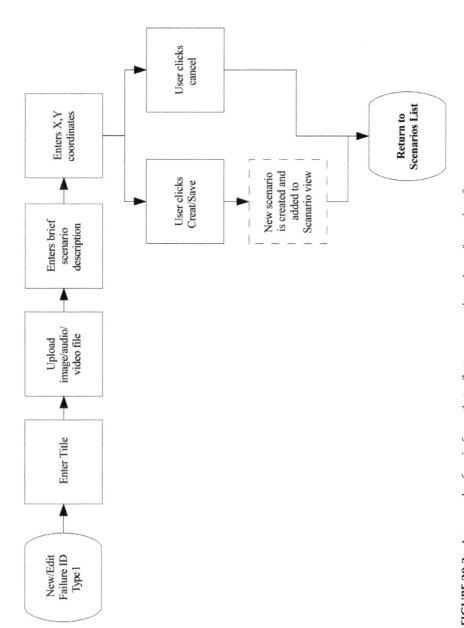

FIGURE 20.3 An example of a wireframe that reflects progress through a software interface.

challenge for the designer is to employ a framework that provides information sufficient to guide the development of an interface that meets the needs of users, taking into account the range of pathways and goals that might be necessary.

20.10.2 WIREFRAMES

Wireframes are schematic diagrams that represent either the features associated with an interface or the pathway that a user might take through a series of interfaces that comprise a piece of software (see Figure 20.3). They are intended to provide an initial representation of an intended design, ensuring a focus and common reference point for users and designers (Silva, Hak, Winckler, & Nicolas, 2017). As schematics, they can be manipulated relatively easily, enabling different options to be evaluated and compared.

20.11 DESIGN THINKING

Sustaining successful performance in an increasingly competitive, dynamic marketplace requires more than simply the provision of goods and services more efficiently. Sustained growth requires a process of continuous innovation and development. Design thinking is an analytical and creative process that is intended to enable innovation to flourish.

Design thinking is an iterative process that comprises reflection, interaction, and critique (Razzoul & Shute, 2012). The goal is to progress from an abstract representation of an issue or problem to be solved, to a product or outcome that is concrete and targeted. This involves the deconstruction of the problem and the progressive creation and analysis of potential solutions or aspects of the solution. It has been described as a nonlinear process where previous stages in the process are revisited as part of a cyclical approach to better inform and further validate subsequent stages (Goldschmidt & Weil, 1998).

An important part of design thinking is to establish the frame within which a problem is perceived. The frame offers a lens through which a solution might be advanced, together with a working principle that provides a focus for the development of the solution. For example, in designing a passenger-initiated device to trigger an alarm on board a train, a working solution might involve a button that can be depressed. With further analysis, it might be established that a button that is not recessed might be activated inadvertently, leading to a parameter to the effect that, if a button is to be used, it must be recessed.

21 Human Factors and Ergonomics

21.1 ERGONOMICS

In addition to the information-processing limitations imposed on human performance, users are also limited in their physical capabilities. Consequently, systems must be designed to take into account both the psychological and the physical limitations of the potential user population. Ergonomics is the discipline that focuses most closely on the relationship between operators and their physical environment (Nath, Akhhavian, & Behzadan, 2017).

Ergonomics is generally regarded as a subdiscipline of human factors, and is applied alongside the discipline of anthropometry to assist the design of systems that meet the physical demands of users (Karwowski, 2005). Anthropometry comprises a set of methodologies that enables the physical dimensions of humans to be recorded and compared within a scientific framework (Kroemer, 1997). The outcome is a set of data that describe a range of physical dimensions such as extended reach, seated eye height, and wrist height. These data provide the basis from which designers can develop systems for use physically by a range of users.

The significance of an adherence to ergonomic and anthropometric principles during the design process is illustrated by the crash of a British Airways Sikorsky S61-N-II helicopter off the Isles of Scilly in 1983. The aircraft was en route to an off-shore oil rig when it impacted the ocean with the loss of 20 of the 26 personnel aboard (Air Accidents Investigation Branch, 1984). According to the Air Accidents Investigation Branch (1984), one of the factors that may have contributed to this accident was the obscuration, by the control column, of a warning light that illuminated when the aircraft descended below 200 feet. Had the pilot seen the light, the collision may have been averted.

This accident illustrates both the consequences associated with inadequate ergonomic design, and the relatively simple design-related issues that can precipitate failures. However, the relationship between the physical environment and the human operator is often more complex, and the remedial action for inadequate design will typically require more than simply moving the position of a light. For example, an Air Inter A-320 suffered a total loss of engine power soon after take-off from Paris Orly after the non-flying pilot aboard the aircraft inadvertently shut down the engines (Sedbon, 1994). The error occurred when the non-flying pilot retracted the leading-edge slats and flaps following departure, but neglected to retract the undercarriage. This resulted in a greater level of aerodynamic noise than was normally the case, and instead of reaching for the undercarriage levers, the pilot grasped the throttles.

Clearly, this sequence of events raises both psychological and ergonomics issues, and it emphasises the difficulties associated with the delineation between human information processing and ergonomic principles. Since systemic errors involve interactions between components, changing one element of the system to the exclusion of changes to other elements will not necessarily improve human performance (Buckle, Clarkson, Coleman, Ward, & Anderson, 2006). Consistent with this perspective, ergonomics and anthropometry must be considered as just two more factors in the range of factors that might be employed as part of an overall strategy to improve the relationship between operators and systems (Hendrick, 2000).

21.2 ERGONOMICS AND THE NORMAL DISTRIBUTION

Despite the best of intentions, it is not always possible to design a system that meets the individual needs of an entire population. For example, a system may have a physical limit to the dimensions of a console so that a proportion of users may not be able to reach a control. To identify the proportion of users whose needs may not be met by such limitations, a designer must be aware of the entire range of physical dimensions that characterise a population. This enables a decision to be made concerning the acceptability or otherwise of the design in meeting the objectives associated with usability.

Within any given population, the features with which they are characterised can typically be distributed as a normal or Gaussian distribution. The normal distribution can be displayed where the vertical axis represents the frequency, and the horizontal access represents units of measurement such as height in centimetres (see Figure 21.1). Typically, the shape of the normal distribution approximates an inverted 'U' and occurs when the mean, median, and the mode coincide (Stanton, & Young, 2003).

From the perspective of the designer, the value of the normal distribution lies in specifying the range of values that might occur for a specific dimension. In the case of workstation design, a designer might be interested in the eye height of the user in a sitting, slumped position to determine the optimal height at which a computer screen might be placed. Since the elevation of the screen should be located so that it is horizontal to the user's eye height, the normal distribution enables the designer to determine the elevation at which the needs of the maximum proportion of users will be met. However, it also allows the designer to determine the proportion of users whose needs may not be met by the design.

Clearly, the optimal dimensions for a design are those that meet the needs of a maximum proportion of users. In the case of a normal distribution, this is the point at which the mean, median, and mode coincide. This is referred to as the 50th percentile, since it is considered the point below which 50 per cent of the values of a particular dimension fall (Porta, Saco-Ledo, & Cabañas, 2019). For example, the

Human Factors and Ergonomics

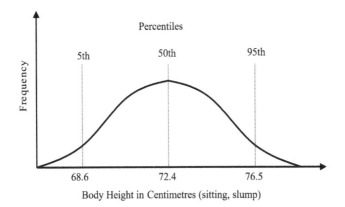

FIGURE 21.1 An illustration of a normal distribution for females for eye height in centimetres in the sitting, slumped position for the 5th, 50th, and the 95th, percentiles. (Adapted from Das, 1998.)

50th percentile for females' eye height in a sitting, slumped position is 72.4 cm from a standard reference point, whereas the 95th percentile is 76.5 cm (see Figure 21.1).

Distributions of physical characteristics provide important information in the context of design, since they enable judgements concerning the range of potential users' needs that may have to be met by a single system. In effect, this becomes something of a cost-benefit analysis, since the commercial environment may dictate that developing solutions for eye heights below 70.9 cm and above 81.3 cm may not be viable, since it would only cater to 10 per cent of the potential population of users.

The calculation of percentiles occurs following the collection of raw scores relating to a particular dimension. To obtain an accurate reflection of the characteristics of the population, a random sample should be sought to ensure that the probability of obtaining a specific value is consistent with the probability with which the value occurs within the broader population. The result is a list of raw scores and associated frequencies as listed in Table 21.1. (Note that only a partial list of the range of possible waist circumferences is listed in Table 21.1.) To determine the percentile rank for a particular score, the cumulative frequency must be calculated, initially by summing the frequencies of all the values up to the value for the percentile rank to be determined. For example, in the case of waist circumference in Table 21.1, the cumulative frequency for a score of 72.5 cm is calculated as:

$$3 + 8 + 46 + 160 + 448 + 699 = 1364$$

Having obtained a cumulative frequency, the percentile rank is calculated by dividing the cumulative frequency by the total number of scores and multiplying by 100. For example, the percentile rank for a waist circumference of 72.5 cm is calculated as:

$$(1364/6682) \times 100 = 20.4$$

This value indicates that a waist circumference of 72.5 cm is located at the 20th percentile where this value and below accounts for 20 per cent of the overall distribution. This value can be used to determine the limits to which a particular garment might be restricted. For example, it might be argued that, since a waist of circumference of 72.5 cm only accounts for 20 per cent of the overall US male population, it may not be cost-effective to develop a garment with these dimensions.

In addition to information pertaining to a single dimension, Table 21.1 also provides an illustration of the difficulties faced by designers in the interaction between design criteria. For example, if a garment was to be designed that took into account the maximum number of individuals based on their waist circumference and their crotch height, the total proportion of the population for whom the garments could fit would amount to only 140 individuals or 2.0 per cent. This illustrates the complexity faced by designers in accounting for the different permutations that may need to occur in meeting the physical requirements of users.

21.3 WORKSTATION DESIGN AND THE APPLICATION OF ERGONOMICS

In the contemporary industrial environment, oversight of complex systems is often exercised by system operators who are based at workstations. The increasing reliance on workstations requires that they are designed appropriately so that they encourage appropriate posture from users and thereby optimise the efficiency with which they undertake physiological tasks (Workineh & Yamaura, 2016). Ensuring the appropriate design of workstations is presumed to result in a reduction in the downtime arising from physiological complaints and improvements in morale and the efficiency with which both physical and cognitive tasks are performed (Woo, White, & Lai, 2016).

Consistent with the principles of systems development, the first stage of workstation design involves an assessment of the need for the system within the operational environment. Ideally, the proposed design should improve on current products, and should embody advantages that will ensure that the workstation is marketable and offers a value proposition to prospective customers.

From an ergonomics perspective, the designer is faced with a decision concerning the type of approach that is likely to yield an optimal outcome for the majority of potential users. In essence, there are three options available:

1. Create a single design that will meet the needs of all users;
2. Create a design that is adjustable; or
3. Create a design for which it is possible to construct several sizes (Chapanis, 1996; Workineh & Yamaura, 2016).

Typically, the option selected will occur as part of the development of the operational concept and will be based on an assessment of the potential market for the product, the financial costs associated with the development of different options, and the consequences associated with limiting the application of the product (Hugo, Kovesdi, & Joe, 2018). According to Haber and Haber (1997), there is always a limit to the extent to which a design can meet the needs of every potential user. However, users will

TABLE 21.1
Partial List of Waist Circumferences and Crotch Heights (cm) for Males

						Crotch Height						
	...	77.5	79.0	80.5	82.0	83.5	85.0	86.5	88.0	89.5	91.0	92.5	...
Waist Circumference	85.0	20	39	49	65	63	60	55	41	32	29	14	503
	82.5	32	37	73	70	97	74	84	49	41	24	16	647
	80.0	38	62	90	119	107	110	94	86	72	36	27	908
	77.5	43	60	122	100	119	140	91	87	57	43	23	937
	75.0	40	82	72	131	139	11	102	89	57	48	23	964
	72.5	38	50	62	84	96	104	80	54	48	18	12	699
	70.0	36	40	41	56	58	63	15	33	17	19	3	448
	67.5	5	10	18	17	27	15	20	13	6	5	4	160
	65.0	3	7	5	6	1	4	2	2	2	5	1	46
	62.5		1		1	3		2					8
	60.0				1	1	1						3
	...	1	2	6	13	21	64	110	174	333	468	645	

Source: Adapted from White & Churchill (1971).

usually tolerate some level of mismatch between their optimal needs and the design of a system. Therefore, one of the challenges for the designer is to determine the degree to which users are willing to tolerate a mismatch and integrate this information as part of a broader cost-benefit analysis. The goal is to limit the number of different systems that need to be constructed, thereby creating a more cost-effective product.

An important part of the operational concept stage of design is the identification of specific objectives that will meet the operational needs of users. In the case of workstation design, Das (1998) advocates a process of knowledge acquisition whereby existing designs are compared, and users are interviewed to determine any difficulties or problems that may exist. Part of this process may involve the distribution of questionnaires to identify the extent to which user performance depends on the perceived characteristics of the workstation. In addition to the provision of pertinent information for the design process, this type of information may also be useful to advocate for changes that might be proposed subsequently.

Complementing the information derived from potential users, there are a series of general ergonomic principles that dictate at least some of the objectives that ought to be included in the design process. These objectives include ensuring that:

- The height of the workstation is appropriate for the range of users;
- Frequently used tools are within normal reach;
- Controls are located in a position whereby they can be easily manipulated if required;
- There is sufficient elbow room and room for movement about the waist; and
- Displays are located at eye height.

Although these are very general rules, they do provide some level of prescription to ensure that workstations and other similar systems are designed with ergonomic principles in mind. However, the extent to which these objectives will be met will be determined during the concept exploration process where the different demands on the design process must be balanced. For example, there may be limitations imposed on the design due to the costs associated with developing or implementing solutions and/or existing regulatory requirements.

The physical dimensions of the proposed workstation are generally specified as part of the concept demonstration phase of the design process. Das (1998) suggests that there are four main envelopes that must be considered, including the height of the workstation, the normal and maximum reaches, lateral clearance, and the angle of vision and eye height. In dealing with each of these components, anthropometric tables must be considered to determine the maximum and minimum dimensions that can be entertained.

Although anthropometric data are very useful at this stage, they must be used with some caution, since the data are generally based on samples that may be atypical of the general population. For example, anthropometric data often arise from analyses of military personnel who will have been preselected on the basis of specific physiological attributes such as a minimum height and weight. Furthermore, anthropometric data may be outdated and therefore, may not reflect the changes that have occurred over recent years in the physical characteristics of the general population.

In addition to anthropometric tables, designers would be expected to consult the international and national standards pertaining to the design of workstations. For example, the Federal Aviation Administration (FAA) specifies that workstations at which operators are seated should be located between 740 mm and 790 mm above the floor (Wagner, Birt, Snyder, & Duncanson, 1996).

Bearing these issues in mind, it is also useful to acquire anthropometric data from the potential users of the product. This ensures that the data are appropriate to the specific population for whom the system is designed. This approach is also consistent with the principles associated with participatory ergonomics, where users are incorporated as an integral aspect of the process of design (Burgess-Limerick, 2018; Wilson & Haines, 1997).

The simulation of the proposed design can be accomplished by creating a physical or computer-generated mock-up to illustrate and test whether the design objectives are likely to be achieved. In the case of computer-generated systems, there are a number of packages available that enable three-dimensional models to be rendered in a range of environments. The advantage of a computer-generated approach is that it allows a greater level of flexibility to change and test designs rapidly as more information becomes available.

Having developed and tested the design, a full-scale construction of the workstation should begin prior to deployment within the operational environment and subsequent evaluation. This does not preclude any changes being made, although it should be noted that the costs associated with changing a system prior to deployment are invariably lower than the costs incurred subsequent to deployment.

An important part of the ergonomic design process involves the development of a series of standards against which future developments can be constructed and evaluated. For example, the Human Factors and Ergonomics Society (1988) specifies that visual display terminals such as those used for computers should be positioned between 0 and 60° below the horizon. Keyboards should be angled towards the user at between 0 and 25°, while the height should be between 58.7 cm and 71 cm from the floor to the row in which the 'home' key is located. This affords maximum efficiency with minimal probability of injury.

21.4 TRAINING AND ERGONOMICS

Despite attempts to optimise the design of a product, training in its use often remains necessary due to compromises that may have been required during the design process. For example, taking into account the broadest range of users may have been possible only by developing a product where adjustments were necessary to meet different needs. Therefore, the utility of the product lies in ensuring that a user has the capability to both determine the appropriate settings and make the adjustments necessary to achieve those settings.

In the absence of training, there is a possibility that a product may be used incorrectly or discarded on the assumption that it fails to meet the perceived needs of users. Therefore, training programs should be designed to ensure the acquisition of appropriate attitudes and knowledge concerning the intended use of products, and the skills necessary to use these products safely and effectively. The nature and scale of this training will depend on the existing knowledge and capabilities of users and the resources available within the organisation.

Training in the use of new products can be particularly challenging for customers, since interaction with systems may be infrequent and require a minimum level of capability. A relevant example can be drawn from the initial introduction of the Automated Teller Machine (ATM), a system designed to increase the flexibility of the services provided to customers, thereby reducing the demand for transactions conducted with a banking representative (Rogers & Fisk, 1997).

Survey research following the introduction of ATMs suggested that approximately 38 per cent of banking customers in the United States described themselves as non-users of ATMs, despite the fact that a significant proportion possessed an ATM card (Rogers, Cabrera, Walker, Gilbert, & Fisk, 1996). Since non-users could generally be described as older, and with limited experience with technology, it suggested that their lack of engagement with ATMs was possibly due to a lack of knowledge and capability in the use of the interface.

According to Rogers, Cabrera, et al. (1996), 21 per cent of the non-users of ATMs reported that they would be willing to undertake training. Rogers, Fisk, Mead, Walker, and Cabrera (1996) suggested that, in this case, the most effective type of training program was one where users received interaction and feedback as they were interacting with the ATM. The aim was to provide specific practice on the various components of the service with assistance if, and when, required.

In developing training for new technologies, it is important to note that training should not occur in isolation of design improvements that might otherwise be initiated. For example, Rogers and Fisk (1997) noted that one of the most significant problems amongst older users of ATMs involved forgetting to take their card and/or the receipt, following the transaction. A simple design measure was subsequently introduced to overcome this problem that involved withholding any cash until the card had been removed from the ATM. Other issues that needed to be considered included improving the safety of the users of ATM machines, and standardising the design of machines across different banks.

In the context of new designs, training offers an opportunity to bridge the metaphorical gap between the designer and the user. This complementary process that integrates design and training can be employed as part of a broad-based macro-ergonomic approach in which both technical issues and organisational factors are examined as part of the process of improving human performance (Joyce, 1998).

21.5 MANUAL HANDLING

Heavy lifting is a task where ergonomics has had a profound influence in the development of both training systems and system aids (Woodhouse, McCoy, Redondo, & Shall, 1995). In many workplaces, training strategies for heavy lifting are a well-established part of broader occupational health and safety initiatives. However, these training strategies have been linked with design strategies including:

- Systems that enable manual lifting without bending or twisting;
- Space around objects to be lifted;

- The provision to slide, rather than lift objects;
- Handholds that enable an appropriate grip; and
- The labelling of heavy objects (adapted from Corlett & Clark, 1995).

Back support devices such as exoskeletons have also been introduced as part of a strategy to increase intra-abdominal pressure, thereby reducing the lifting loads imposed on the lower back area (Theurel, Desbrosses, Roux, & Savescu, 2018). While the effectiveness of many of these design initiatives has yet to be established conclusively, there is a growing body of evidence to suggest that the combination of training and physical support devices reduces the physical stressors imposed by manual lifting (Lavender, Thomas, Chang & Andersson, 1995; Huysamen, de Looze, Bosch, Ortiz, Toxiri, & O'Sullivan, 2018; Walsh & Schwartz, 1990).

Beyond back injuries that are due to manual handling tasks, there is evidence to suggest that back ailments among industrialised populations are becoming an increasingly serious issue (Deyo, Mirza, & Martin, 2006). For example, up to 51 per cent of school students report back problems by the age of 16 (Troussier, Davoine, De Gaudemaris, Fauconnier, & Phelip, 1994). However, while students are often given extensive training to encourage correct posture for sitting, many continue to revert to an incorrect posture in which they hunch over the desk (Mandal, 1997).

The inclination towards poor posture amongst school students reflects the broader challenge in relying on training and encouragement as the basis for changes in behaviour. Like the situation involving heavy lifting, the optimal solution appears to lie in a combination of training and changes in design. For example, Mandal (1997) proposed a solution based on the design of school chairs and tables, where they were capable of being tilted with an angular posture at 10° from the vertical. In this case, the natural inclination of students to lean back on their chairs provided the basis for an optimal ergonomic design.

21.6 COMMAND AND CONTROL SYSTEMS EVALUATION

Macro-ergonomics is a holistic approach to the measurement and assessment of a system (Luczak, Krings, Gryglewski, & Stawowy, 1998). It incorporates some of the principles of Total Quality Management (TQM) where the focus is not simply directed towards the quality of a product but jointly to the quality of the systems through which the product was developed. Consequently, macro-ergonomics has important implications for the assessment of broad-based systems, including those that exercise command and control over a number of subsystems (Hendrick, 2007; Kleiner, 2006).

Command and control structures arose from the military environment where there are very clear chains of authority. According to Metcalf (1986), a structured approach to command and control is essential when the success of an operation depends on a number of interdependent units. Deficiencies in military command and control structures are evident in slow response times, an inefficient or ineffective decision-making process, and/or a lack of continuity of responsibility (Crecine, 1986; Roberts, Stanton, & Fay, 2017).

In the ergonomics environment, command and control structures are typically computer-based systems that provide information pertaining to a range of subsystems, and on an ongoing basis (Kleiner, 2002; Taylor, Charlton, & Canham, 1996). However, the principles of operational control and the outcomes of an inappropriate structure are entirely consistent with the experience within the military environment where a central command structure is relatively well established.

21.7 A TOTAL QUALITY MANAGEMENT APPROACH TO ERGONOMICS

Given the interdependence that exists between the components of command-and-control structures and the factors that are likely to impact system performance, it would be reasonable to consider TQM as one of a number of mechanisms to identify areas for potential improvement. Specifically, TQM advocates an approach to management where users have the capacity to adapt their environment, as and when required, to fulfil personal and organisational objectives (Pambreni, Khatibi, Azam, & Tham, 2019).

In the context of command and control structures, this may include the capability to change an interface to improve performance, the flexibility to change the level of control that is exercised over the system, and/or the capacity to access additional information when required (Luczak et al., 1998). This concept has a number of parallels with the principle of adaptive task allocation where the technology supports, rather than directs, the system (Parasuraman, Mouloua, & Molloy, 1996). In addition, there is an intrinsic principle of trust in the capability of the operator to respond appropriately to changes within the system.

Despite the emphasis on the decentralisation of responsibility, the adaptive changes that are initiated by users must occur within a prescribed organisational framework. Therefore, a TQM approach requires a continuous process of planning and assessment where potential system failures can be identified and remediated at the earliest opportunity (Luczak et al., 1998). Specifically, the process involves the:

- Assessment and evaluation of the risks and opportunities;
- Instigation of appropriate remedial procedures if necessary;
- Development and implementation of a monitoring process; and the
- Implementation of a strategy to foster adjustments where necessary.

Part VI

Human Factors in Context

22 Human Factors and Automation

22.1 AUTOMATED SYSTEMS AND HUMAN PERFORMANCE

Improvements in technology within complex socio-technological environments have generally resulted in an increase in the reliance on automation to manage complex tasks. In the contemporary workplace, automated systems are now able to exercise control over multiple systems, assess and respond to uncertainty, and initiate responses to safeguard a system (Endsley, 2017). In introducing automated systems, the aim generally is to relieve the operator of otherwise complex or onerous tasks, thereby reducing costs while improving the overall performance of a system (Parasuraman & Riley, 1997).

Despite the goals associated with the development of automated systems, the transition from 'human as controller' to 'human as systems manager' has not been without difficulty. For example, there have been a number of cases where automated systems have been implicated in systems failures (Cook & Woods, 1996; Funk, Lyall, Wilson, Vint, Niemczyk, Suroteguh, & Owen, 1999). However, it should be noted that examples of uncommanded automation-induced failures are relatively uncommon. Rather, automation-induced failures tend occur most frequently due to differences between the operator's expectations of the automated system and the subsequent behaviour of the system (Lee, Abe, Sato, & Itoh, 2021).

Referred to as mode error, differences between operator expectations and the behaviour of a system are typically explained by reference to inadequate design or training. For example, amongst some drivers of Tesla motor vehicle drivers, there appears to be an expectation that the 'autopilot' system enables the vehicle to be operated in the absence of driver oversight. However, a number of crashes have indicated that the system is not infallible, and that drivers need to continue to monitor the system (Tenhundfeld, de Visser, Haring, Ries, Finomore, & Tossell, 2019).

22.1.1 Automated Systems and Interface Design

The design of appropriate interfaces between the operator and automated systems is the key to improving human performance within the operational environment (Jaimieson & Vicente, 2005; Woods, Johannesen, Cook, & Sarter, 1994). This requires an interface design that enables the observation of automated systems readily, and in a form that is meaningful for the operator. Where information pertaining to the function of automated systems is not directly apparent to the operator, errors are likely to be remain undetected.

In addition to the detection of errors, Woods et al. (1994) advocate an approach to interface design that enables error recovery where the potential for negative consequences is minimised. This facility requires that the behaviour of automated systems is monitored at various stages throughout an automated process. The intention is to provide an opportunity for the performance of the system to be compared against the operator's expectations. Where differences emerge, a timely recovery mechanism can be initiated.

Error recovery is often complicated by the introduction of advanced technology, since it may increase the range of strategies available to recover from an error. This situation can be likened to the development of a road transportation network within a growing city, whereby a destination can be reached via a number of different, yet equally advantageous, routes. The choice between a range of strategies, each of which targets the goal, epitomises the flexibility of advanced systems to adapt to meet the specific needs of the user (Parasuraman & Riley, 1997). However, in doing so, it also increases the number of choices available for the operator, and may actually impede the rate at which responses are initiated to identify and resolve system failures.

22.1.2 The 'Out of the Loop' Syndrome

The 'out of the loop' syndrome is said to occur when an automated system performs functions that are not anticipated by the operator (Merat, Seppelt, Louw, Engström, Lee, Johansson et al., 2019). This tends to be the most common error that occurs as a result of interaction with advanced technology (Endsley, 2017; Sarter & Woods, 1995). Part of the difficulty appears to lie in both the accuracy and the reliability of automated systems, to the extent that operators may become complacent regarding the potential system failures that can occur (Singh, Molloy, & Parasuraman, 1993)

From the perspective of human reliability, the likelihood that a system will perform functions that are unanticipated by the operator is related to both the inherent behaviour of the automated system and the performance shaping factors (PSF) that impact the operator (see Figure 22.1). Where a system is relatively unreliable, operators tend to maintain a comparatively higher level of vigilance, thereby reducing the reaction time in response to an unexpected change in the system state (Foroughi, Sibley, Brown, Rovira, Pak, & Coyne, 2019). However, where a system is relatively reliable, operators may develop a degree of trust in the system, the consequence of which may be an increase in the reaction time in response to an unexpected change in the system state (Chavaillaz, Wastell, & Sauer, 2016).

The complexity of the interrelationship that exists between operators and automated systems underscores the requirement that analyses of human behaviour account for more than simply system design and operator training. Issues associated

Human Factors and Automation

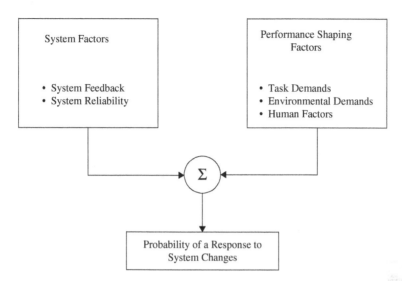

FIGURE 22.1 An illustration of the interaction between system factors and performance shaping factors and their impact upon the probability of detecting a change in the system state. (Adapted from Kantowitz and Campbell, 1996.)

with trust, and the development and maintenance of a dynamic mental model are equally important issues that have the potential to impact human performance in an automated environment (Schwarz, Gaspar, & Brown, 2019; Singh et al., 1993). Moreover, they reflect problems associated with automation that are more fundamental than might be implied by referring simply to the need for improvements in training and design.

Muir (1994) suggests that, irrespective of issues such as design and training, the notion of automation itself has implications for human performance, especially in the context of failure detection and diagnosis. For example, evidence arising from empirical analyses suggests that a lack of direct involvement in the performance of a task increases the time required to establish control over a system in the event of failure (Endsley & Kiris, 1995; Kessel & Wickens, 1982). Therefore, the difficulty associated with automation might arise due to the lack of cognitive involvement in the performance of a task. In the absence of such involvement, the cues arising from changes that occur within the operational environment are no longer immediately evident, other than through secondary sources such as instrumentation.

In addition to a reliance on secondary sources of information, the introduction of automated systems has resulted in a shift in the cognitive and perceptual demands on the performance of the operator. Where non-automated systems tend to be reliant on active participation and adherence to standard operating procedures, automated systems tend to require a more supervisory approach involving skills such as diagnosis, troubleshooting, and planning (Hollnagel, 1995; Kummetha, Kondyli, & Devos, 2021). As a consequence, the demand for predominantly skill-based and rule-based behaviour in non-automated systems has been replaced by a demand for knowledge-based behaviour in automated systems (Li & Burns, 2017).

22.1.3 Cooperative Automated Systems

One of the strategies that might be employed to overcome the lack of cognitive involvement in an automated task involves a cooperative approach where some tasks remain the responsibility of the operator, while other tasks are managed through automated system aids (Wei et al., 1998). However, the success of this type of strategy depends on a number of factors, the most significant of which involves the process by which tasks are delegated to the operator and/or the system. Where an automated system functions independently of the context or the needs of the operator, cognitive involvement in the task is likely to be lower than if the operator has the flexibility to select a level of automated assistance as required.

The flexibility to select a level of automated assistance is intended to enable the operator to retain a level of cognitive involvement in the performance of a task. Nevertheless, there is no guarantee that an automated system will be activated in a particular situation. Rather, automated systems have, in some cases, been deactivated by operators due to a perceived increase in the workload required to program and reprogram the system, particularly under time-constrained and rapidly changing conditions (Naweed, Chapman, Allan, & Trigg, 2017).

Ultimately, the nature of the design of an automated system aid will determine the extent to which it is used appropriately within the operational environment (Wiegmann, Rich, & Zhang, 2001). Where automated systems are perceived as too difficult to manage or program, and/or the time required is significant, then it is less likely that a system aid will be employed in the management of a task. However, the extent to which an automated system will be employed will also depend on the capabilities and perceptions of the individual user.

22.1.4 Adaptable Systems

As technology has advanced within complex socio-technological environments, so has the potential for the system interface to adapt to meet the needs of individual operators. For example, advanced technology motor vehicles have the capacity to display information in a range of sequences, in different locations, and at different frequencies (Seppelt & Lee, 2019). Similarly, the information on a desktop computer screen can be altered to the extent that the information displayed to one user may be markedly different to the information displayed to another.

Although contemporary operators have an opportunity to interact with interfaces that can be adapted to suit their individual needs, the extent to which these changes impact human performance is a matter of some concern. For example, in a study of five different computer interface designs, Keyson and Parsons (1990) observed that the interface that resulted in the highest level of performance also received the worst ratings in its preference amongst potential users. Similarly, users appear to prefer the use of an iconic interface, rather than an alphanumeric interface, despite the fact that the latter tends to yield fewer errors (Kacmar & Carey, 1991).

22.2 AUTOMATED SYSTEMS AND RELIABILITY

The introduction of automated systems has tended to increase the complexity required for analyses of human reliability. Specifically, the range of behaviour that must be anticipated increases markedly, due to the flexibility of automated systems and the extent to which their introduction impacts the information-processing strategies that might be employed by the operator. For example, knowledge-based behaviour, such as diagnosis and planning, is likely to be more apparent in an automated environment than in a non-automated system (Khastgir, Birrell, Dhadyalla, & Jennings, 2018).

The adaptability of system interfaces also presents significant difficulties in establishing the likelihood of human error. Analyses need to incorporate a far greater range of options, unless standard operating procedures prescribe a particular style or format of the interface. Nevertheless, the opportunity for the adaptation of a system interface must be taken into account to ensure a valid and reliable account of human performance.

Although the likelihood of error recovery is difficult to estimate, it remains a significant aspect of contemporary automated systems. However, the issue is complicated by the relationship that exists between error detection and recovery, and the design of the system. In some cases, the probability of error detection may be reduced, not simply by the lack of awareness on the part of the operator, but by an inappropriate design that makes the probability of detection difficult, if not impossible.

The interaction between system design and human performance is likely to differ between individuals, and will typically be influenced by performance shaping factors. However, there have been some successful attempts to model aspects of automated system design and human performance. For example, Wei et al. (1998) sought to develop a model that explained the relationship between the allocation of tasks to human operators or automated systems, and subsequent human and system performance. The relationship can be encapsulated by the Degree of Automation (DofA) and is expressed as:

$$DofA = 1 - \frac{\sum_{i=1}^{N} t_i w_i}{\sum_{i=1}^{N} w_i}$$

where t_i refers to the type of task (automated or manual), N denotes the total number of tasks, and w_i represents the weighting attached to task i. If the aim is to determine the relative Task Mental Load (TML) that is imposed by an automated system, the TML for the task needs to be calculated initially in a non-automated condition. The resulting value represents the weighting (w) that is attached to the task. Where an automated system is concerned, the resultant $DofA^{TML}$ can be interpreted as the mental load that is experienced by the operator in an automated system, against the mental load imposed by the task in a non-automated condition (Wei et al., 1998). For example, in the case of a combination of one automated and one manual switching system, it might be calculated that the mental workload involved in the non-automated condition is 0.2 on a scale from 0 to 1. Substituting, where an automated condition = 1 and a non-automated condition = 0:

$$DofA^{TML} = 1 - \frac{(1 \times 0.2) + (0 \times 0.2)}{0.2 + 0.2}$$

$$= 1 - \frac{0.2}{0.4} = 0.5$$

This result indicates that the automation of one of two tasks reduces the TML involved by 0.5 or 50 per cent. Having established the basic principle, it is also possible to calculate the predicted TML for different levels of automation and for a wide range of tasks.

The calculation of TML enables the quantification of the influence of different levels of automation on aspects of human performance. These values can be integrated subsequently into fault tree or event tree analyses to establish the overall likelihood of an occurrence. However, it should be noted that the TML is based on an assumption that automated systems generally reduce the mental workload involved in a task. It does not take into account qualitative changes in workload that may occur as a result of automation. An increasing reliance on cognitive skills, as a result of automation, may not result in a reduction in overall mental load. Rather, it may simply shift the load from one feature of information processing to another.

22.3 AUTOMATION AND COGNITIVE WORK ANALYSIS

The complexity and diversity of automated systems means that a fully quantitative analysis is unlikely to yield an accurate indication of the likelihood of human performance failures within the operational environment. Rather, a combination of qualitative and quantitative approaches is likely to be most effective where the qualitative evidence arising from case studies is integrated with a broader, quantitative examination of the characteristics of human performance. This type of mixed-methods approach to the analysis of human behaviour is likely to result in information of greater utility across a range of domains and tasks.

Although knowledge acquisition strategies such as Cognitive Task Analysis (CTA) have the capability to integrate qualitative and quantitative data, this type of methodology has generally been applied in the investigation of specific cases. Therefore, there is often a significant investment of both time and resources, the results of which may only be applicable in a relatively narrow range of tasks. While it represents a useful approach, the extent to which it represents the most efficient means of knowledge acquisition is a matter of some debate (Vicente, 1999; Naikar, 2017). For example, the outcomes of CTA need to be considered in the context of a cost-benefit analysis. Where the outcomes of a knowledge elicitation technique are limited to the immediate environment, the potential costs associated with the strategy may outweigh the benefits involved.

Vicente (1999) regards CTA and other forms of task analysis as normative approaches to the investigation of human performance in the workplace. Normative approaches can be characterised by a level of prescription concerning the type of activities that need to be performed to achieve a goal. For example, Redding and Cannon (1991) examined the nature of expertise in an air traffic control environment to determine those capabilities that were required for skilled performance in the management of airspace. Similarly, Roth and Mumaw (1995) recommend the application of CTA as a basis for establishing the optimal design specifications for a human–machine interface.

Cognitive Work Analysis (CWA) is a flexible, efficient knowledge elicitation technique that offers an alternative to CTA (Vicente, 1999). Where CTA is designed to focus on the performance of the operator, CWA is designed to focus on the object with which the operator is interacting (Lundberg, Arvola, Westin, Holmlid, Nordvall, & Josefsson, 2018). Therefore, it represents a fundamental shift in the process of knowledge elicitation within a complex operational environment. CWA can be conceptualised as a logical progression from the highly contextualised approaches that have characterised the initial attempts to examine human performance, to one that is more general and attempts to encapsulate the broader nature of human performance.

CWA is a formative model of knowledge elicitation, since the process targets the identification of those factors (human and machine) that need to be satisfied to ensure that a system functions as intended (Naikar & Elix, 2016). In this respect, it has some similarities to a competency-based approach to training, whereby the analysis focuses on the identification of the task goal, rather than the behaviour necessary to achieve the goal (Schneider, 1985). This contrasts with a normative approach where there is an emphasis on the performance of the operator, rather than the achievement of the task.

According to Rasmussen, Pejtersen, and Goodstein (1994), CWA comprises a five-phase approach to knowledge acquisition, including:

1. Work Domain Analysis
2. Control Task Analysis
3. Strategies Analysis
4. Social Organisation and Cooperation Analysis
5. Worker Competencies Analysis.

The aim of this process is to progress from a consideration of the task in respect of the broader context within which it occurs, through to an analysis of the competencies that are required amongst operators. This increasing level of specificity is designed to enable the acquisition of information pertaining to a case while maintaining an understanding of the operational environment within which a task occurs (Rasmussen et al., 1994).

22.4 WORK DOMAIN ANALYSIS

Work domain analysis begins with a relatively detailed description of the operational environment, independent of a task or operator. By understanding the nature of the environment, it becomes possible to consider a wide range of activities and behaviours and the interactions that may exist. This involves a description of the environment and the identification of some of the broader concepts that appear to be significant (Vicente, 1999). For example, in railway operations, concepts such efficiency, automation, safety, cost-effectiveness, customer relations, and comfort might be identified as significant aspects of the analysis.

Following the field description, constraints will be identified that impose limits on human and system behaviour (Vicente, 1999). These may include regulations and conventions that prescribe human behaviour in particular situations, or operational limitations that constrain the use of equipment to specific types of operating environments.

22.4.1 Control Task Analysis

Having identified the nature of the operating environment, a more specific analysis is required to identify the goals that are likely to be sought in various situations. Vicente (1999) refers to this type of strategy as control task analysis, since it identifies the tasks that need to be 'controlled' to achieve the goal. It is important to note that the achievement of a goal can occur in a number of different ways, and each of these strategies needs to be considered as part of a control task analysis. This is especially the case where automated systems may be employed.

22.4.2 Strategies Analysis

Where a control task analysis is designed to identify the tasks that need to be accomplished to achieve a goal, a strategies analysis is intended to detail how a particular task is to be accomplished (Morineau & Flach, 2019). These strategies would normally comprise both a condition statement and an action statement in the form or a rule or production. In addition, part of this process might involve the development of an event tree analysis or some other representation such as an information flow map of the relationship between the initial state and the goal state (Rasmussen et al., 1994).

The identification of task-related strategies can occur through a verbal protocol analysis, process tracing, and/or through the observation of performance in the operational environment. However, as part of this process, it is important to establish the range of strategies that might be employed and the context in which they tend to occur. In most socio-technological environments, there will be more than one strategy available to solve a problem. The frequency with which different strategies occur will have significant implications for subsequent assessments of performance.

22.4.3 Social Organisation and Cooperation Analysis

The social organisation and cooperation analysis is intended to establish the social and technological aspects of the system that contribute to effective performance (Vicente, 1999). From the perspective of automated systems, this type of analysis might be used to distinguish those functions that are automated from those that require input from an operator. More importantly, the aim is to describe the nature of the relationship that exists between the various components of the system.

A social organisation and cooperation analysis also provides an opportunity to examine the nature of the system and the relationship that exists between various elements within the organisational hierarchy. Therefore, there is an assumption that decisions at a higher level within an organisation have an impact on the performance of operators at lower levels of the organisation. This is consistent with the systemic model of accident causation in which failures associated with high-level decision-makers may create a latent condition in which active (operator) failures have the potential to occur (Reason, 1997).

22.4.4 WORKER COMPETENCIES ANALYSIS

Rather than prescribe a set of operator competencies at the outset, CWA provides a methodology whereby competencies are identified as a product of a detailed analysis of the requirements of the system. These competencies are considered with reference to the Skills-Rule-Knowledge (SRK) framework, to ensure that any competencies that are developed are consistent with the limitations imposed by human information processing (Naikar, 2017; Rasmussen et al., 1994). This is the point during CWA at which quantitative assessments of human performance can be incorporated into the model of the system.

22.5 IMPLICATIONS OF COGNITIVE WORK ANALYSIS

CWA provides an alternative to cognitive task analysis in which human performance is integrated within an organisation structure. This type of methodology has significant implications for automated systems, since it encapsulates the relationships that may exist between human operator and the automated system. In addition, it is possible to capture the broad range of problem-solving strategies that might be employed at any given time.

Although CWA provides useful additional information for the purposes of human performance, it should be noted that estimates of human performance remain, to some extent, subjective. For example, although there is evidence to suggest that three different strategies might be employed to resolve a specific problem, the probability that any one of these strategies will be employed must be determined if predictions are to be made concerning human behaviour. In many cases, these estimates will be made subjectively, in the absence of empirical data. As a result, some of the inherent limitations associated with human reliability may not be resolved simply through the application of a CWA.

23 Human Factors and Aviation Systems

23.1 THE SIGNIFICANCE OF SURVIVABILITY

In the absence of any acceleration, the human body experiences a *g* loading of + 1.0*g*. This denotes a force equivalent to the normal weight of an object at sea level. However, in the case of rapid accelerations or decelerations, *g* loading may alter substantially, with a consequent impact on human performance. Acceleration forces in excess of +10.0*g* can cause significant damage to the human body (Fahlstedt, Halldin, & Kleiven, 2016). This represents a weight equivalent to ten times the normal weight of the object at sea level.

From an operational perspective, a positive *g* load, such as that experienced during an acceleration or deceleration, may result in the 'pooling' of blood in the lower extremities, so that the brain becomes starved of blood and, therefore, of oxygen (Ernsting & King, 1988). In contrast, a negative *g* load, such as that experienced when driving quickly over the top of a steep hill may, in extreme situations, result in the accumulation of blood in the upper extremities, resulting in what is referred to as 'red-out'.

During a collision, the impact forces sustained may be significant, depending on the trajectory and velocity of the vehicle during the impact sequence, and the rate of deceleration. In the case of the crash of an Avianca Boeing 707 near New York in 1990, Wilson (1991) estimates that the cockpit crew would have been subjected to a loading in excess of +70.0*g* for up to 0.099 seconds. The aircraft was estimated to have descended at a pitch-up attitude of +5.5 degrees while the angle of the terrain was -23.75 degrees. This resulted in a net 'crush angle' of approximately -18.25 degrees.

Although the aircraft impacted the terrain at approximately 117 knots (approximately 216 km/h) equivalent airspeed, the deceleration was such that the forward section of the aircraft came to rest only 90 feet from the initial impact point. Nevertheless, despite the significant *g* forces that were imposed on the occupants during the impact sequence, 54 per cent survived with either serious or minor injuries (National Transportation Safety Board, 1990c). Combined with the survival rates associated with other aircraft accidents, this suggests that even in instances of major structural damage to a vehicle, at least a proportion of the occupants are likely to survive.

DOI: 10.1201/9781003229858-29

The extent to which occupants survive an accident depends on a number of factors including the structural integrity of the vehicle, the number and type of exits available, the number of occupants on board, the types of restraints, and the nature of the impact. Therefore, in improving survivability, a combination of strategies is likely to be most effective.

23.2 CRASH SURVIVABILITY

The crash of Flight 92, a British Midlands Boeing 737-400, in 1989 highlighted the significance of the physical and ergonomic characteristics of the aircraft cabin in maximising the survivability of occupants following an aircraft accident. The aircraft lost power during an emergency approach to East Midlands Airport and came to rest along an embankment to the west of the M1 motorway, approximately 900 metres from the threshold of Runway 27.

The initial impact occurred to the east of the M1 on level ground. During the second impact sequence, the aircraft sustained two major structural failures, the first of which occurred forward of the wing section and resulted in severe disruption to the forward passenger cabin. The second failure occurred aft of the wing section and forced the tail section into a position so that it was almost inverted (Air Accidents Investigation Branch, 1990).

The aircraft was operated by British Midlands and was equipped with a total of 156 passenger seats configured three abreast, either side of the centre aisle (Carter, 1991). Apart from the seats adjacent to the overwing exit, the seat pitch (defined as the horizontal distance between two corresponding points on seats in adjacent rows) ranged from 30 to 32 inches, and the seat backs were capable of pivoting forwards to facilitate evacuation. Consistent with most aircraft, the seats were attached at four points along two tracks that ran the length of the aircraft.

The seats were also designed to withstand a level of dynamic loading, similar to that expected during an impact sequence (Carter, 1991). For example, in the case of Flight 92, the majority of seats remained at least partially attached to the seating track, despite the major disruptions that had occurred. However, the seating track itself separated from the cabin floor structure. The cabin floor remained intact for 11 of the 27 rows of seats. The other areas suffered major disruptions, and it was within these sections that the majority of fatalities occurred.

A total of 118 passengers were aboard Flight 92, distributed throughout the cabin in a single-class configuration. The six cabin crew were located at positions fore and aft of the main cabin during the impact. Although all eight crew members survived the impact sequence, the majority sustained severe injuries and were incapable of assisting with the evacuation.

The nature of the injuries sustained was consistent with a rapid deceleration along the longitudinal axis and this was particularly evident in rows one through nine. The Air Accidents Investigation Branch (1990) determined that, in cases where the seat separated from the cabin floor, secondary impacts occurred with other debris that compounded the initial injuries that were sustained. Evidence to support this assertion was derived from survival rates amongst passengers seated in the overwing area where the seating and seat track remained fixed to the cabin floor. However, there were abdominal injuries sustained as a result of the lap belt, and this might have

TABLE 23.1
Number and Proportion of Survivors and Non-Survivors Who Suffered Injuries Sustained during the Crash of Flight 92

Type of Injury	Survivors	Non-Survivors
Head Injury	74 (85%)	38 (97%)
Neck	6 (7%)	21 (54%)
Upper Limb	28 (32%)	19 (49%)
Chest	18 (21%)	38 (97%)
Abdominal	2 (2.2%)	36 (92%)
Lower Limb	13 (15%)	22 (56%)
Total at the Scene	87	39

Source: Adapted from Air Accidents Investigation Branch (1990).

created significant difficulty had there been a requirement to evacuate the aircraft in the case of fire.

In comparison to survivors, non-survivors of the crash of Flight 92 were considerably more likely to have sustained head and/or neck injuries (see Table 23.1). This was probably due to a combination of the concertina effect which was evident between rows one and nine, and the major disruptions that occurred to the cabin flooring (White, Rowles, Mumford, & Firth, 1990). In addition, there was some evidence to suggest that the contents of the overhead lockers may have become dislodged and may have contributed to the prevalence of head injuries.

Although the impact forces were greater than those that the seating and aircraft structure were designed to withstand, the nature of the injuries suggests that a number of alternatives might have improved the survival rate and minimised the types of injuries sustained. For example, the rearward-facing seats occupied by the flight attendants provided significantly greater protection than the forward-facing seats occupied by nearby passengers (Carter, 1991).

Although common in military transport aircraft, the use of rearward-facing seating in commercial aviation is a matter of some debate, since aft-facing seats are likely to impose a greater load on the cabin floor structure than forward-facing seats. There is also a perception that commercial passengers may feel less comfortable in rearward-facing seats. Finally, rearward-facing seats risk potential exposure of the face to the debris that may detach during rapid deceleration.

Following the analysis of the crash of Flight 92, the Air Accidents Investigation Branch (1990) made a number of recommendations that were designed to enhance the survivability of occupants following an aircraft accident. These included:

1. The design of seats so that they minimise injury during aircraft accidents;
2. The modification of existing cabin flooring to prevent distortion during dynamic loads such as that experienced by Flight 92; and
3. The investigation of an upper torso restraint system to reduce abdominal injuries due to lap straps.

23.3 CABIN SAFETY HAZARDS

During both normal and non-normal operations, the aircraft cabin is an area where hazards have the potential to impact safety. These hazards include luggage dislodging from overhead lockers, cuts and abrasions from galley equipment, and impact due to turbulence. Remedial strategies may involve more effective locking mechanisms and/or the design of equipment so that structural failures are minimised. However, while these types of solutions may be relatively unproblematic in principle, the implementation of remedial strategies may be more difficult to achieve in practice.

In the case of in-flight turbulence, the main factor that contributes to passenger injuries appears to be lack of compliance amongst passengers to fasten their seatbelts while seated during flight (Chang & Liao, 2009). This is likely due to a number of factors, including flight attendant workload that prevents reminders, the unwillingness of passengers to adhere to the advice, and the clarity and context of the cabin briefing prior to departure (Molesworth & Burgess, 2013). It also illustrates the competing goals faced by flight attendants to both maintain customer satisfaction and ensure an optimal level of safety.

23.3.1 OVERHEAD LOCKERS

The challenges faced by flight attendants in managing the behaviour of passengers is illustrated in the use of overhead lockers. Despite guidelines that specify the maximum dimensions for carry-on luggage, passengers regularly possess carry-on luggage that exceeds the dimensions specified (Rhoden, Ralston, & Ineson, 2008). In response, flight attendants are required to stow the luggage elsewhere which, in turn, has an impact on customer satisfaction.

In most jurisdictions, carry-on luggage is restricted according to both its weight and dimensions. To reinforce these limits, airlines provide templates for measuring carry-on baggage prior to boarding. Luggage that exceeds these requirements must be stowed in the hold. The intention is to minimise the physical stress on overhead lockers, particularly in the case of an accident or incident where dislodged luggage/lockers may cause further injuries and impede the evacuation of the aircraft.

The crash of a Scandinavian Airline Systems DC-9-81 (McDonnell Douglas MD-80) at Gottröra, Sweden, in 1991 illustrates the physical demands that are imposed on overhead lockers during an accident sequence. Flight SK751 was conducting a scheduled flight to Warsaw via Copenhagen. The aircraft lost power shortly after take-off from Gottröra, and collided with terrain shortly thereafter. The fuselage separated into three sections and this enabled the evacuation of the aircraft. There were no fatalities and there was no fire (Swedish Board of Accident Investigation, 1992).

The subsequent investigation revealed that all of the overhead lockers in the forward section of the aircraft (rows one to seven) had separated from the ceiling and the majority were resting on the seat-backs. In the mid-section of the aircraft, most of the overhead lockers had remained attached, although many of the Passengers Service Units (PSU), which are located beneath the lockers, had become dislodged or had separated completely. The PSU is normally located above the passenger seating, and contains the oxygen generator, oxygen masks, and loudspeakers.

The overhead lockers were manufactured to cope effectively with a 9g static load. There was no provision for dynamic loading and the latches for the overhead lockers consisted of a single locking mechanism. The Swedish Board of Accident Investigation (1992) noted that the certification requirements for overhead lockers were below those required for other cabin furnishings such as seating. Therefore, it was recommended that overhead lockers be strengthened to withstand dynamic loads equivalent to other cabin furnishings. In addition, it was recommended that passengers be encouraged to place heavy items beneath the seat, rather than in the overhead lockers, as a means of reducing the loads imposed during an impact. This has since become a recommendation employed by a number of major airlines.

23.3.2 Fire Hazards

In most aircraft accidents where fire is a major factor, the occupants of the aircraft will more than likely die of asphyxiation and/or the inhalation of toxic gases, rather than immediate contact with the fire itself (Birch, 1988; Ekman & De Backer, 2018; Hsu & Liu, 2012). This reflects the level of thermal toxicity of the materials that can be used in aircraft furnishings. The most common gases arising from aircraft fires include hydrogen cyanide, nitrogen dioxide, and hydrogen chloride (Chaturvedi, 2010; Chaturvedi, Smith, & Canfield, 2001). Each of these gases has the potential to impact respiratory and nervous system functioning.

A collision between a USAir Boeing 737-300 and Skywest Metroliner at Los Angeles International Airport in 1991 illustrated the significance of smoke inhalation following aircraft accidents. The USAir aircraft had been cleared to land on Runway 24 Left, with the Skywest aircraft still on the runway. The collision caused both aircraft to leave the runway and impact an unoccupied fire station (National Transportation Safety Board, 1991b).

While the initial impact resulted in fatal, traumatic injuries to the majority of occupants of the Skywest aircraft, the fatalities aboard the USAir aircraft were primarily the result of asphyxiation due to smoke inhalation (National Transportation Safety Board, 1991b). In addition, the majority of the fatalities aboard the USAir aircraft occurred in the forward section between rows two and eight.

According to reports from passengers aboard the USAir aircraft, thick black smoke filled the cabin shortly after the collision. Moreover, the National Transportation Safety Board (1991b) suggested that oxygen released by the flight-crew oxygen system may have accelerated the combustion of materials and the release of toxic gases. Since the aircraft involved in the crash was manufactured prior to 1985, it was not subject to a regulation issued by the Federal Aviation Administration (Federal Aviation Regulation 25.853) that mandated the installation of fire-retardant cabin materials.

According to the National Transportation Safety Board (1991b), the airline was only obligated to utilise fire-retardant cabin furnishings if the aircraft was to undergo a 'general retrofit'. Minor refits were not subject to the Federal Aviation Administration (FAA) regulation. It was argued that any general requirement for a major retrofit would have economic implications for the airline industry.

The argument for the installation of fire-retardant materials is often based on the premise that a relationship exists between the ignition of a substance and the release of toxic materials. However, in many cases, the release of toxic gases occurs both prior to, and following the ignition stage. Consequently, the installation of fire-retardant materials may not in itself reduce fatalities, and other mechanisms ought to be explored as a means of reducing fatalities.

Possibly the most controversial suggestion to mitigate the impact of toxic gases during evacuation concerns the installation of Passenger Protective Breathing Equipment (PPBE) such as smoke hoods or masks. Nader and Smith (1993) argued that smoke hoods would provide a further five minutes to evacuate an aircraft following the release of toxic gases. However, opponents of the installation of PPBEs argue that the time required to secure the masks would counteract any advantage they may pose in aiding an evacuation (Birch, 1988; Braithwaite, 2001).

23.3.3 Exits and Evacuation

There are five standard types of exits that may be fitted to commercial jet aircraft. These range from the Type A exit, the minimum height and width for which are 183 and 106.7 cm respectively, to the Type IV exit, the minimum height and width for which are 66 and 73.7 cm respectively (see Figure 23.1) (Martinez-Val & Hedo, 2000).

The number and the types of exits with which an aircraft is equipped are determined by both the number of occupants and the speed with which they are able to evacuate the aircraft during a simulated emergency. The nominal duration within which an aircraft must be evacuated is 90 seconds, using exits on only one side of the aircraft

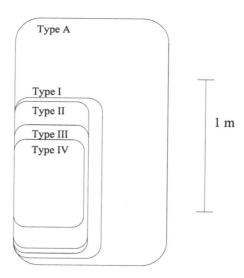

FIGURE 23.1 An illustration of the relative dimensions of standard exits for commercial aircraft. (Adapted from Edwards and Edwards, 1990, p. 140.)

(Wilson & Muir, 2010). This is designed to simulate the possibility that an obstacle (such as fire) may restrict egress.

While there has been considerable interest in the development of suitable exits to enable the evacuation of aircraft, the exit that has received most attention is the Type III, overwing exit. The Type III exit has featured prominently in a number of aircraft accidents, and it is the only exit which is permanently obstructed, usually by seating arrangements. While the actual pitch or distance between the seats depends on the operator, the pitch between the two seats adjacent to the overwing exit is normally 32 inches (Air Accidents Investigation Branch, 1988).

The minimum height for the Type III exit is 91.4 cm, while the minimum width is 50.8 cm (Wilson & Muir, 2010). The exit acts as a window during normal operations and is normally activated by pulling a latch at the top, and using a handle at the base to lift the hatch free of the fuselage.

Concern over the position and size of the Type III exit arose primarily as a result of an aircraft accident involving a Boeing 737-236 aircraft, operating as British Airtours Flight 28M, on a charter flight from Manchester to Corfu. The aircraft had just begun its take-off roll along Runway 24 when a 'thud' was heard, and the captain ordered an aborted take-off. A fire had ignited in the number one (left) engine and the flight crew turned off the runway. The aircraft came to a stop approximately 14 seconds later (Air Accidents Investigation Branch, 1988).

As the aircraft came to a halt, an evacuation was ordered on the right side of the aircraft, as the flames were being fanned across the left side of the fuselage. Of the 131 passengers and six crew members on board the aircraft, 82 passengers and four crew members survived. A total of 27 passengers escaped via the overwing exit.

In the case of the fire involving Flight 28M, a number of difficulties were noted with the Type III exit and the extent to which it might have been used effectively during the evacuation. For example, investigators noted that the passenger briefing card depicted a considerably larger area than was the case in reality where movement could be exercised in removing the hatch. Further, the figure depicted on the passenger briefing card was similar to those figures depicted at other evacuation stations where flight attendants would be expected to open exits and deploy the safety equipment. In reality, the Type III exit is positioned at the centre of the cabin, and it would be highly unlikely that a flight attendant would be in a position to open this hatch. The space available within which to remove the hatch is also restricted.

In an effort to determine the optimal aperture necessary to enable an evacuation through the Type III exit, a number of research programs were initiated, both in the United Kingdom and in the United States. In the United Kingdom, a Hawker Siddeley Trident aircraft was used to examine various aspects of cabin evacuation, including modifications in the aperture between the galley and the central bulkhead along the centre aisle (Muir, Bottomley, & Marrison, 1996). Muir et al. (1996) utilised a competitive environment where the first 50 per cent of passengers to leave the aircraft received payments in addition to their payments for participation. This strategy was designed to simulate the level of competition that is likely to exist during an actual evacuation.

For the Type III exit, the distance between the foremost and rearmost positions of the adjacent seats (also referred to as vertical projection) was varied systematically from 3 inches to 34 inches (where the row of seating is removed). The data were based on the time taken for the 30th person to egress the aircraft and the results indicated that the most rapid egress occurred with a vertical projection of 25 inches. Increasing the distance to 34 inches created more direct competition for the exit, and limited the extent to which the flow of passengers could be streamed in an orderly manner (Muir et al., 1996).

In the United States, a similar research effort was undertaken to compare the egress times for:

(a) A single Type III exit with 20 inch vertical projection;
(b) A single Type III exit where the seat back forward of the exit breaks forward 15°;
(c) A single Type III exit where a 10 inch unobstructed path is provided; and
(d) Dual Type III exits positioned adjacent to one another.

According to McLean, Chittum, Funkhouser, Fairlie, and Folk (1992), the results revealed that the average time taken for all participants to egress the aircraft was least for the dual Type III exit and was greatest for a single Type III exit where the seatback breaks forward 15 degrees.

In addition to the physical characteristics associated with the Type III exit, there were a number of additional factors that were likely to impact the rate of passenger egress. For example, the rate of egress for older participants was slower than younger participants and the egress rate decreased linearly, as a function of increasing weight and waist size (McLean & George, 1995; McLean, George, Chittum, & Funkhouser, 1995).

23.4 PROCESS CONTROL

In continuous systems such as rail and air traffic control, operators exercise a degree of oversight to ensure that the system functions within prescribed tolerances. Referred to as process control, it is a cognitive activity that ranges from monitoring automated systems and rarely intervening, as in the case of factory production; to semi-automated systems, such as rail control where there is a requirement for intervention on an intermittent basis; to non-automated systems, such as air traffic control, where continuous intervention is required.

Different forms of process control have been designed for different purposes. For example, Air Traffic Control (ATC) is a service that maintains the separation of aircraft both on the ground and during flight. It is intended to ensure the safe and expeditious movement of aircraft from departure to destination. In the contemporary aviation environment, aircraft operate along prescribed routes, travelling at different altitudes and in different directions. Under air traffic control, the separation between aircraft is normally managed through interfaces that depict their location.

Rail control tends to function slightly differently, depending on the location of the network. For example, some networks are highly automated, operate at high speed, and require minimal intervention on the part of controllers. On other networks, the movement of trains might be automated according to a timetable, but intersections

Human Factors and Aviation Systems 283

with other parts of the network necessitate intervention by rail controllers to ensure both the separation of trains and the prioritisation of movements.

Where rail networks are shared by both passenger and freight rail lines, conflicts must be negotiated, bearing in mind the priority for passenger services operating to a timetable and the acceleration and deceleration capabilities of heavier trains. Negotiating conflicts is also necessary in remote locations where trains often travel on the same line in opposite directions, passing only when one train is located in a railway siding and the controller grants permission for the transition to occur. This coordination is critical in process control settings to ensure that potential conflicts are avoided and depends on clear procedures and communication.

Poor communication during conflict management was at the centre of an accident involving a British European Airways (BEA) Trident and an Inex-Adria Douglas DC-9 over Yugoslavia on September 10, 1976. The Trident had departed London Heathrow at 0832 en route to Istanbul with 54 passengers and nine crew aboard. Approximately one hour later, the aircraft approached the Zagreb navigational beacon at 33,000ft. At the same time, an Inex Adria DC-9-31, with 108 passengers and five crew members on board also approached the Zagreb beacon, climbing from 26,000ft to 35,000ft (see Figure 23.2). On observing the likelihood of a collision, Zagreb Air Traffic Control requested that the Inex Adria aircraft maintain its present level. This request was issued in Serbo-Croatian and the flight crew replied that they were maintaining 33,000ft. Three seconds after the request from Zagreb ATC, the two

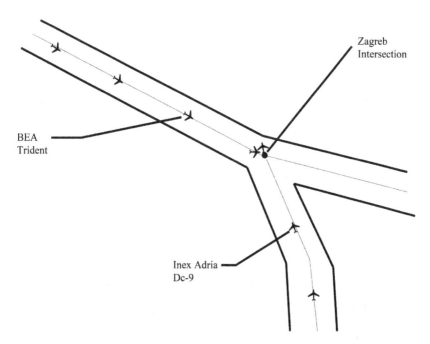

FIGURE 23.2 An illustration of the relative tracks and positions of a BEA Trident and Inex Adria DC-9 that collided over Zagreb in 1976. (Adapted from Gero, 1993, p. 135.)

aircraft collided and the explosive decompression resulted in the immediate disintegration of the two aircraft (Cookson, 2009).

The subsequent investigation indicated that primary responsibility for the collision lay with Zagreb ATC and the communication between the middle and upper sector controllers. It appeared that the upper sector controller was unaware that the middle-level controller had given permission for the Inex Adria DC-9 to climb to 35,000ft (Kletz, 1994). A number of other significant factors were also implicated in the collision including the lack of redundancy, inadequate procedures, and the impact of stress and fatigue (Cookson, 2009).

It was also noted that, in the final moments before the collision between the Trident and the DC-9, the air traffic controller had resorted to communicating in his more familiar Serbo-Croatian, rather than English, as the international language of aviation. This was likely a response to his anxiety in recognising the potential for an imminent collision, and a reversion to a less demanding and more efficient means of communication, particularly given the time constraint.

Like air traffic control, the effective and efficient control of rail networks is also impacted by an adherence to standard language that affords an accurate, joint mental model amongst stakeholders. For example, on May 19, 2004, two freight trains operated by the BNSF Railway Company collided near Gunter in Texas, while travelling in opposite directions on a single railway line. The collision was due to a communication failure where the crew of the southbound train had assumed that the northbound train was already in a siding, having vacated the main line. In fact, the train that the southbound crew had seen was a second northbound train and the rail controller had authorised the southbound train to continue only after the second of two northbound trains had been directed to a siding.

While the complexity of the situation faced by crews at Gunter likely exacerbated the difficulty in coordinating the movement of the trains, it was also noted that the use of non-standard phraseology and colloquial expressions in communicating instructions to the crews probably led to the misunderstanding. Together with the lack of precision in the application of procedures, the confluence of factors created a confusing situation that was difficult to control remotely.

The complexity in managing process control systems has been further complicated by the introduction of advanced technology systems that are intended to operate in parallel with human controllers, typically as defences against error. The impact of this complication has been especially evident in air traffic control where communication from controllers can coincide and contradict the advice from advisory systems on board the aircraft.

The Cockpit Display of Traffic Information, also referred to as the Traffic Collision and Advoidance System (TCAS) is designed to provide an aircraft flight crew with a display of information that assists in the management of aircraft separation and collision avoidance. Catalysts for the introduction of TCAS included the substantial increase in traffic density in the vicinity of major airports, the reduction in the size of flight crews, and its potential as a back-up for air traffic control (De & Sahu, 2018). The TCAS also enables a tighter separation between aircraft, increasing the efficiency of traffic flows (Stokes & Wickens, 1988).

Where potential conflicts are detected, a traffic advisory alert is issued to the flight crew, usually as an auditory and/or visual warning. In more advanced systems, these advisory alerts are accompanied by instructions that are issued based on calculations that are intended to maximise the separation between the aircraft. The difficulty lies in situations where TCAS and air traffic controllers issue conflicting instructions.

The conflict between technical and human advice was illustrated in a collision that occurred mid-air on July 1, 2002 between a Bashkirian Airlines Tupolev 154 and DHL Boeing 757 over Überlingen in Southern Germany (Bennett, 2004). The aircraft were travelling at 36,000ft on a collision course and were under the control of an air traffic controller who was managing two workstations at the time. Together with delayed radar information, the controller became aware of the impending collision only moments before the conflict occurred.

Immediately prior to the activation of the TCAS systems, the controller advised the flight crew aboard the Tupolev aircraft to descend to avoid the Boeing. Both the Tupolev and the Boeing 757 were equipped with TCAS systems and both operated correctly, advising the Tupolev to climb and Boeing to descend to avoid the conflict (Weyer, 2006). However, the Tupolev flight crew had already initiated a descent and the aircraft collided with the loss of all passengers and crew aboard both aircraft (Brooker, 2005).

23.4.1 THE ORIGINS OF AIR TRAFFIC CONTROL

Rudimentary air traffic control was introduced during World War II due to requirements for operations in poor visibility and at night, and to coordinate the activities of defending aircraft. The safe, orderly, and expeditious flow of traffic has since been established as the primary objective of air traffic control. Safety was, and still remains, the critical consideration in this process. The intention is to ensure that aircraft are safely separated at all times, and are provided with safe separation from the ground and other obstacles.

The safe separation of aircraft requires the successful management of three spatial dimensions, namely longitudinal separation, vertical separation, and lateral separation (Durso & Manning, 2008). Longitudinal separation involves a consideration of the distance between aircraft that are positioned in sequence along a particular route. This can be defined according to time or distance. Vertical separation refers to the vertical distance between two aircraft and depends on the altitude of the aircraft at the time. Lateral separation involves a consideration of the distance between aircraft, other than those in sequence along an identical track.

The development of radar systems substantially improved the quality and the reliability of the information relating to the position of aircraft. This enabled the development of air traffic control procedures based on visual representations of the location and speed of aircraft, where a greater number of aircraft could be managed within a given airspace without impacting safety margins.

Beyond radar coverage, procedural control remains a requirement, and this involves the provision of far greater safety margins than would otherwise be the case. Under procedural control, the controller must rely on the accuracy of both the navigational information associated with the aircraft and the reliability of the flight crew to report this information accurately, to determine the precise position of aircraft (Sandilands & Pascoe, 1997).

23.4.2 THE ROLE OF AIR TRAFFIC CONTROL

The role of civilian air traffic control differs with both the phase of flight and the type of flight being conducted. The most significant distinction occurs in controlled and uncontrolled airspace. In controlled airspace, aircraft are required to establish and maintain contact with an air traffic control facility and respond to directives and requests. In uncontrolled airspace, air traffic control generally acts as an advisory service (Cardosi & Murphy, 1995).

Within controlled airspace, specific areas of responsibility are defined into sectors for each controller/s according to spatial dimensions. In addition, distinct sectors require different roles in the provision of air traffic services. For example, Surface Movement Coordination (SMC) is a sector that generally involves the control of movement around the airport apron, including both vehicular and aircraft traffic. Approach and departure sectors involve the management and sequencing of aircraft in the immediate vicinity of the airport (terminal area). En route sectors may be either radar equipped or operate under procedural control, and involve the management of aircraft traffic between terminal areas.

In contrast to civilian air traffic control, military air traffic controllers are typically required to manage a relatively smaller number of aircraft in a confined area. Advice may be given concerning the position and strength of adversaries and the controller may be responsible for the same aircraft throughout the entire flight. While separation remains a significant part of the role of the military air traffic controller, the separation standards are less restrictive than those imposed on their civilian counterparts due to operational requirements.

Irrespective of the type of air traffic control, controllers are expected to develop a three-dimensional mental representation of the location and movement of aircraft. This occurs in a context where different aircraft might be capable of travelling at different speeds, at different altitudes, and will be operating to different destinations (Callari, McDonald, Kirwan, & Cartmale, 2019). The result is a complex series of decisions and judgements that must be made quickly and accurately to ensure that an optimal balance is maintained between aircraft separation and operational efficiency.

In addition to more immediate decisions, air traffic control is an operational context where some decisions and associated activities need to be deferred. This reliance on prospective memory reflects the dynamic nature of the environment where responses may be necessary only when an aircraft reaches a particular location or after a specified period (Wilson, Farrell, Visser, & Loft, 2018). Prospective memory is especially vulnerable to forgetting, particularly when interruptions occur or where there are few cues that might act as reminders.

23.4.3 THE FLIGHT PROGRESS STRIP: A HUMAN FACTORS CASE

The flight progress strip is the primary tool used by air traffic controllers to manage the position of aircraft within a sector. In some centres, these are completed by hand

and are arranged on a flight progress board where they can be referred to easily by the air traffic controller. The intention in arranging the strips is to enable the controller to represent the relative locations and priority of aircraft. This is especially important in a procedural environment where there may be no radar coverage available to confirm the location of the aircraft.

Typically, a flight progress strip will be prepared in advance of the aircraft reaching a sector (usually when the flight plan is submitted) to enable the controller to plan the most efficient sequencing of aircraft within the sector. On transiting a sector, flight progress strips are passed from one sector controller to another, denoting the handing over of control of an aircraft from one controller to another (MacKay, 1999).

Improvements in air traffic control systems have led to the development of electronic flight progress strips that have since obviated the completion of flight progress strips by hand. However, Isaac and Guselli (1996) questioned whether reading and entering information using a keyboard involves the same cognitive structures that are used when completing the flight progress strip by hand. It was argued that completing the flight strip by hand: maintained the situational awareness of the controller through the physical positioning and movement of the strips; reinforced memory traces in physically writing material onto the strip; developed strong memory traces through physical contact with the strip; and maintained prospective memory in having to actively prepare for the task by completing a flight progress strip.

While electronic flight strips have increased the efficiency of air traffic controllers' management of documentation, adaptations in behaviour have also emerged, including the use of alternative cues that act as reminders or trigger responses in the absence of written flight strips (Huber, Gramlich, & Grundgeiger, 2020). This adaptation to electronic flight strips in the context of air traffic control reflects a similar response that has occurred in the rail industry following the transition from paper-based train graph recording systems to electronic train graphs. Train graphs are used by railway controllers to map the current and future location of trains, thereby enabling the identification of potential sources of conflict (Caimi, Chudak, Fuchsberger, Laumanns, & Zenklusen, 2011).

23.5 PROCESS CONTROL AND SITUATIONAL AWARENESS

The successful management of air traffic, like other complex systems, requires a capacity for situational awareness. In an initial analysis of 33 incidents involving flight crew and air traffic controllers, extracted from the Aviation Safety Report System (ASRS), 69 per cent involved Level 1 situational awareness errors, while 19 per cent and 12 per cent of incidents involved Level 2 and Level 3 situational awareness errors respectively (Jones & Endsley, 1996). Further analysis indicated that the primary Level 1 errors related to the failure to monitor or observe data (51.5 per cent), due in part, to distractions. Although these data provided a useful foundation for the investigation of operational errors in air traffic control, they also illustrated the challenge in isolating patterns of error causation in complex, interdependent systems. Endsley and Rodgers (1997) approached the problem from a more rigorous perspective using the Systematic Air Traffic Operations Research Initiative (SATORI) to identify the foundations of successful performance. Developed by Rodgers and Duke (1993),

SATORI was designed to provide a simulated air traffic control environment using actual data, implemented within a prescribed context.

Three incidents were used to examine the situational awareness of 20 experienced air traffic controllers. At a specific point during the simulation, the screen was 'frozen' and the participants were asked to recall information such as location of aircraft within the sector, aircraft call signs, aircraft altitudes, aircraft ground speeds, and aircraft headings. Overall, the participants identified 67.1 per cent of the aircraft correctly, with a mean position error of 9.6 miles. In addition, 79.9 per cent of alphabetic call signs were recalled correctly, although this degree of accuracy decreased to 38.4 per cent with numeric call signs (Endsley & Rodgers, 1997). Finally, participants recalled the correct altitude (±300ft) for 59.7 per cent of aircraft, the correct groundspeed (±10 knots) for 28 per cent of aircraft, and the correct heading (±15°) for 48.4 per cent of aircraft. These results suggest that the information on which air traffic controllers are managing traffic is likely to be tacit, rather than explicit, and may be highly contextualised, consistent with expertise.

Using the probe technique, Friedrich, Biermann, Gontar, Biella, & Bengler (2018) were able demonstrate that, as the demands of the air traffic environment increased, the situational awareness of controllers tended to decrease. In response, controllers adopted compensatory strategies to ensure that the impact on their operational performance was minimised. Differences in the extent to which these strategies were applied suggests that an opportunity exists to develop interventions to identify and improve controllers' capacity to recognise and respond to features that are likely to impact situational awareness, and develop strategies that mitigate the impact of these losses (Falkland & Wiggins, 2019; Jipp & Ackerman, 2016).

23.6 WORKLOAD MANAGEMENT AND PROCESS CONTROL

In process control environments such as air traffic control, changes in operational demands are dynamic and are difficult to predict with certainty. Therefore, amongst controllers, the capacity to prioritise functions and thereby manage high levels of workload is critical in maintaining operational performance. In air traffic control, the number of aircraft under the direct control of a controller is generally associated with the frequency of errors that are likely to occur. Operational errors (OE) in this context occur when an air traffic controller allows the separation between aircraft, or between an aircraft and obstacle, to decrease below the minimum separation standard (Endsley & Rodgers, 1997).

Rodgers, Mogford, and Mogford (1998) suggest that OEs reflect the frequency of features to be managed, the area over which they are to managed, and the extent to which any unpredictable activities may influence the management process. This is consistent with research where a relationship is evident between the complexity of a sector under control and the performance of the controller (Endsley & Rodgers, 1997). It also suggests that workload is a major determinant in the frequency and the severity of OEs within process control environments (Shorrock & Kirwan, 2002).

Although the design of a system is likely to represent an important determinant of the workload experienced by a controller, training offers another mechanism to improve the management of workload (Kluge, Silbert, Wiemers, Frank, & Wolf,

2019). For example, Gronlund, Dougherty, Ohrt, Thomson, Bleckley, Bain, Arnell, and Manning (1997) sought to determine the extent to which the air traffic controllers' memory of aircraft positions could be manipulated through a combination of design changes and training. However, the results failed to reveal any significant changes in controllers' memory for aircraft, despite the application of a number of strategies to increase the relative significance of aircraft features such as ground speed and altitude.

Gronlund et al. (1997) suggest that memory for flight details in the context of air traffic control may not be a particularly appropriate strategy in maintaining situational assessment, since this type of information may interfere with competing tasks. For example, since air traffic control is dynamic and subject to rapid change, a vast repertoire of stored flight details may detract from, rather than enhance, the performance of controllers. This suggests that operational performance is likely to depend on skills and capabilities other than the capacity to recall features such as the ground speed or altitude of an aircraft. It also suggests that measures of performance in process control environments need to be considered carefully to ensure that they constitute a valid and reliable measure of performance.

Opportunities to support and enhance the performance of air traffic control lie in both redesigning the work environment to ensure that workload is better distributed, and in the development and introduction of training initiatives. As a means of examining the training requirements of air traffic controllers, Thompson, Agen, and Broach (1998) examined the perceptions of experienced controllers located at various towers throughout the United States. The aim was to determine whether differences existed between the perceptions of controllers and the type of training that they considered necessary for each of three different terminal control environments.

Terminal control facilities were divided on the basis of the average density of air traffic, with Level III terminal facilities experiencing an average of 20 to 59.99 operations per hour, while Level IV and Level V terminal facilities experienced 60 to 99.99 and greater than 100 operations per hour respectively. Participants were asked to rate a number of statements relating to Knowledge, Skills, and Attitudes (KSA) and consider whether they were important in operating successfully at Level III, Level IV, or Level V terminal facilities. The frequency of responses for each level of terminal facility is listed in Table 23.2.

The results indicated that experienced controllers considered human factors more important for high-density terminal facilities than for those facilities with relatively lower traffic densities. This contrasts with perceptions of communication strategies where the reverse was evident.

On the basis of this evidence, it might be concluded that the ability to adapt to changing conditions is perceived to be more important in high workload operating environments than it is in relatively low workload conditions. Consequently, the training requirements for higher levels of workload may require the development of a broad base of skills that can be applied depending on the particular features of the process control environment.

The adaptive approach to training opportunities and initiatives reflects a similar approach in managing and distributing the workload of air traffic controllers (Boag, Neal, Loft, & Halford, 2006). As workload increases, predictive modelling can be

TABLE 23.2
Frequency of Responses for KSA Dimensions Required to Function at Level III, Level IV, or Level V Towers

KSA Dimension	Terminal Facility		
	III	IV	V
National Certifications	77%	70%	78%
Human Factors	85%	83%	94%
Communication	94%	95%	89%
Weather	88%	93%	94%
Unusual Situations	77%	69%	79%

Source: Adapted from Thompson et al. (1998, p. 3).

employed to identify areas of high demand, enabling interventions such as redistributing some of the load, and/or providing technical support through automated or decision-support systems (Loft, Sanderson, Neal, & Mooij, 2007).

An adaptive approach to the management of workload typically depends on the operator signalling that the workload is increasing beyond a manageable level. However, controllers may be unwilling to acknowledge the increase in workload or may signal the increase at a point where the workload has already exceeded capacity, increasing the likelihood of operational errors. Therefore, models that predict excessive levels of workload, based on the characteristics of the operational environment, offer a useful alternative since they enable interventions in advance of extreme levels of workload (Langan-Fox, Sankey, & Canty, 2009).

As is the case in any adaptive system, the challenge lies in ensuring that the transition to adaptive control is seamless and assists, rather than exacerbates, the demands on controllers (Edwards, Homola, Mercer, & Claudatos, 2017). This includes maintaining an awareness of the distribution of responsibilities to ensure that the likelihood of operational errors is minimised to the greatest extent possible. The process may be aided in the future with the introduction of artificially intelligent systems and free-flight, with aircraft ostensibly negotiating separation in the absence of direct involvement by an air traffic controller, the role of whom will be to monitor the separation of aircraft using displays.

23.7   THE SIGNIFICANCE OF DISPLAYS

Irrespective of their role, displays are the principal means through which operators interact with the aviation environment. Therefore, optimal system performance depends on the accuracy and the reliability of the information presented, together with the accurate and timely interpretation of the information amongst operators. This requires that the design of displays accounts for different users, operating under a range of conditions, and potentially engaging with a number of displays that convey different types of information relating to the health of the system.

At its heart, system design involves the integration of a number of principles and concepts, each of which may be applicable within a particular domain. The design and development of displays involves a similar process of integration to determine:

- The optimal size of a display;
- The optimal representation of system dynamics;
- The number and types of modes involved;
- The extent to which the display conforms to the principle of pictorial realism;
- The extent to which the display conforms to the compatibility of movement;
- The optimal level of luminance and contrast;
- The optimal viewing distance; and
- The optimal symbols and colour.

The importance of display design was highlighted by a number of aircraft incidents and accidents, one of the most significant of which involved British Midlands, Flight 92, where the flight crew inadvertently shut down the wrong engine during the initial stages of an in-flight emergency. The pilots of Flight 92 had recently undergone conversion training from the Boeing 737-200 to the Boeing 737-400. The former was equipped with an electro-mechanical-style instrument display, while the latter was equipped with an electronic instrument display. One of the instruments included on both aircraft was a vibration indicator to indicate the level of vibration associated with each of the engines (Air Accidents Investigation Branch, 1990). The instrument was approximately 5 cm in diameter and had a reputation amongst B737-200 crews as being extremely unreliable. However, in the later version of the aircraft, the electronic instrument was reliable and, had it been interrogated, might have provided the critical information necessary to determine the performance of the engines.

23.8 THE CDTI: A CASE STUDY IN DISPLAYS

Although it is now ubiquitous in commercial and some general aviation aircraft, the Traffic Collision and Avoidance System (TCAS), originally referred to as the Cockpit Display of Traffic Information (CDTI), was the subject of considerable analysis to determine the optimal representation of the information to flight crew. Catalysts for the introduction of CDTI included the substantial increase in traffic density in the vicinity of major airports, the reduction in the size of flight crews, and the potential opportunities offered by free-flight where aircraft self-separate in the absence of air traffic control. It also enabled closer separation between aircraft, increasing the efficiency of traffic flows.

Irrespective of the mode of information presentation, air traffic avoidance depends on the timely detection of an intruding aircraft, the accurate probability of the estimation of the conflict, and appropriate decisions and actions concerning evasive manoeuvres (Andre, Wickens, Moorman, & Boschelli, 1991). Consequently, the success of the CDTI could be determined by the effectiveness with which flight crew detected and responded appropriately to conflicting aircraft.

23.8.1 Pictorial Representation

Originally, the CDTI was conceived as a planar-view orientation where a map display moved around a stationary, symbolic representation of the host aircraft (Palmer, Jago, & Baty, 1980). However, through experimentation, it became evident that pilots were more likely to perform horizontal (two-dimensional), rather than vertical, evasive manoeuvres when presented with a planar-view of conflicting traffic (see Figure 23.3) (Smith, Ellis, & Lee, 1984). Flight crew also demonstrated a strong tendency to turn towards, rather than away from, the conflicting traffic during situations of low or moderate danger. It was surmised that these responses might arise due to the difficulty in utilising the supplementary information on a plan-view display to mentally construct a three-dimensional representation of the interaction (Ellis, McGreevey, & Hitchcock, 1987).

Unlike the plan-view format, a perspective display of traffic information incorporates a pictorial representation of vertical separation (see Figure 23.4). Arguably, it provides a more accurate reflection of the three-dimensional space, enabling the visualisation of air traffic separation. The relative advantages associated with this design were demonstrated in a comparison between perspective and plan-view displays of traffic information, with a perspective display resulting in a relatively rapid decision time, and a lesser inclination to engage in purely horizontal manoeuvres (Ellis et al., 1987).

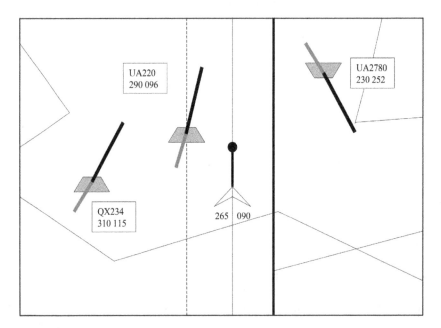

FIGURE 23.3 A sample planar view of the cockpit display of traffic information. (Adapted from Ellis et al., 1987, p. 61.)

Human Factors and Aviation Systems

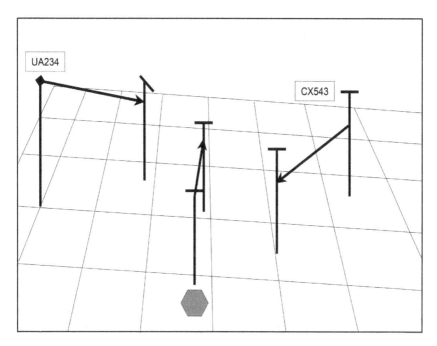

FIGURE 23.4 A sample perspective-view of the cockpit display of traffic information. (Adapted from Ellis et al., 1987, p. 61.)

The outcomes of the comparative displays of air traffic concur with the principle of pictorial realism, and the importance of matching the user's mental model (Chechile, Eggleston, Fleischman, & Sasseville, 1989; Sanders & McCormick, 1993). However, the difficulty associated with the perspective display was the ambiguity of distance information along the line of sight. To alleviate this deficiency, depth cues were incorporated into the display, including interposition, to ensure a degree of functional realism.

23.8.2 Display Size and Position

The optimal display size of an instrument depends both the physical dimensions of the instrument and the distance from which it is viewed. According to a survey of airline pilots during the development of the CDTI, 95 per cent of respondents preferred a plan-view of at least 17.5 cm or larger (Hart & Loomis, 1980). An alternative mechanism for determining the size of displays is to use the optimal display size for the legibility of a single letter at a viewing distance of one metre. This type of information provides a useful rule of thumb to determine the optimal size of a display.

23.8.3 Compatibility of Movement

The degree to which the motion of the display conforms to human expectations impacts subsequent performance. In addition to improvements in reaction time, the rate of learning tends to increase, and there are fewer errors when a display conforms to the principle of compatible motion. For example, Roscoe and Williges (1975) examined the three motion relationships that are used to depict the attitude of an aircraft in relation to the horizon. The first presented a stationary symbol against a dynamic horizon, and resulted in the poorest performance during recovery from an unusual situation (such as inverted flight). In comparison, a moving symbol against a stationary background resulted in relatively greater performance with fewer control reversals.

The most effective display involved a frequency-separated relationship where both the aircraft and the horizon move in response to the change in angular acceleration. As the aircraft is turned, there is an immediate shift in the angle of the aircraft with reference to the horizon (Roscoe, Johnson, & Williges, 1980). As the aircraft becomes established in the turn, both the horizon and the aircraft rotate slowly in the opposite direction to the turn, so that the final outcome is a horizontal aircraft displayed against a tilted background.

The compatibility of movement is applicable beyond aircraft attitude displays and is a necessary consideration during the design of any instrument that provides a dynamic, graphical depiction against an extraneous feature such as the horizon or a target aircraft. A three-dimensional CDTI typically incorporates both a reference to the horizon, and a depiction of aircraft with reference to the host aircraft.

23.8.4 Luminance and Contrast

One of the challenges associated with airborne displays is the inability of the human visual system to adapt quickly to different levels of luminance (Rogers, Spiker, & Cicinelli, 1985). The principal source of luminance within the cockpit is the glare through the windscreen, and this contrasts with the relatively low luminance of electronic displays (Leber, Roscoe, & Southward, 1986). Further, prolonged exposure to high luminance contrasts has been associated with a number of eye complaints and temporary changes in visual accommodation that may impede visual perception. Increasing the screen luminance, using negative polarity displays, and increasing the overall level of illumination, are potential methods to reduce the contrast associated with high luminance.

Improvements in response time tend to be associated with increases in the contrast ratio and background luminance of displays. Therefore, increases in luminance contrast facilitates the detection and discrimination between stimuli against a background. The challenge of luminance has become especially important with the introduction of Electronic Flight Bags (EFB) where tablets are fitted to cockpits to obviate the need for paper documentation. In many cases, the levels of luminance and contrast afforded by these devices have not been designed for high-glare environments, with pilots complaining that critical information becomes imperceptible or is misinterpreted (Cahill & McDonald, 2006).

The introduction of Head-Up Displays (HUD) also raised issues concerning visual accommodation and luminance, particularly in the context of avoidance manoeuvres (Begault, 1993). In practice, the luminance level has inevitably been a compromise between the relative benefits associated with increased discriminability, and the costs involved in visual discomfort and problems with adaptation. A general recommendation places the luminance ratio between a task and its surrounds as no greater than 1:20, although this will be difficult in the cockpit or air traffic control environment due to the requirement for external visibility (Sanders & McCormick, 1993).

23.8.5 CONCEPTUAL COMPATIBILITY

Conceptual compatibility refers to the degree to which the colours, codes, and symbols portrayed on a display are meaningful to the user (Haskell & Wickens, 1993). Generally, the greater the conceptual compatibility, the more rapid the interpretation of the information by the user. Therefore, symbolic representations should correspond, as closely as possible, to symbols that are typically associated with features in practice.

Prior to the introduction of the CDTI into practice, Hart and Loomis (1980) conducted an extensive review of pilots' opinions concerning the optimal display format and symbology for a plan-view CDTI. Consistent with the principle of conceptual compatibility, most pilots selected representations that closely resembled aircraft shapes. Further, all of the respondents expressed the opinion that the symbol for the host aircraft should be clearly differentiated from other aircraft, either by the size, shape, and/or colour.

23.8.6 COLOUR

According to Christ (1975), colour coding of displays may be the single most effective coding dimension, being superior in both discrimination and reaction time to size, shape, and/or luminance. Further, colour is a feature that enables information processing, reducing clutter on a display, and improving performance during periods of high workload (Stokes & Wickens, 1988). However, the impact of colour tends to be most prominent when it is used as a mechanism to reorientate a display that would otherwise be difficult to interpret. For example, colour might be used on a two-dimensional, planar display of terrain to differentiate areas based on height. The effect of colour on a three-dimensional, perspective display would be less significant, since the different heights would clearly be visible. The use of colour has become an important consideration in the transition to synthetic representations of terrain.

23.8.7 DISPLAY MODE

Consistent with other instruments on the flight deck, the CDTI was expected to provide both qualitative and quantitative information concerning the position of the host aircraft relative to other aircraft in the vicinity. During situations where the probability of traffic conflict is low, flight crew tend to prefer a qualitative representation of the proximate position of other aircraft. This enables attention to be directed to other tasks with greater priority. During periods of high traffic density, there is a greater reliance on quantitative information such as the precise heading, altitude, and airspeed of aircraft that may conflict with the host aircraft.

Effective displays of qualitative information involve optimising the ability to locate and observe trends both efficiently and accurately. Consequently, a qualitative representation should be devoid of extraneous information to enable the rapid recognition of critical features. The use of coding mechanisms may be appropriate in this type of situation, since distinctions between shapes and colours facilitate the location and discrimination of stimuli (Lintern, Roscoe, & Sivier, 1990). Similarly, performance on tracking tasks can be improved through the application of one or more coding dimensions (Lintern & Garrison, 1992).

Once a potential conflict becomes evident, quantitative information is necessary to estimate accurately the probability of a conflict and to conduct appropriate evasive manoeuvres. Therefore, quantitative displays are generally designed to provide users with discrete information concerning the precise status of system components (e.g. a digital speedometer). In the case of the CDTI, discrete data are usually presented in the form of a text box linked to an aircraft symbol. A similar approach is used in air traffic control displays to highlight discrete information. Typically, the pertinent information includes an aircraft identifier, the current altitude of the aircraft, the aircraft track, the aircraft airspeed, and whether the aircraft is climbing, descending, or maintaining an altitude.

23.8.8 PROBABILITY ESTIMATION

One of the principal requirements in successfully maintaining aircraft separation is the ability to accurately predict the future position of aircraft. This is true of both CDTI and air traffic control displays, and ensuring the accuracy of these estimates facilitates the decision-making process in maintaining optimal separation standards. Typically, pilots perform poorly on air traffic conflict tasks where either aircraft is turning, or when the relative heading of an intruder is greater than 45° (Palmer et al., 1980; Smith et al., 1984). These results suggest that pilots may require additional information when predicting the future position of aircraft.

Predictive displays offer an advantage in assisting decision-making, since both the present and the future states of the system can be portrayed simultaneously (Jensen, 1981). Predictions are based on algorithms that provide the operator with the capacity to formulate responses in advance of any conflict. The effectiveness of predictive displays has been established in a number of environments, including planar displays of air traffic information. In the context of the CDTI, the evidence suggests that predictive displays improve performance during air traffic conflict scenarios where one or both aircraft are turning (Palmer et al., 1980).

Despite the opportunities afforded by predictive displays, some of the disadvantages include a potential increase in display-clutter, and the possibility of increased visual workload to recognise and respond to additional task-related information (Yin, Wickens, Helander, & Laberge, 2015)

23.8.9 DECISION-MAKING

Increasing the certainty with which information is available and interpreted is likely to result in a more rapid and accurate response. However, the challenge associated with the design of the CDTI is the extent to which the flight crew are directly responsible for the decision-making process and therefore, the outcome, particularly as aircraft are now capable of executing manoeuvres to avoid a threat, independent of pilot input. In the case of contemporary systems, it is possible that:

- The flight crew take sole responsibility for the decision, based upon information derived from the display;
- The aircraft systems take sole responsibility for the decision, based upon the information derived from the display; and/or
- The flight crew *and* the aircraft systems bear joint responsibility for the decision-making process.

This is a situation consistent with the challenges associated with adaptive automation. While there is evidence to suggest that error detection and response is significantly more efficient if operators are in direct control of a task, their effectiveness depends on sustained attention and remaining 'within the loop' (Molloy & Parasuraman, 1996). Distraction due to competing priorities and/or a need to disengage with an effortful activity potentially degrade performance. The collision between the two aircraft over Öberlingen highlighted the complexity where air traffic control is also engaged in the management of aircraft separation. In this case, the instruction of the air traffic controller to the Tupolev conflicted with the advisory issued by the traffic collision and avoidance system, with the pilots responding to the 'human' air traffic controller (Weyer, 2006).

24 Human Factors and Energy

24.1 ENERGY GENERATION AND TRANSMISSION

Since the development of nuclear energy as a reliable power source, society has been faced with a difficult and, at times, controversial dilemma. On one hand, nuclear energy has the capability to provide a relatively cost-effective form of energy (Zinkle & Was, 2013). On the other hand, the potential consequences associated with a system failure can be significant (Stack & Thomas, 1991). Therefore, there is a need to consider the balance between the potential advantages afforded by nuclear energy against both the potential for a system failure and the resultant costs that might emerge as a consequence (Teperi, Puro, & Ratilainen, 2017).

The generation of electricity from nuclear energy involves a process known as fission, during which the nucleus of an atomic particle is split (Hyde, 1993). Splitting an atomic particle produces significant amounts of radiated heat that can be used to increase the temperature of water and thereby creating pressure through steam. The steam is used to drive a series of turbines and ultimately creates electrical energy.

The fuel for a nuclear reactor comprises rods of enriched uranium and one of the by-products of the generation of nuclear energy is radioactivity (Champlin, Kastenberg, & Gale, 1988). Radioactive material can be conceptualised as nuclear particles that are energised. The three main classes of radiation are ionising radiation, non-ionising radiation, and electromagnetic radiation. By far the most lethal is ionising radiation which can shatter soft tissue cells (Kempf, Azimzadeh, Atkinson, & Tapio, 2013).

Ionising radiation includes gamma rays and X-rays that penetrate bodily organs and tissue, and exposure to these is a significant concern in healthcare (McColl, Auvinen, Kesminiene, Espina, Erdmann, de Vries, Greinert, Harrison, & Schüz, 2015). Non-ionising radiation includes ultraviolet, infrared, microwave, and radio activity, where the effects on the human body appear to be less dramatic. Electromagnetic radiation is produced by high-tension power lines and a wide variety of appliances and is possibly the most common form of radiation experienced in day-to-day life.

System failures associated with high-reliability, high-consequence operating environments will always attract a high public profile. In the nuclear industry, the most catastrophic of these failures is the 'nuclear meltdown'. A nuclear meltdown occurs when the heat generated through fission is not dissipated and the temperature of the rods reaches a level at which the nuclear core begins to melt. Unless the heat is contained, it is possible that the radiation produced may be vented into the atmosphere through cracks or fires that emerge as part of the process.

Given the potentially severe consequences associated with a meltdown, the nuclear energy industry has been at the forefront in developing proactive strategies for the management of system safety. These have included generating estimates of the likelihood of errors that might be associated with system failures, the identification and analyses of hazards, and the development and implementation of strategies that are intended to control risk (Kim, Park, Kim, Kim, & Seong, 2017).

Like many industries, the design of early generation nuclear power plants was heavily influenced by technological limitations that required oversight over a large array of indicators and associated controls. This led to the duplication of controls that were similar in form but which differed in function. To avoid errors, operators installed visual and tactile reminders, in one case using 'beer taps' as a means of differentiating levers (Norman, 1988).

The challenges associated with the design of early generation nuclear control rooms was illustrated in 1961 when three military servicemen were killed while conducting maintenance on a prototype nuclear facility located at the National Reactor Testing Station at Idaho Falls in the United States (Casey, 1998). The three servicemen were tasked with reassembling part of the reactor in preparation for a test the following day. While the instructions were relatively detailed for the disassembly of the components, they were limited to the following for the process of reassembling the components: 'Assembly of the rod drive mechanism, replacement of concrete blocks and installations of motor clutch assembly are *the reverse of disassembly*' (Casey, 1998, p. 126).

While attempting to replace one of the rods that became stuck, the servicemen attempted to remove the obstruction by pulling the rod vertically. In doing so, the rod was dislodged, but it caused a chain reaction that increased the temperature in the core to 3,740° F, releasing significant levels of radiation. The ambiguity in the instructions, and the lack of knowledge concerning the consequence of particular behaviours, resulted in a meltdown at the site. Six days after the accident, the dose of ionising radiation in the body of one of the servicemen was still measuring 1,000 Rem per hour, where the maximum recommended level of exposure to ionising radiation is normally limited to 3 Rem over three months (Casey, 1998; Goetsch, 1993).

The accident at the National Reactor Testing Station illustrated the importance of a systemic approach to the assessment of safety in nuclear systems where a complex interrelationship often exists between components. A systemic analysis allows the identification of hazards, the calculation of risks, and the implementation of risk controls. As is the case in other systems, the goal is to minimise, to the greatest extent possible, the likelihood of a system failure, together with the associated consequences.

In the absence of a comprehensive, systemic analysis, failures are likely that may emerge from otherwise innocuous activities. For example, the behaviour of operators that preceded the explosion of the Number 4 reactor at the Chernobyl Nuclear Power Plant on April 26, 1986 involved what was considered a routine test (Beresford, Fesenko, Konoplev, Skuterud, Smith, & Voigt, 2016). The goal was to take advantage of a scheduled shutdown and determine the period during which the turbines would continue to supply power to the circulating pumps following the loss of electrical supply (Kortov & Ustyantsev, 2013). A similar test had been carried out the previous year.

Under normal conditions, water was used to regulate the heat generated by the nuclear reaction. This process was supported by an automated reactor core cooling system which was disengaged for the purposes of the test. Due to the design of the reactor, slowing the turbines resulted in a reduced flowrate and the entry of water warmer than was necessary to maintain the core temperature (Salge & Milling, 2006). This resulted in a rapid and unregulated increase in the heat generated by the nuclear reaction, causing ruptures and an increase in the generation of steam, the pressure from which caused the first of two explosions.

The events at the Chernobyl Nuclear Powerplant were due, in part, to a series of systemic failures on the part of both management and operational personnel. For example, the testing plan in use at the time was flawed, and the operators were not sufficiently familiar with the systems and the potential hazards that were involved. There is also some evidence to suggest that the design of the reactor itself was flawed, with the type of reaction employed leading to an increase in reactivity in cases of malfunction (Salge & Milling, 2006).

24.2 ACCIDENT CAUSATION IN NUCLEAR SYSTEMS

According to Green (1988), accident causation in complex systems can be conceptualised as an unintended transfer of energy between different subsystems, due to an inability to contain the elements within specified limits. This model of accident causation is based on the assumption that there are three main stages during which a loss of control of the system occurs (see Figure 24.1). At the initial stage, the process is operating at a normal level, and the transition to the abnormal state only occurs through a mistake on the part of the operator, equipment failure, or an activity that enables a latent condition to become manifest. More importantly, Green (1988) notes that the progression from the abnormal state to the final loss of energy and the resulting accident is prevented either through direct human intervention on the part of the operator, or through the intervention of an automated, fail-safe system.

In cases where a failure is recognised by the operator, direct intervention can be very useful to prevent the release of energy. However, where the failure is unexpected, the response of the human operator is less reliable than the intervention of an automated system. Where operators have attempted to exercise control over a system that has failed unexpectedly, there have been instances where the intervention has impeded the otherwise normal intervention of the automated system.

The main implication associated with this type of model of accident causation is that the three stages of accident causation involve independent systems, the probability of failure of which can be estimated relatively accurately. For example, the progression from normal to abnormal operations might be dependent on the probability that the cooling system in the plant will fail. However, this failure is independent of other failures that might occur, and that could otherwise have prevented radiation (energy) from being released into the atmosphere. The two elements are

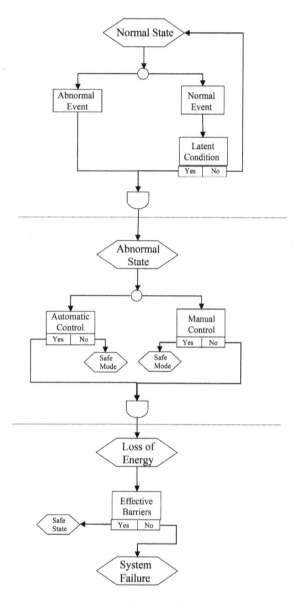

FIGURE 24.1 A diagram of an energy-based accident causation model. (Adapted from Green, 1988.)

quite separate systems and, therefore, it might be assumed that the risk associated with the failure of each of these systems is largely independent.

Although this type of model enables the quantification of risk, it is also based on the assumption that the prevention of the loss of energy, rather than the prevention of the initial error, is the issue of primary importance. Therefore, Green (1988) recognises that errors and failures are inevitable, but that the consequences must be prevented from culminating in an accident or release of energy. This principle is broadly consistent with the notion of error management where the occurrences of errors in an organisation is accepted, but that they need to be acknowledged, examined, and learning outcomes devised to prevent a similar occurrence in the future (van Dyck, Frese, Baer, & Sonnentag, 2005).

Despite the intuitive nature of the energy-based model of accident causation, the success of this model depends on the development of accurate estimates of the probability with which a particular system or piece of equipment will fail. According to Singleton (1984), this can be even more problematic when reliable estimates of human performance are required. Therefore, the aim is to develop a mechanism whereby human performance can be included as simply another component of a detailed risk analysis of a system.

Swain and Guttman (1983) suggest that one of the most effective strategies to develop an estimate of human performance is to divide a task into its constituent components and obtain estimates of error rates based on the simplified tasks. These error rates can be combined with a performance factor, such as stress, to produce an overall estimate of the rates of error associated with complex human behaviour. For example, driving a car involves a number of different perceptual, cognitive, and psychomotor elements, each of which can be isolated and examined.

The outcome may include a number of estimates of the rate at which errors are made for tasks, such as reading the speedometer, or making a left-hand turn using the indicator. By creating a composite figure, the overall rate at which errors are likely to be made can be estimated and thereby included in a global assessment of the reliability of a system. However, it should be noted that estimates of the probability of a failure can only be validated once a failure has actually occurred. The implication is that estimates of reliability and risk, and particularly those associated with human performance, need to be considered with caution, since they may be unreliable.

24.3 HUMAN ERROR IN NUCLEAR SYSTEMS

Where human error has been targeted retrospectively in many high-consequence environments, the approach in the nuclear power industry has tended to focus on estimates of the likelihood of error as a basis for broader assessments of vulnerability (Kim et al., 2017; Preischl & Hellmich, 2013). This has enabled comparative analyses and led to evidence-based decisions drawing on differences in performance. For example, the frequency of errors associated with the interpretation of information from one of two similar displays positioned adjacent to one another tends to be greater than the interpretation of same information from displays positioned some distance apart (Dougherty & Fragola, 1988). These results led to a general principle concerning the proximity of similar displays. However, the results also suggest that the design of displays with similar physical features is problematic, especially when the displays refer to different types of information.

In addition to the frequency of human error, the nuclear industry has also targeted probabilistic assessments of error, particularly where performance depends on the context in which behaviours occur. For example, oral communication is generally associated with a probability or error of 0.03 or 3 in 100 trials, whereas typing performance for a touch-typist has an associated error probability of 0.01 or 1 in 100 trials (Kirwan, 1994). Consequently, it might be concluded that written forms of communication are less prone to error than oral communication.

In considering the benefits of one form of communication over another, it should be noted that the probability of errors associated with typing performance is based on 'touch typing', a skill that must be acquired and maintained. Not only does the requirement for this type of training represent an added expense to a system, but it also increases the range of elements on which the success of the system is dependent, including the availability of sufficiently skilled operators.

Valid comparisons between probability estimates depend on an equivalent base rate and the similarities that exist between the environments within which the information was derived (Park, Kim, & Jung, 2018). In many cases, the error probabilities reported relate to specific energy-related contexts, including energy production, transmission, or distribution. Differences in the level of technical support available, the operational demands, and opportunities for training and development will have an impact on the probability that errors will occur in these contexts (Kim, Park & Jung, 2017).

Where probabilistic assessments of human error are calculated independently of a specific context, the difficulty lies in determining the extent to which a specific context will influence the outcomes. For example, the probability of error associated with a complex, non-routine event that must be accomplished under some level of stress is estimated at 0.3, or 3 in 10 trials (Kirwan, 1994). From a systems design perspective, it becomes necessary to determine the extent to which a novel system is likely to require this type of activity. Part of this process may involve some consideration of the nature of the stressors, and the impact of different stressors on the performance of complex, non-routine tasks.

The reliability of non-routine or knowledge-based tasks can also be problematic if time constraints are imposed during the performance of the task (Hoc & Amalberti, 2007). However, like other task-related features, time constraints can be perceived differently, depending on the context. For example, in a busy accountancy firm, a time constraint might constitute three days to determine the viability of a business proposition. In the case of an impending nuclear meltdown, an accurate response might be required within minutes.

In establishing accurate estimates of the probability of error, Rasmussen (1982) suggests that assessments should be based on the cognitive resources that are required to perform a task. Drawing on the SRK model of human performance, knowledge-based behaviours involve higher-order cognitive skills such as problem-solving, where a solution may be required in response to a novel task. This is an activity where the imposition of a time constraint is likely to impede performance due to the reliance on working memory, thereby increasing the likelihood of error. Occupying the other end of the continuum, skill-based behaviour demands relatively few cognitive resources, since responses are made automatically in response to a stimulus. Therefore, the imposition of a time constraint in this case is less likely to impede performance.

In estimating the probability of human error, the SRK model has been employed in the nuclear power industry as a 'step-ladder' approach to decision-making where different types and frequencies of errors are expected to occur depending on the

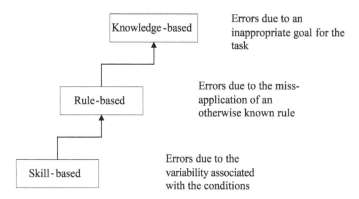

FIGURE 24.2 Characteristics of errors associated with knowledge-based, rule-based and skill-based behaviour.

type of behaviour that is likely to be engaged (Cacciabue and Hollnagel, 1995) (see Figure 24.2). For example, rule-based reasoning would be more likely to involve errors such as the misapplication of a known rule or a failure to recall an appropriate procedure (Rasmussen, 1982). In identifying the behaviours that are likely to be engaged, together with the errors that are likely to emerge, appropriate defences can be prioritised and implemented.

Although the SRK approach is a useful model of human performance, it is an example of micro-cognition, where the aim is to build an accurate theoretical model and identify the specific components of human performance. This type of approach is focused more on the relationship between the theoretical model and empirical evidence, rather than an evaluation of cognitive performance in realistic tasks (Klein & Wright, 2016; Roberts & Stanton, 2018). Therefore, a model such as SRK needs to be considered together with macro-cognitive strategies where the aim is to examine human performance within a realistic operational context.

24.4 HAZARD IDENTIFICATION IN NUCLEAR SYSTEMS

Hazard identification and risk analysis are examples of macro-cognitive strategies, and are important initiatives that provide appropriate and timely information pertaining to the safety of a specific system (Yin et al., 2021). In the nuclear industry, both hazard identification and risk analysis have well-established principles and have been used extensively to identify potential sources of failure (Jeong, Lee, & Lim, 2010).

The Hazard Identification Technique (HIT) is a strategy that can be employed both to facilitate systems design and to identify potential sources of system failure (Howard & Faust, 1991; Moreno, Guglielmi, & Cozzani, 2018). The principles are similar to other forms of hazard identification and analysis, and include:

1. Listing all of the components that comprise the system;
2. The identification of all of the potential system interfaces; and
3. An examination of the interfaces for potential sources of error.

To enable the identification of potential hazards, the components of a system should be organised into discrete elements such as equipment, personnel, and material. These categories are similar to the components that comprise the conceptual, SHEL (Software, Hardware, Environment, Liveware) model of system resources (Edwards, 1988). Having isolated the various components of a system, the second stage of the technique requires that interfaces between components be identified. This process should not normally involve any interpretation on the part of the analyst, and even the most benign interface should be included (Howard & Faust, 1991).

The final stage of the HIT is the most complex and detailed. In the first instance, it requires that each element of the interface is labelled as either 'active' or 'passive'. The aim of this process is to ascertain the extent to which a change in the state of one component can cause a change of state in the other. Where a change is anticipated, there is a need to consider whether new, second-generation interfaces emerge. This might be the case where an abnormal system state triggers an interaction with an emergency override system that might not be used during normal operations.

When considering the relationship between system components, the aim is to determine whether there is a potential for a mismatch between components, so that a change in one component has the capacity to impact adversely on another. For example, if it can be determined that the inadvertent activation of a specific subsystem can disrupt other components, it might be reasoned that some form of control mechanism might be required to prevent the activation of the subsystem. This is referred to as a critical control point where the active interface is the (H)azard, the consequence is the (T)arget and the relationship between the two is referred to as the (I)nterface.

An understanding of the relationship between system components is an important requirement for complex systems such as nuclear reactors. Part of the difficulty faced by the operators at the Chernobyl nuclear reactor related to the sheer complexity of the system. The operators were required to rapidly interpret information from a number of distinct sources to develop a coherent mental model of the situation (Norman, 1988).

As a means of resolving the difficulties faced by operators in advanced nuclear power plants, a number of training initiatives have been developed that emphasise the development of accurate and reliable mental models of the system state (Crichton & Flin, 2004; Takano, Sasou, & Yoshimura, 1997). In broad terms, the strategy emphasises:

1. The development of predictive scenarios in which operators anticipate changes and develop preventative measures as appropriate;
2. Investigations of the causes of changes in the system state and the development of appropriate remedial responses; and the
3. Development of strategies whereby changes in the system state can trigger immediate responses.

24.5 PREDICTIVE STRATEGIES

With experience, operators responsible for managing large arrays of information often develop strategies that enable assessments of the system state based on a limited number of critical indicators (Sharma, Bhavsar, Srinivasan, & Srinivasan, 2016). This strategy reduces the demands on information processing, since it reduces the number of features that need to be monitored during normal operations (Sturman et al., 2019).

24.5.1 INVESTIGATION OF CAUSES

Since critical indicators are selected on the basis that they represent the state of the system, they draw on a mental representation that can be used quickly and accurately for the purposes of fault diagnosis (Kim & Seong, 2019). While rapid responses may be necessary in some cases, the ideal approach in high-consequence, proceduralised environments such as nuclear power control is one that draws on System 2 reasoning where an operator generates a hypothesis based on an understanding of the relationship between the symptoms and causes (Wang, Gao, Li, Song, & Ma, 2017). This hypothesis is then investigated using information derived from the system, eventually enabling the identification of the cause of a system fault (Smith & Borgonovo, 2007). Once the cause has been identified, the appropriate response can be implemented according to published procedures.

24.5.2 IMMEDIATE RESPONSES

In highly time-constrained situations, operators rely on well-established procedures that have been practised regularly to initiate a response and prevent a system failure. These procedures are normally specified as part of a manual that should be accessible and easily interpretable under different cognitive demands (Kim & Seong, 2019). In some contexts, a quick reference handbook may be used alongside a more detailed procedures manual to enable rapid responses to emergencies. However, like manuals in general, their utility depends on a design that enables the identification of the appropriate fault to be addressed and the specification of procedures in a form that enables their interpretation and implementation (Ahmed, 2019).

24.6 REQUIREMENTS FOR HAZARDOUS FACILITIES

The requirements for the control of hazardous facilities are normally prescribed by a code of practice established by the relevant jurisdiction. Hazardous facilities are defined on the basis of the amount of toxic material held on site as a proportion of a threshold quantity (Khakzad, Reniers, & van Gelder, 2017). The threshold quantities of toxic substances should be consistent with those specified by the United Nations for the purposes of classifying dangerous goods.

As a part of the code of practice for the development of hazardous facilities, operators should identify each of the hazards that are likely to impact the normal operation of the system (Purba Lu, Zhang, & Ruan, 2011). In this case, a hazard analysis begins with the identification of the potential causes of system failures. This process is consistent with a fault tree analysis whereby failures are used as the starting point from which causal factors associated with the events are identified as target hazards (Kang, Kim, Lee, Lee, Eom, Choi, & Janget, 2009). The outcomes of this

TABLE 24.1
Hazard Identification Diagram for the Storage of Flammable Liquids in External Tanks

Functional Area	Possible Initiating Events	Possible Consequences	Prevention Measures
Tanks containing flammable liquids	Tank roof collapse	Tank fire	Regular maintenance Foam injection systems
	Ignition during maintenance	Possible escalation to other tanks	External water cooling systems
	Lightning	Explosion of vapours in tanks	Flame arresters on vents

Source: Adapted from National Occupational Health and Safety Commission (1996).

analysis should be presented in a tabular form so that the relationships between events can be identified (see Table 24.1).

Failing to accurately estimate the interactions between the different features that comprise a high consequence environment can lead to significant consequences. For example, the meltdown of the Fukushima-I nuclear power plant in Okuma, Japan, on March 11, 2011 was due to a combination of environmental factors and decisions that were made by both managers and operators (Khan, Jasan, & Sarkar, 2018). Initially, a magnitude 9.0 earthquake off the coast of Japan caused the reactors in three units to stop automatically, while the reactors in the remaining three units had already been shutdown for maintenance. The earthquake also caused the supply of electricity to fail, necessitating back-up generation to ensure the circulation of coolant to control the nuclear reactions.

The earthquake generated a tsunami that flooded the reactors, causing a loss of power to the pumps that were circulating the coolant. The loss of coolant resulted in an increase in temperature and the eventual release of ionising radiation. The meltdown required the evacuation of approximately 154,000 people. Although there were no deaths that were directly attributable to the meltdown, concerns remained for the population living in the vicinity of power plant.

Despite the relatively unusual confluence of events that led to the meltdown at the Fukushima-I power plant, subsequent investigations revealed that the owners of the facility had been warned of the possibility of a tsunami and the possible consequences (Kim, 2018). However, the validity of the analyses on which the assessments were based was questioned and concern was expressed at the time that the publication of the material might cause unnecessary alarm. In effect, the hazards that had been identified were ignored (Hollnagel & Fujita, 2013).

Human Factors and Energy

In high-consequence environments, the likelihood of a system failure must be calculated once hazards have been identified as part of a probabilistic risk assessment (Kang et al., 2009). In addition, estimates are necessary to quantify the level of toxic material that is likely to be released, the dispersion of the material, and the impact of this material on the population and the environment in the event that a system failure occurs (Mancuso, Campare, Salo, & Zio, 2017). This enables the development of strategies that are intended to prevent a system failure and/or manage the outcomes of a system failure.

In addition to the identification and assessment of hazards, safety management systems are necessary to monitor, identify, and respond to situations where the safety of the system may be compromised (Wahlström, 2018). Amongst regulatory authorities, this constitutes a recognition of the role of management in facilitating an appropriate response to the potential for a system failure (Hollnagel & Fujita, 2013). In particular, there is an emphasis on the development of safety objectives and performance measures against which the safety management system can be evaluated.

24.7 MANAGEMENT AND NUCLEAR POWER

Consistent with a growing awareness in other high-consequence environments, the nuclear industry is recognising the significance of management factors in precipitating system failures. For example, Marcus and Nichols (1991) argue that high-profile nuclear accidents such as those that occurred at Three Mile Island in 1979 and Chernobyl in 1986 were as much a function of inadequate management as they were a function of errors on the part of operators. They identified three elements that appeared to contribute to the safety performance of a nuclear power plant. These included:

- Problem identification
- Financial performance
- Power plant experience

Therefore, the safety of a power plant can be conceptualised as a combination of the frequency of significant incidents, the frequency with which system safety initiatives are activated, the frequency of system safety failures, and the level of exposure to radiation (Marcus & Nichols, 1991). However, it should be noted that the application of resources to specific elements, rather than the general availability of resources, tends to be most significant in predicting safety performance.

Consistent with the intention of Marcus and Nichols (1991), Wahlström (1991) sought to identify those areas of management that were associated with high-performing nuclear power stations. In this case, a comparative analysis was employed whereby similarities were sought between various high-performing systems, and these were presumed to be indicative of an effective approach towards the management of nuclear power stations. The results of this analysis revealed the following factors as characteristic of high-performing organisations:

- Well-defined goals that are communicated and understood by staff;
- A clearly defined organisational structure;
- Interaction between staff members and management;
- The application of quality assurance principles; and
- The maintenance of staff proficiency.

25 Human Factors and Marine Operations

25.1 THE MARINE ENVIRONMENT

Activities in the marine environment range from the small-scale recreational to large-scale military and commercial operations. Consequently, approaches to safety in the marine environment differ, depending on the context (Gretch, Horberry, & Koester, 2008). Amongst recreational users, the emphasis tends to target personal responsibility and the capability to identify and assess risks (Wiggins, Griffin, & Brouwers, 2019). For larger military and commercial activities, there has been a focus on systemic analyses, triggered largely by system failures.

Like other technical domains, there have been a number of serious marine accidents, resulting in a significant loss of life and substantial environmental damage. For example, on March 16, 1978, the *Amoco Cadiz*, a supertanker operating under a Liberian flag, discharged 70 million gallons of oil after it ran aground off the coast of France (Lukaszewski & Gmeiner, 1993). The vessel was heading for the United Kingdom when the rudder jammed after it had turned sharply to avoid another ship. Repairs to the rudder were initiated once the engine had been shut down. However, an onshore wind caused the *Amoco Cadiz* to drift towards the coast. Despite the efforts of a tugboat and other ships to draw the vessel away from the coast, it ran aground, breaking apart before the oil could be removed (Vidmar & Perkovič, 2018).

On March 24, 1989, the *Exxon Valdez* lost 11 million gallons of oil after running aground shortly after departing Prince William Sound in Alaska. The vessel had departed shortly before midnight and was en route to Long Beach, California. At the helm was the relatively inexperienced third mate, who was negotiating the departure through an ice field that had detached from Columbia Glacier. The vessel was equipped with a collision avoidance radar, but the system was inoperative and had yet to be repaired (Håvold, 2010)

Travelling outside the normal sea lane that would have enabled safe passage, the vessel ran aground on Bligh Reef at 12.04 am. Despite significant efforts to contain the oil spill, it resulted in catastrophic environmental damage over the immediate and longer term (Bragg, Prince, Harner, & Atlas, 1994). The associated economic loss following the grounding of the *Exxon Valdez* was estimated at approximately $3 billion (Harrald, Marcus, & Wallace, 1990).

Accidents at sea, like other industrial accidents, involve a number of causal factors, each of which contributes to the occurrence. For example, an analysis of 100 marine accidents by Wagenaar and Groeneweg (1988) identified between 7 and 58 causal factors associated with individual marine events. In most cases, combinations of environmental, mechanical, and human-related factors were likely to have been involved.

Human error tends to be the feature most evident in marine accidents and incidents where the consequences are groundings and/or collisions (Department of Transport and Regional Development, 1996). In an analysis of 24 cases that occurred between 1991 and 1995 around Australia, the main issues included:

- A lack of planning;
- Poor communication;
- A lack of position monitoring (situational awareness); and
- A failure to adhere to standard procedures.

Overshadowing these failures is the role of the organisation and the lack of an appropriate safety management system that, in many cases, might have enabled the identification of these issues as risks that needed to be addressed (Corrigan, Kay, Ryan, Ward, & Brazil, 2019). Like other complex environments, it can be difficult to identify latent conditions as they emerge. Even in the aftermath of an accident or incident, establishing with certainty a relationship between organisational expectations and the behaviour of operational personnel can be difficult. This is especially the case where the events that precede an adverse event are complex and may have occurred over an extended period.

The relationship between organisational expectations and the behaviour of operational personnel was illustrated in an accident involving the merchant vessel *Anne Holly* in St Louis Harbour, Missouri in 1998. The vessel was owned by American Milling LP, and was towing 14 barges along the Mississippi River on April 4, when a number of barges collided with a pier of the centre span of the Eads Bridge (National Transportation Safety Board, 2000). The impact caused eight of the barges to break away from the *Anne Holly* and drift back down the river towards the *President Casino on the Admiral* (*Admiral*). The *Admiral* was permanently moored below the Eads Bridge and was struck by three of the drifting barges. The total cost arising from the collisions was $11 million.

According to the National Transportation Safety Board (2000), the Mississippi River was effectively in flood on April 4, and the initial collision occurred at about 7.50 pm local time. The flood conditions had resulted in an increase in the river flow, such that the captain of the *Anne Holly* had requested the assistance of another vessel in the fleet to navigate through four bridges, including the Eads Bridge. However, the captain was informed that no vessel was available to assist the *Anne Holly*, and he continued to make his approach to the bridges unaided.

The decision to continue the journey in darkness, with a fast-running river and in the absence of a vessel to assist with the navigation of the bridges, was not considered prudent by the National Transportation Safety Board (2000). Further, the investigators suggested that the accident might have been prevented had American Milling LP developed and implemented an effective safety management system. In the case of the *Anne Holly*, the lack of a safety management system created an environment where the captain of the vessel was faced with a difficult decision either to wait until the conditions improved, or take the risk of navigating through the bridges.

The National Transportation Safety (2000) noted that American Milling LP had not established any guidelines concerning the management of safety during high-water situations, nor where situations required the assistance of other vessels. In addition, there were no guidelines provided to indicate the point at which operations should cease due to safety concerns. It reflects a latent condition that only impacted performance under particular operational circumstances.

The complex relationship that exists between latent conditions and system failures is also evident in the design of vessels, particularly where there are significant demands that prevent a considered assessment of the risks. This was a situation illustrated in the design and development of a fleet of cargo ships during World War II. Referred to as the 'Liberty Ships', they were designed using an all-welded technique, rather than the riveted technique that had previously been in use (Petroski, 2012). The reduction in the requirement for riveting reduced both the costs of labour and the time required to manufacture the ships (Puryear, Ramirez, Botard, & Kenady, 2018).

The difficulty arose when one of the ships, lying at anchor in Portland, Oregon, spontaneously broke apart and sank almost immediately. Initially, it was assumed that the failures were due to the inexperience of the ship builders and the use of an inappropriate welding technique. However, a subsequent investigation revealed that it was the type of construction, rather than the quality of manufacturing, that lay at the heart of the problem. The design was inappropriate for this type of vessel, given the types of conditions under which these ships were expected to operate (Puryear et al., 2018). It was established that steel is susceptible to a phenomenon known as 'brittle fracture', where low water temperatures and irregular loading on the seams can result in cracks that travel quickly along the plates comprising the base of the vessel.

This case of the Liberty Ships illustrates the challenge in establishing the relationship between latent conditions and outcomes, particularly over extended periods. Nevertheless, improvements in system safety and performance require that these relationships are established in the case of an incident or accident, so that appropriate remedial strategies can be implemented. To assist with this process, a causal network model can be employed where the relationships between significant factors and system failures are depicted schematically through the use of AND and OR gates.

25.2 CAUSAL NETWORK ANALYSIS

A causal network model of accident causation is an approach to the analysis of system failure in which the outcome (accident or incident) is the starting point for the process. The ideal application of this methodology involves an anticipation of an undesired event, and the identification of remedial strategies to prevent the occurrence. For example, Harrald, Mazzuchi, Spahn, Van Dorp, Merrick, Shrestha, & Grabowski (1998) attempted to develop a model of the risks associated with the transportation of oil through Prince William Sound, Alaska. The probabilities associated with specific events were estimated on the basis of interviews with Subject-Matter Experts (SME). This information was used subsequently to develop an overall probability of system failure, and the relative impact of risk reduction measures.

Wagenaar and Groeneweg (1988) adopted a slightly different approach in which the causal network methodology was used to identify patterns across a series of marine accidents. A causal network model typically consists of a decision tree in which each event has a series of preceding factors (Ren, Jenkinson, Wang, Xu, & Yang, 2008). For example, successfully stopping at a traffic light depends on both the driver seeing that the light is red AND the capability of the vehicle to stop within the distance available. Neither of these factors alone will ensure that the event occurs. This relationship might be depicted as in Figure 25.1.

Where there are multiple factors required before an event occurs, the complexity of the diagram increases. Figure 25.2 illustrates an AND gate where the ability of the driver to observe the traffic light is predicated on the driver attending to the road ahead and the traffic light being visible. Therefore, seeing the traffic light becomes an intermediate factor and the root factors are the traffic light being visible, the driver attending to the road ahead, and the capability of the vehicle to stop within the distance available.

Where there are a number of factors, any of which may trigger an outcome, an OR gate is employed, as depicted in Figure 25.3. In this case, it is assumed that the driver seeing the traffic light is predicated on the light being visible and the driver OR the passenger attending to the road ahead.

Botting and Johnson (1998) suggest that the use of accident reports for the development of causal network models can be problematic, due to the potential lack of information arising from reports. Moreover, the accuracy of the model depends

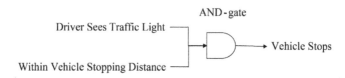

FIGURE 25.1 An illustration of the application of an AND gate in a causal network model for stopping a vehicle before a red traffic light.

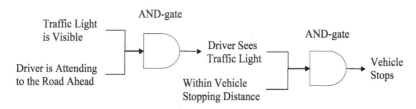

FIGURE 25.2 An illustration of the application of multiple AND gates in a causal network model for stopping a vehicle before a red traffic light.

Human Factors and Marine Operations

on the accuracy of the causal relationships identified by the accident investigators. Therefore, the outcomes of causal network models need to be interpreted with a degree of caution.

Causal network models also fail to convey the precise sequence during which events may have occurred. Neither do they necessarily differentiate human from other sources of error. Nevertheless, causal network models can be useful where the accident is otherwise inexplicable. For example, the causal network model in Figure 25.4 has been developed to partly explain an accident aboard the *Farmsum* in 1982, the circumstances of which were initially considered inconceivable (Wagenaar & Groeneweg, 1987).

The *Farmsum* had been sailing using water ballast in hold 4, when poor weather resulted in a leak that allowed some of this water to accumulate in the adjacent hold 5. Although the missing water was noted, the first mate explained the discrepancy as water that had been pumped to clean hold 6. As a result, hold 4 was refilled. In doing so, the water in hold 5 also increased to the extent that the load on the partition between holds 5 and 6 was too great and the partition collapsed, releasing approximately 6,000 tons of water. At the time, four sailors were cleaning hold 6, and three were drowned.

Based on Figure 25.4, it should be evident that causal network analysis has the capability to differentiate the various elements that comprise an event. It can be a very useful, initial step in the process of explanation. However, it should also be evident

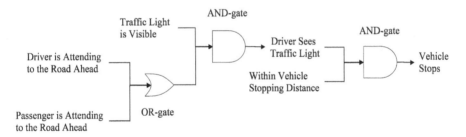

FIGURE 25.3 An illustration of the application of an OR gate in causal network model for stopping a vehicle before a red traffic light.

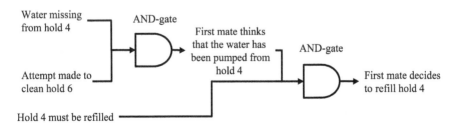

FIGURE 25.4 A causal network model to explain the factors associated with the decision by the first mate to refill hold 4 aboard the *Farmsum*: a decision that ultimately led to the loss of three crew members. (Adapted from Wagenaar and Groeneweg, 1988.)

that the duration across which these events occurred is difficult to determine on the basis of the information available. It is also difficult to differentiate those events that were related to human error, from those that were related to other factors, such as the environment or the organisation.

25.3 OCCUPATIONAL HEALTH IN THE MARINE ENVIRONMENT

In addition to the potential for large-scale accidents and incidents, the marine environment is also particularly susceptible to the influence of occupational health and safety hazards. Some of these hazards include unsecured loads, extreme environmental conditions, irregular work shifts, fatigue, slippery work surfaces, and confined workspaces (Aziz, Ahmed, Khan, Stack, & Lind, 2019).

The impact of hazards in the marine environment can be exacerbated by the environmental conditions under which operations are expected to occur. For example, the transfer of marine pilots is often necessary when vessels approach a port. In comparison to the captain, the marine pilot has greater experience with the idiosyncratic conditions that are likely to be experienced within a port and therefore this strategy mitigates the risks associated with berthing a vessel.

The hazards to which marine pilots are exposed relate primarily to the transfer from a tender to a ship (Weigall, 2006). This transfer typically occurs in open water where the tender is positioned alongside the vessel while the pilot reaches for either a ladder or ramp that has been lowered for the purpose. Climbing the ladder or ramp can occur in large swells, causing pilots to lose their footing, with the possibility that they will fall into the sea. In some ports, the risk is so high that port authorities use helicopters rather than tenders to transfer pilots, except in those conditions where helicopters are unable to operate.

In an analysis of the causes of death associated with the transportation of dangerous goods, Romer, Haastrup, and Petersen (1995) compared the number of fatalities in the marine environment with those that are associated with other forms of transportation. The results suggested that, where a toxic emission occurs, the affected population is much greater in the marine environment than in other forms of transportation. This is probably explained by the relative proximity of operators within the marine environment.

25.4 HUMAN FACTORS INITIATIVES

In response to the observation that the majority of human error-related marine accidents involve collisions, regulatory authorities have implemented a Vessel Traffic Service (VTS), especially for areas of high traffic density (Praetorius, Hollnagel, & Dahlman, 2015). The aim of VTS is to provide traffic information so that the captains of vessels can anticipate difficulties and take appropriate corrective measures.

The validity of the VTS in preventing collisions was established following an analysis of 936 accidents that were classified into one of four groups according to the factors involved (Le Blanc & Rucks, 1996). The results indicated that those accidents where the crew had never used the VTS were more likely to be associated with navigation conditions considered reasonable for the purposes of traffic avoidance. By comparison, those accidents where the VTS had been used regularly by crew were more likely to be associated with navigation conditions considered difficult for the purposes of traffic avoidance. This suggests that the availability of VTS may provide captains with an assurance of safety that they may be unwilling to accept in the absence of VTS.

Goossens and Glansdorp (1998) note that the availability of VTS does not eliminate accidents from occurring. While there has been a reduction in the frequency of collisions in those areas where VTS has been introduced, errors remain and collisions do occur from time to time (Puisa, Lin, Bolbot, & Vassalos, 2018). However, the introduction of VTS, like advanced technology systems in other contexts, has changed the nature of the errors that occur. Therefore, VTS is one of a number of strategies that need to be considered in the prevention of marine collisions. These include the application of appropriate error management procedures and the introduction of management training referred to as Bridge Resource Management (BRM) (Yıldırım, Başar, & Uğurlu, 2019).

In addition to improvements in the performance of the operator, human factors principles have also been introduced as part of the process of marine design. According to Pan & Hildre (2018), human factors principles are an integral aspect of the design of both contemporary vessels and the legislative and regulatory context within which they function. Similarly, human factors principles have been applied within the submarine environment, especially in the development of contemporary command and control systems (Roberts, Stanton, & Fay, 2017).

25.5 HUMAN PERFORMANCE AND OFF-SHORE PLATFORMS

Off-shore oil platforms are subject to many of the human performance issues that are evident in other areas of the marine industry. For example, they are often located in hostile environments, some distance from a shoreline, with operators spending extended periods on board, unable to disengage from the workplace. The activities that they undertake can also be physically demanding, impacted by poor sleep patterns due to long hours of shift work (Hudson, 2007).

According to Mansfield, Michell, and Finucane (1991), the risk of serious injury aboard off-shore oil platforms is exacerbated by the proximity of drilling stations and production facilities to crew living quarters. This tight coupling between system components typically results in a rapid escalation of the consequences of a system failure. Therefore, the capability to contain the adverse effects of system failures and prevent causalities can be limited.

The rapid escalation of system failures was highlighted in an explosion aboard the Piper Alpha oil and gas platform in the North Sea on July 6, 1988 (Swuste, Groeneweg, Van Gulijk, Zwaard, & Lemkowitz, 2018). Located 273 kilometres from Aberdeen, Piper Alpha was a fixed platform designed to pump crude oil and gas. The accident was triggered when members of the night shift attempted to start a condensate pump used in gas processing that had previously been shut down for maintenance (Paté-Cornell, 1993). Due to a breakdown in communication with the preceding shift, the night shift was unaware that a safety relief valve had been removed and that the replacement seal was not secure.

The platform had originally been designed so that the most dangerous operations were located some distance from the areas in which crews were operating. However, a design change to enable gas production resulted in the more consequential gas compression facility being located adjacent to the control room (Flin, Slaven, & Stewart, 1996). Immediately after the condensate pump was switched on, the missing safety valve caused an increase in gas pressure that eventually escaped and ignited. The explosion blew through the firewall and destroyed the control room. Despite the best efforts of the crew and associated first-responders, the platform was destroyed completely within three hours of the initial explosion. Of the 226 crew members aboard the platform, 165 were killed.

Given the difficulty associated with containing a system failure, an alternative approach in the context of off-shore oil and gas processing has focused on the root causes of human error. Gordon (1998) describes an approach in which 25 accidents were examined based on the information submitted as part of an accident reporting system. The causal factors associated with accidents were classified as either immediate causes (unsafe acts), or underlying causes (latent conditions). Examples of the former included:

- Operating without authority
- The use of defective equipment
- The improper use of equipment
- Horseplay
- Improper lifting
- Lack of attention

Underlying factors were further classified into personal and job factors and included:

- Inadequate training
- Lack of education
- Fatigue
- Inadequate planning
- Inadequate discipline
- Poor job descriptions

Both the immediate and underlying factors associated with accidents aboard offshore oil platforms are consistent with the types of causal factors that occur in other industrial domains including aviation (Gerbert & Kemmler, 1986). The difference lies in the speed with which these failures can escalate to interact with other systems. Mansfield et al. (1991) suggest that strategies such as fault tree analysis may not be sufficient to account for the complex relationships that exist between system components.

In addition to the application of fault tree analysis, Mansfield et al. (1991) advocate a strategic approach to the management of oil platform safety in which potential hazards are identified and SME are asked:

(a) To assess the extent to which the consequences of a system failure can be contained; and
(b) Whether protective measures would be sufficient to prevent the consequences of a system failure from interacting with other systems.

Consistent with an energy-based model, this type of strategy acknowledges the inevitability of system failures, but is designed to identify strategies that might be employed to contain the consequences of such failures. The strategy proposed by Mansfield et al. (1991) requires estimates of both the probability of an occurrence and the duration over which the events are likely to occur. Consequently, the accuracy of this type of model is likely to be dependent on the skills and knowledge of the SMEs concerned.

26 Human Factors and Healthcare

26.1 HUMAN FACTORS AND HEALTHCARE

Like aviation and marine operations, healthcare is a high-consequence environment where successful performance depends on a combination of skilled practitioners operating as an efficient and effective team, using technology and systems that are well designed, usable, and that reduce the likelihood of error. However, where other operational contexts are characterised by routine activities punctuated by limited periods of non-normal operations, healthcare tends to be more variable, with situations escalating quickly, demanding interventions from a range of specialists who are operating under uncertainty (Carayon, 2006).

The role of human factors in healthcare likely first emerged following concerns associated with usability of medical devices and the frequency of errors that were occurring. Of particular concern was the intravenous (IV) pump, and the challenges in ensuring the accuracy in the rate at which a drug was administered (Schaeffer, 2012). Like many systems, early designs of IV machines lacked user involvement, particularly in complex, time-constrained environments, where accuracy and efficiency were necessary.

Beyond the design of devices, early interest in the role of human factors in healthcare was advanced by anaesthetists in particular, who identified parallels between operations in the aircraft cockpit and the management of teams during surgery (Toff, 2010). The result was a particular interest in resource management and the application of these principles in the surgical context.

Like the experience in aviation, initial approaches to introduce resource management into healthcare were designed to generate interest, largely supported by non-jeopardy training courses and workshops (Jones, Fawker-Corbett, Groom, Morton, Lister, & Mercer, 2018). At this initial stage, the goal was to improve knowledge and awareness amongst practitioners, and courses included topics such as communication, leadership, and situational awareness. The content of these initial courses drew heavily on the experience in aviation, including case examples.

Among anaesthetists, the initial introduction to resource management led to a broader concern for human factors, including the development and introduction of checklists, system design, and the management of errors (Mercer Whittle, Siggers, & Frazer, 2010). This is an approach that continues, with detailed analyses of occurrences supported by the application of system-related taxonomies. As in other operational contexts, it reflects a progression towards understanding human error as an outcome of failures that may have occurred at the system level.

26.2 RESOURCE MANAGEMENT AND TEAMS

Capitalising on the early experience of resource management initiatives in healthcare, training programmes and initiatives have since been implemented across a broad range of healthcare professions, from nursing staff to surgeons (Oriol, 2006). Depending on the experience of the cohort, there may be an emphasis on awareness and knowledge as a precursor to the acquisition and maintenance of skills. These capabilities are now reinforced through simulation, drawing together specialist teams that can practise and further develop the components of resource management (Cheng, Donoghue, Gilfoyle, & Eppich, 2012).

With experience and understanding that the medical context differs from other high-consequence environments, examples and cases tend no longer to be drawn from the aviation or other external contexts, but are sourced from occurrences that are health-related or that relate directly to the specialisation. Employing relevant cases creates a degree of engagement and arguably facilitates the transfer of training into practice.

Resource management initiatives in healthcare have been further supported through the development of an inventory of non-technical skills that was originally developed for the aviation environment but was subsequently adapted for health specialisations (Mishra, Catchpole, & McCulloch, 2009). Rather than a set of universal non-technical behaviours, editions have been developed that are directed specifically to specialisations, including surgery. They are typically administered by an observer, focusing on team behaviours, rather than the behaviour of a particular individual within the team (Pradarelli, George, Kavanagh, Sonnay, Khoon, & Havens, 2021).

In developing the capacity afforded by effective and efficient teams in healthcare, learning outcomes have been drawn from the coordination evident among Formula One pit-crews (Catchpole, Sellers, Goldman, McCulloch, & Hignett, 2010). These crews are charged with the responsibility to refuel and service racing cars safely and efficiently within very short periods. Their effectiveness depends on highly defined roles and responsibilities, the activities associated with which are well-coordinated and well-rehearsed. Parallels can be drawn with emergency medicine, where accurate, coordinated responses are also necessary within very short periods. Establishing and understanding roles in this context appears to be associated with improved patient outcomes.

While highly coordinated activities have proven successful in the context of emergency medicine where behaviour is centred around a singular event, less time-critical situations may require the coordination of healthcare resources over an extended period, involving a number of medical specialities. For example, a cancer diagnosis will normally require staging, during which radiologists will identify the course of the malignancy throughout the body. This information will be used by surgeons and oncologists to determine whether surgery, chemotherapy, and/or radiation therapy constitutes the most appropriate treatment.

In determining a course of treatment under conditions of uncertainty, multidisciplinary teams will normally be involved and will meet to discuss the various options, together with the risks for the patient. The success of these teams depends on a range of collaborative skills, including clinical planning, executive tasks, and relations among team members (Sutton, Liao, Jimmieson, & Restubog, 2011). The identification and categorisation of these skills has led to the development of the Team Functioning Assessment Tool (TFAT), an instrument that can be employed to evaluate the non-technical skills of multidisciplinary teams in healthcare, the outcomes of which are sensitive to group processes (Sutton, Liao, Jimmieson, & Restubog, 2013).

26.3 ORGANISATIONAL FACTORS AND HEALTHCARE

As has occurred in other complex industrial systems, there has been a significant interest in the role of organisational factors in precipitating failures in healthcare (Hudson, 2003). These generally relate to the availability of resources, the quality of leadership, and expectations of performance. Since healthcare is a labour-intensive context, one of the most significant issues that concerns nurses in particular is the level of workload imposed during shifts, together with the distribution of this load.

The successful management of the patients within a ward depends on an accurate expectation of patient needs. Establishing the anticipated needs of patients enables judgements concerning the level of nursing support required, including the availability of specialist nurses. To manage costs, nurses with a range of skills and capabilities might be rostered for a shift, potentially creating constraints that limit the types of care that can be administered to patients (Westbrook, Rob, Woods, & Parry, 2011). This creates a dynamic situation where there may be periods of a shift where there are increases in the demands on a limited number of nurses due to the particular and unanticipated needs of patients.

The reliance on a limited number of nurses to undertake highly skilled activities potentially leads to instances where a higher workload, together with increasing demands from patients, creates the conditions for adverse events, including medication errors (Reason, 2000). Medication errors involve unintended variations in the process of treatment that prove hazardous for the patient. Medication errors have been implicated in a greater number of deaths in the United States than car accidents (Fry & Dacey, 2007).

Using a Hierarchical Task Analysis (HTA), Lane, Stanton, & Harrison (2006) were able to demonstrate that the precursors to medication error occurred at one or more of five stages, including during prescription, the documentation of the medication, the dispensation of the medication, or finally, the administration and monitoring of the effects of the medication. Remedial strategies emerging from the HTA included electronic tagging of medication, clearer labelling, and limiting and standardising equipment to minimise uncertainty.

Employing the Systematic Human Error Reduction and Prediction Approach (SHERPA), it was possible to estimate the likelihood of different types of error, including action errors, checking errors, and communication errors. However, it is important to note that the incidence of medication errors can be further exacerbated when nurses are required to undertake tasks in addition to the immediate care of the patient, such as transporting patients and general housekeeping (Kang, Kim, & Lee, 2016).

In anaesthetics, the rate of medication errors has been estimated at approximately 1 in 13,000 administrations, with the poor labelling of syringes contributing to many of these occurrences. While national and international standards have been introduced to ensure a degree of consistency in the labelling of syringes, there are provisions for situations that do not require labelling, especially where the same person has drawn up the dose of medication and administers the medication immediately. This might occur in emergency situations where the time required to label a syringe potentially impacts the treatment of the patient. Nevertheless, Merry, Shipp, & Lowinger (2011) recount a situation in which a trainee physician drew up and administered what he believed to be magnesium to treat the elevated blood pressure of a patient. In fact, the medication administered was dopamine which caused a further increase in the patient's blood pressure. The error occurred due to the similarity in appearance between the ampoules containing magnesium and dopamine. Merry et al. (2011) argue that pausing to label the syringe would have potentially identified the error since the mismatch between the label on the syringe and the label on the ampoule would have become evident.

The challenges associated with medication errors raise a debate that exists in healthcare concerning the role of systemic and organisational factors and the responsibilities of healthcare providers. Violations of best practice or accepted practice constitute a challenge in healthcare, although the distinction is made between work-as-imagined and work-as-done (Clay-Williams, Hounsgaard, & Hollnagel, 2015). In many cases, the development of procedures and practices that are designed to reduce the probably of adverse outcomes are so cumbersome that, inevitably, violations emerge.

In the medical context, a clinical guideline comprises a series of tasks that is intended to ensure that a task is completed safety. However, guidelines are often based on assumptions that may not be valid under particular circumstances. For example, a guideline might be predicated on the availability of existing materials or resources. Similarly, a lack of specificity may lead to differences in interpretation, which might explain differences in the degree of compliance. This was evident in the administering guidelines associated with the management of infections in intensive care units, where ambiguity was evident in the tasks, responsibilities, expectations, and exceptions (Gurses, Seidl, Vaidya, Bochicchio, Harris, Hebden, & Xiao, 2008).

Employing the Functional Resonance Analysis Method (FRAM), Clay-Williams et al. (2015) compared Work-as-Done with the guidelines associated with the transfer of critically ill patients within a hospital and noted a number of inconsistencies, including the expectation that doctors and nurses would be available simultaneously for a joint meeting prior to the transfer. In addition to the difficulties associated with the coordination of personnel, the meeting was predicated on the availability of a range of data that would normally take time to compile accurately. The outcomes highlighted the role of organisational factors in the administration of healthcare to patients.

26.4 RESILIENCE AND HEALTHCARE

Healthcare, like other complex, high-consequence environments, engages a process of investigation in the context of an adverse outcome. An important part of this process involves the Root-Cause-Analysis (RCA) methodology, on the basis that a limited number of components of the system can be targeted and then redressed to prevent a reoccurrence (Peerally, Carr, Waring, & Dixon-Woods, 2017). However, despite the prevalence of RCAs as a safety management strategy in health, there is also a recognition that investigations in response to isolated occurrences do not necessarily improve the capability to respond to changes in system demands in the future (Latino, 2015).

Capitalising on the capabilities of resilience engineering, approaches to safety management are complemented by strategies that are designed to ensure that healthcare systems are capable of withstanding dramatic changes in demands (Smith & Plunkett, 2019). This can involve simulations that are intended to establish the fragility of the system. For example, simulations in an emergency department might reveal a capacity to cope with a large number of patients with non-life-threatening complaints. However, introducing a patient with a viral infection might reveal a lack of personal and protective equipment, and an inability to isolate the patient adequately, thereby reducing the capacity of the department to admit other patients. The consequence would be a requirement to transfer patients to other facilities, which may involve additional travel and the associated risks to the patient, together with increases in the demands on other hospitals.

Like other proactive approaches to system safety, one of the most significant challenges associated with a resilience approach to healthcare is the investment necessary in advance of an occurrence. These represent direct costs to the healthcare system with the promise of identifying weaknesses that, when resolved, will improve the capacity of a hospital or department. These can be difficult to accrue when the healthcare system may already be impacted by fiscal constraints.

Identifying opportunities for large-scale simulations or tests of the system also present barriers to a resilience or Safety II approach to safety management in healthcare (Braithwaite, Wears, & Hollnagel, 2015). While these simulations can be table-top exercises where participants mimic the roles that they might play during an event, greater fidelity is achieved in simulations that represent reality as closely as possible. Allowing an event to 'play out' as it would in the operational environment enables the emergence of dependencies that might otherwise remain unobserved (Sujan, Furniss, Anderson, & Braithwaite, 2019). It is precisely these 'hidden' dependencies that can cause fragility or brittleness in a system, since they are unmonitored.

As occurs following 'Black Swan' scenarios, it is rarely the case that a particular simulation will account for all the interactions and permutations that may impact a system over an extended period. Therefore, periodic risk assessments are necessary to identify potential or emerging threats that might be employed as triggers for a simulation. For example, winter is typically associated with an increase in respiratory illness, particularly amongst the elderly. Elderly patients also present with significant co-morbidities, drawing on resources from multiple health specialties. Consequently, a risk assessment might lead to the conclusion that an assessment of the resilience of the healthcare system is necessary prior to the onset of winter.

26.5 DECISIONS AND REASONING

One of the most important roles amongst physicians is the capacity to identify symptoms of a condition and generate diagnoses that might lead to effective and timely treatment. For physicians, the challenge lies in discerning relevant from less relevant features, the acquisition of information from a range of sources and in a range of forms, the integration of this information, and then matching this information with existing representations in memory to draw a conclusion. To further complicate the situation, this complex process of diagnosis may need to occur quickly, and in an emotionally charged environment where the consequences of a misdiagnosis may be significant.

Like other forms of human reasoning, the diagnostic process engaged by physicians is liable to error (Croskerry, 2013). These errors may be due to an incomplete mental model, the failure to identify a critical feature or symptom, and/or the inability to integrate the information available into a coherent and meaningful frame. Opportunities for error tend to emerge in situations where symptoms emerge slowly over time, where multiple medical specialties might be engaged, and/or when there are environmental or organisational demands that might prevent a detailed analysis.

The role of heuristics and biases has emerged as a significant issue in the context of medical misdiagnosis, since there is evidence to suggest that physicians are likely to be susceptible to confirmation bias where pre-existing information might be available and that might lead to a preconceived expectation concerning the patient's condition (Croskerry, 2017). Interventions that are intended to improve performance generally consist of training initiatives where physicians are encouraged to adopt System 2 thinking during diagnoses. In practice, this is referred to as differential diagnosis, and involves a process whereby the physician considers an initial diagnosis, together with alternative explanations for a complaint. Systematically, each alternative explanation is examined against the pattern of symptoms until the most likely explanation remains.

Simulation training has proven effective in raising awareness of the potential for bias amongst clinicians. However, it appears that remedial strategies are most effective where clinicians experience the bias personally and participate subsequently in a process of debriefing (Altabbaa, Raven, & Laberge, 2019). Debriefing enables the provision of feedback and provides an opportunity for reflection.

Amongst nurses, reflection has been associated with improvements in decision-making in intensive care (Razieh, Somayeh, & Fariba, 2018). It is a process that encourages the use of open-ended questions to engage in thoughtful analysis, drawing attention to context-related information, and considering the appropriateness of interpretations and responses. The intention is to enable the development of strategies necessary for self-directed feedback that can be implemented in practice.

Like other initiatives, improvements in clinical decision-making need to account for the complexities associated with healthcare, including the availability of appropriate diagnostic tools. However, there are individual differences in diagnostic capability that are evident, irrespective of broad measures of experience. These individual differences likely emerge due to the nature of the health conditions to which clinicians are exposed. For example, clinicians operating is some jurisdictions may become more familiar with particular conditions due to the nature of population whom they are treating. This potentially leads to a prevalence bias where clinicians inadvertently bias their diagnoses based on perceived base-rates.

Overcoming prevalence bias requires a recalibration where clinicians are exposed to features that are intended to correct a tendency towards a particular hypothesis. This begins with an assessment of clinicians' performance that enables the provision of feedback, ideally accounting for experience. Feedback that enables the identification of areas of development provides the motivation for engagement in learning and provides the basis for the development of strategies to improve performance in the operational environment.

Clinical decision-making in the future is likely to be informed by Artificially Intelligent (AI) systems operating alongside clinicians. AI systems are already demonstrating high levels of accuracy, particularly in diagnosis (Loh, 2018). However, this cooperative endeavour between diagnosticians and automated systems is raising issues consistent with adaptive automation in driving. These include the reliability of advanced technology diagnostic tools, the extent to which clinicians transition from diagnosticians to monitors of the diagnostic process, and the resulting impact on the skills of clinicians over the longer term.

Together with AI, robotic systems are already in use in surgery, often in the absence of direct engagement by the surgeon. The challenges associated with robotic surgery include the costs of systems, their specificity, and the training needs of surgeons.

Part VII

Assessment and Report Writing

27 Human Factors Testing Methodology

27.1 HUMAN FACTORS TESTING

Testing and evaluation in the context of human factors is a process similar to empirical research. The aim is to acquire data that will assist in the development of an understanding of the existing state of the environment. Consequently, the principles that underscore appropriate empirical research also apply in human factors testing.

The process of research is based on a number of premises, one of the most important of which is that the method of data acquisition and analysis yields outcomes that accurately reflect the nature of the environment. As a result, the methodology employed must be appropriate for the particular environment, and must be transparent, so that other researchers can employ the same methodology and thereby arrive at the same or similar outcomes.

The selection of an appropriate methodology is one of the most important decisions faced by the human factors specialist, since it must be appropriate for an operational environment where there may be a limited degree of experimental control that can be exercised over the variables. For example, if a workplace injury is reported, it may be difficult to identify the causal factors simply on the basis of a retrospective analysis. For a number or reasons, including a potential lack of information, significant factors may be overlooked, resulting in an inaccurate assessment and conclusion.

27.1.1 Reliability

Given the lack of experimental control that often characterises the operational environment, it is important to consider the extent to which more than one methodology might be employed as a means of establishing the reliability of the research outcomes. For example, where subjective perceptions of ergonomic issues are required, a questionnaire may be distributed initially, and the results might be compared against the outcomes of focus groups or other similar methods of data acquisition. This process of triangulation, where the extent to which the perceptions expressed in the questionnaire are consistent with those that arise from other forms of data acquisition, can be considered an indication of the overall reliability of the outcomes.

The principle of reliability has important implications for the assessment of human factors initiatives, since it enables comparisons between the outcomes of evaluations prior to, and following, the instigation of remedial measures. From an organisational perspective, a judgement can be made concerning the extent to which a particular initiative has been successful in either mitigating the potential for error and/or improving productivity.

27.1.2 Validity

In addition to reliability, the validity of the research outcomes must be established to ensure that the data acquired are an accurate reflection of the issues under consideration. For example, an assessment of the frequency with which accidents or incidents occur is not necessarily a valid reflection of the level of safety within an organisation since the rate at which accidents or incidents are reported in one organisation may be lower than in a comparable organisation. Consequently, valid assessments of the level of safety are likely to be based on factors in addition to the frequency of accidents or incidents.

The principal means of assessing the validity of a particular measure is to compare the outcomes to other measures with which a relationship is presumed to exist (Kanis, 2000). Therefore, if a workplace assessment technique is presumed to reflect the level of occupational health and safety, it might be assumed that a relationship would exist between a poor workplace assessment and the rate at which worker's compensation claims are sought. The strength of this relationship will determine the confidence with which subsequent conclusions can be drawn on the basis of the outcomes of workplace assessments.

Although it may be possible to establish the validity of a technique in a particular domain, the extent to which the results can be transferred to other domains is difficult to determine in the absence of additional research (Stanton, Salmon, & Rafferty, 2013). Organisational factors, such as leadership, motivation, morale, and command and control structures, all have the potential to impact the reliability and validity of assessment techniques. As a result, a great deal of caution needs to exercised when drawing comparisons across different organisations and across different domains.

27.2 SUBJECTIVE VERSUS OBJECTIVE DATA

Although there are a number of measures that might be employed as part of a human factors testing methodology, the range of strategies available can typically be divided on the basis of whether they target subjective or objective forms of data. Subjective data consist of information including perceptions or attitudes where participants are asked to provide information based on their individual experience and expertise within a particular context. The measures that are used to acquire this type of information include questionnaires, interviews, rating scales, and checklists.

The main advantage associated with subjective measures of data acquisition is their ease of administration. Furthermore, they tend to be more cost-effective and less time-consuming than objective methods. Objective data comprise information such as response latency, accuracy, and choice decisions. The main advantage associated with the application of objective measures of data acquisition is the objectivity that is afforded in the responses and the interpretation of the outcomes.

Human Factors Testing Methodology

The choice between subjective and objective measures of data acquisition will be governed by a number of factors including the availability of the information, the costs associated with the process of data acquisition, and the particular constructs under investigation. For example, in evaluations of operator workload, Yeh and Wickens (1988) established that subjective measures of data acquisition are less effective than objective measures in measuring changes at very high and very low levels of demand. Consequently, the choice between measures of data acquisition will depend, in part, on expectations concerning the types of data that are likely to emerge and the conditions under which they emerge.

In addition to the nature of the data, the appropriateness of a methodology will also be influenced by the political and social environment within which an assessment takes place. Organisational morale, the level of organisational structure and control, and the level of cooperation, all have the potential to influence the outcome of both subjective and objective measures of human performance. Therefore, the methodology employed as part of a human factors assessment must be sufficiently robust to yield valid and reliable data, irrespective of the nature of the organisational culture or industrial situation at the time.

A combination of both subjective and objectives measures is one of the obvious strategies to ensure that the data arising from an evaluation are both valid and reliable. However, it should be noted that the costs involved tend to increase when different forms of data are acquired. Consequently, a balance must be struck between the costs associated with the process of evaluation, the potential for unreliable and/or invalid data, and the potential outcomes that might emerge.

27.3 HUMAN FACTORS METHODS

As a discipline, human factors comprises a broad range of methodologies that might be applied for the purposes of data acquisition. What follows is an examination of some of the more common approaches that have been applied in the context of human factors. In most cases, they represent relatively well-established mechanisms to acquire data that are relevant to the performance of a task.

27.3.1 OPERATIONAL ANALYSIS

Operational analysis is generally conducted during the design phase of a system and involves an analysis of the operations that are likely to occur as part of a system. The aim is to identify, during the design stage, any difficulties or problems that might be associated with the function of the system within the operational environment (Saad, Abdel-Aty, Lee, & Wang, 2019).

Information pertaining to the intended operation of the system can be obtained from a number of sources, including interviews with experts and the planning documentation employed as part of the development of the system. The intention is to identify any sources of problems that might arise in the function of the system, and any limitations or failures that might occur.

Operational analysis is typically conducted as part of the operational concept phase of systems development. From a human factors perspective, it offers an opportunity to conduct an evaluation of the system prior to development and production. Therefore, it represents a proactive approach to the identification and management of human performance.

27.3.2 FUNCTIONAL FLOW ANALYSIS

Functional flow analysis is a methodology based on the principles of engineering in which a system is divided into subcomponents so that the relationships can be isolated and observed. The outcome is a functional flow diagram in which the inputs and outputs are depicted in addition to the position of the subcomponent within the function of the system (see Figure 27.1).

Figure 27.1 illustrates a functional flow diagram for a system intended to progress from the initiation of a flight to departure. It comprises two levels, the first of which is referred to a zero-order function, as it represents the broad categories that enable the achievement of the system goal. The other level is referred to as a first-order function, since it constitutes the deconstruction of zero-order functions.

From a human factors perspective, one of the main advantages associated with functional flow analysis is that it determines the allocation of responsibility for the various aspects of the system. This is most important when automated systems are involved, since there is a requirement for a clear delineation between the role of the operator and the role of the automated system, particularly within a complex and uncertain operating environment (Parasuraman, Mouloua, & Molloy, 1996).

Functional flow analysis also provides the basis for the functional allocation of responsibility across multiple operators, and the isolation of potential flaws within the system (Adriaensen, Patriarca, Smoker, & Bergström, 2019). In the case illustrated in Figure 27.1, the preparation of the aircraft might be a responsibility undertaken by a crew member other than the crew member whose responsibility involves checking the flight plan. Therefore, the potential flaw in the system becomes evident in failing to incorporate a review of the preparation of the aircraft prior to departure.

27.3.3 CRITICAL INCIDENT ANALYSIS

Critical incident analysis is a methodology in which a critical incident is employed to trigger the cognitive, perceptual, and behavioural strategies that would normally be applied to cope with a situation had it occurred within the operational

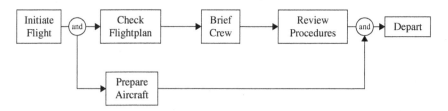

FIGURE 27.1 An example of a first-order functional flow diagram for a zero-order function. (Adapted from Chapanis, 1996.)

environment (Seamster, Redding, & Kaempf, 1997). The incident trigger may be simulated or actual, and the data may be acquired concurrently or retrospectively, depending on the nature of the environment. The aim is to provide as many of the operational cues as necessary to ensure that the data are a valid reflection of the strategies that are actually engaged within the operational domain (Hoffman, Crandall, & Shadbolt, 1998).

Combined with a structured interview, critical incident analysis has the potential to yield information pertaining to the operator's understanding of the environment, the extent to which procedures are applied consistently, and the extent to which a system copes with critical situations that may occur (Arora, Johnson, Lovinger, Humphrey, & Meltzer, 2005; Crandall & Getchell-Reiter, 1993). Therefore, the methodology is designed to enable the acquisition of information beyond simply the behaviour of operators in response to changes in the environment (Flanagan, 1954). Rather, it is designed to assist in obtaining an understanding of the nature of the psychological dimensions of human performance, including decision-making, situational awareness, and stress management.

From a theoretical perspective, the critical nature of an incident trigger is presumed to test the underlying cognitive and perceptual structures that an operator engages to perform a particular task. The successful resolution of this task will depend on an accurate mental model, and the knowledge and skills necessary to respond to changes in the demands of the task (Thordsen, 1991). In the absence of these skills and knowledge, errors are expected to occur and deficiencies within the system can be identified.

Although there are a range of approaches that can be regarded collectively as forms of critical incident analysis, the Critical Decision Method (CDM) is the most formalised of these approaches and has been applied successfully in domains including nursing (Coté, Notterman, Karl, Weinberg, & McCloskey, 2000; Crandall & Getchell-Reiter, 1993), and air traffic control (Redding & Cannon, 1991). CDM requires that participants recall a non-routine event that has occurred recently within the operational environment. This provides a trigger for the application of a semi-structured interview protocol designed to elicit information including the decision goals, the various options that may have been generated, and any analogous situations that may have been recalled during the decision-making process (Kaempf, Klein, Thordsen, & Wolf, 1996; Klein, Calderwood, & Macgregor, 1989).

Typically, CDM has been applied in situations where it has been necessary to develop an understanding of successful decision-making. For example, Leas and Chi (1993) used CDM to isolate the characteristics of expertise amongst swimming coaches. Similarly, Kaempf et al. (1996) adopted CDM to establish an understanding of the decision-making processes of experienced personnel involved in a military command and control environment.

Given the focus of CDM approaches in defining the characteristics of successful performance, participants tend to be subject-matter experts. However, the methodology is sufficiently robust to be applied successfully to non-experts and to provide comparative data between various systems and subsystems. It has the potential to provide information concerning the accuracy of mental models, the management of information in a time-constrained environment, and the application of procedures to resolve uncertainties. Therefore, it represents a practicable approach for the acquisition of data in applied industrial settings.

The main disadvantage associated with the critical incident technique in general, and CDM in particular, is the amount of data that can emerge as part of the process. At a practical level, this often results in a restriction in the sample size and/or in the scope of the investigation.

27.3.4 TASK ANALYSIS

Task analysis comprises a number of strategies that, combined, are designed to provide information pertaining to the nature and function of a particular task. Generally, the aim of task analysis is to provide sufficient information to enable the development of either appropriate training systems or systems that have the potential to augment human performance (Means, 1993; Rose & Bearman, 2012). However, the process of task analysis is also applicable as a mechanism for human factors testing, since it enables judgements concerning both the task and the operator (Fastenmeier & Gstalter, 2007).

Task analysis involves a process of structured deconstruction where a complex task is divided into its subcomponents (Schaafstal, Schraagen, & Van Berl, 2000). In the case of behavioural task analysis, the division and isolation of subcomponents is based on performance that can be observed and isolated (Seamster et al., 1997; Thordsen, Militello, & Klein, 1993). For example, the complex task of starting a manual motor vehicle begins by checking that the handbrake is retarded, depressing the clutch, and placing the gear lever into the neutral position. Once these subtasks have been completed, the brake lever is depressed and the key is turned. This description is an example of a behavioural task analysis that can be depicted as a linear process as in Figure 27.2.

In contrast to behavioural task analysis, cognitive task analysis targets cognitive and perceptual factors in addition to the behavioural elements that comprise the performance of a task (Schaafstal et al., 2000; Seamster et al., 1997). The shift from behavioural to cognitive task analyses occurred through a realisation that human information processing involves more than simply the activation of behavioural responses. Rather, the perception of task-related information and the capacity to formulate decisions also has a significant impact on human performance, especially within complex dynamic environments where behaviour cannot always be prescribed (Roth & Woods, 1989; Wei & Salvendy, 2004).

Task analysis is very similar to functional flow analysis, since both methodologies are designed to identify the subcomponents of a system and the mechanisms through which they interrelate. However, where functional flow analysis is a broad approach based on the performance of the system as a whole, task analysis is a specialised approach with human performance as the primary focus of interest.

Typically, the data necessary for task analyses are obtained from operators who have experience in the performance of the task. The process of acquiring this type of information is referred to as knowledge elicitation and it includes methods such as interviews, questionnaires, protocol analysis, observation, and process tracing

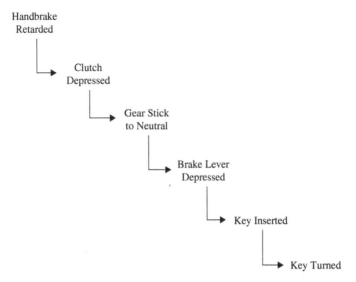

FIGURE 27.2 An illustration of the outcome of a behavioural task analysis for starting a manual vehicle.

(Hoffman, Shadbolt, Burton, & Klein, 1995). Each method provides a different perspective concerning the performance of a task, and they are normally applied in combination to ensure that valid and reliable data can be obtained through triangulation.

Consistent with the variety of methods of data acquisition, there are a number of strategies available to display and analyse data including cognitive graph analysis, conceptual modelling, and verbal protocol analysis. These strategies are designed to generate graphic displays of task-related information that illustrate the sequence of activities in combination with the cognitive and perceptual requirements. This process requires a level of interpretation and should be applied with caution. However, one of the most effective mechanisms to ensure the reliability of these types of research outcomes is to involve subject-matter experts during the analysis. In this case, the main role of SMEs is to comment on the accuracy of the conclusions derived as a valid reflection of the task (Militello & Hutton, 1998; Shadbolt & Burton, 1995). Where differences exist, the SME has the opportunity to advise the researcher concerning changes that may need to be made, prior to the application of the information within the operational environment.

The main advantage associated with the application of a task analytic procedure is the amount of information that can be acquired concerning the performance of complex tasks. However, it is precisely the amount of information acquired through task analysis that tends to result in difficulties, particularly in data processing and analysis. As a result, task analysis can be both time-consuming and costly, and needs to be applied judiciously to ensure that sufficient returns are likely from the anticipated improvements in the system.

27.3.5 Fault Tree Analysis

Where critical incident analysis and task analysis are generally associated with exemplary human performance, fault tree analysis is a methodology that focuses, almost exclusively, on the identification and anticipation of human error. The aim of fault tree analysis is to determine the range of errors that may occur within a system, the sequence of events that may lead to an occurrence, and the frequency with which system failures are likely to occur.

Ideally, fault tree analysis should be applied prior to the development of a system as a means of identifying remedial strategies or redesigning a system to prevent an unintended occurrence. For example, Kirwan (1987) employed fault tree analysis during the design phase of an off-shore oil platform to identify the potential for human error. The results led to a significant review of the design based on the analysis of five anticipated scenarios that might lead to accidents.

In each of the scenarios, Kirwan (1987) created a sequence of operational events, anticipating the point at which the system would fail. For each of the events, a description of the task was generated and the possible human responses were identified. The probability that a human response would be erroneous was estimated on the basis of existing data arising from nuclear power plants. These data were aggregated to produce an overall probability of human error for each of the scenarios. The results led to a number of improvements in the design of the off-shore oil rig, thereby minimising the costs associated with system changes following construction.

27.3.6 Physical Assessment

Although there are a range of methodologies that can be applied within the cognitive and perceptual domains, the assessment of human performance from a physiological perspective is generally limited to a relatively limited number of standardised strategies. These are generally field-based and involve an initial assessment based on a job analysis (Louhevara, Smolander, Aminoff, & Ilmarinen, 1995).

A job analysis is very similar to a task analysis, although the former is less specific, and incorporates physiological requirements in addition to a job description and a list of required functions (Ogle, Rutland, Fedotova, Morrow, Barker, & Mason-Coyner, 2019). For example, a job analysis for a baggage handler might include the capability to lift up to 30 kilograms in a confined space under extreme temperatures. Therefore, it is as much a prescription of the tasks required as it is a description of the tasks that are actually involved in a specific job.

Having completed a job analysis, an analysis of the localised hazards can be completed to ensure that the requirements for the job are suitable, given the environment within which the employee is expected to function. An assessment of the environment is normally made against recognised standards to ensure that the exposure to hazards remains consistent with the principles of best practice (Zimolong, 1997). In some cases, this strategy may also include an analysis of the workplace from an ergonomic perspective, including an assessment of posture and manual handling requirements.

A significant requirement of assessments of the physical load associated with a job involves an analysis of the physiological requirements based on indices such as oxygen consumption, heart rate, systolic blood pressure, and/or energy expenditure. The aim is to ensure that the requirements for the task do not exceed the capabilities of individuals, resulting in stress and strain. Furthermore, the information derived from an assessment of physiological requirements also enables the development of training systems to better manage the physical loads that are imposed.

This combination of job analysis, hazard analysis, and physiological evaluation is a relatively well-accepted approach to the assessment of the physical requirements of a task. As a result, there are a number of assessment tools that are available to assist with the implementation of the process. Nevertheless, it should be noted that workplace assessment from a physiological perspective remains subject to the political and industrial conditions within which it is applied. For example, establishing minimum physiological requirements for a particular occupation may discriminate against a proportion of the population. Therefore, the process needs to be applied with caution and with the view that improvements in performance can also be achieved through changes to the design of a system.

27.4 ETHICS AND HUMAN FACTORS TESTING

Since the aim of human factors testing and evaluation is often to improve human performance, there is an inherent assumption that existing performance may be inadequate. As a result, some of the strategies that are employed may be distressing for participants, and may lead to unreliable information and/or a lack of information pertaining to the performance of a task. Therefore, having regard for the needs of participants is an important part of the process of human factors testing and evaluation.

As is the case for all psychologists, there is a requirement to ensure that individual data emerging from any investigative process remain confidential, and that the overall conclusions are both reliable and valid. Furthermore, it is important at the outset that the purpose of the evaluation, the precise nature of the tasks to be undertaken, and the types of data that are being acquired are communicated clearly to prospective participants. Participants also ought to be given the right to withhold their participation or withdraw at any time during the process of evaluation.

Ideally, the goal of human factors testing should be to ensure that participants regard themselves as an integral part of the process and as catalysts for improvements within the operating environment (Ferguson, Crist, & Moffatt, 2017). This ensures that participants are committed to the provision of accurate and reliable information. The principle is also consistent with the goals associated with participatory ergonomics where potential users are involved from system design through to evaluation (Burgess-Limerick, 2018).

A participatory approach to human factors testing requires initial investments in both time and resources. However, the advantages associated with involving participants include the rapid identification of problem areas, assistance with the selection of appropriate methodologies, and the provision of solutions to task-oriented problems. A participatory approach is also likely to lead to a ready willingness to test potential solutions to problems once they have been developed.

As a mechanism to overcome any ethical implications that may emerge as part of organisational research, some organisations have developed ethics committees to consider research proposals. These committees may comprise representatives from throughout and/or outside an organisation and their role is to consider the demands on participants against the anticipated outcomes of the analysis. The goal is to ensure that participation is voluntary, the goals of the study are clearly articulated, and that the outcomes of the research will be made available to participants.

27.5 THE PROCESS OF DATA ACQUISITION

In general, the extent to which participants will volunteer to participate in applied research will depend on the extent to which they can identify with the potential outcomes. For example, if operators perceive a testing strategy as an opportunity to generate improvements in their own operating environment, there is a greater likelihood of involvement. Conversely, if an approach appears esoteric and/or the outcomes of previous investigations have failed to improve the operating environment, voluntary participation is less likely to occur.

From an industrial perspective, an important prerequisite for involvement in research strategies involves the provision of sufficient time for participation during the course of a normal shift or work period. Generally, operational personnel are less likely to participate in organisational research in the absence of some form of compensation from the employer. However, the provision of compensation also needs to be exercised cautiously, as there are ethical implications where potential participants feel coerced to participate in testing.

Where video or audio recorders are used to collect testing-related data, participants may be unwilling to reveal contentious information that may be of use to the investigative process. Therefore, good practice dictates that participants are informed that all recordings will be erased once the data have been transcribed for subsequent analysis. This is a particularly important principle where the research process involves the observation of behaviour, since participants may fear that the information will be used subsequently for punitive action.

28 Human Factors Assessments

28.1 HUMAN FACTORS ASSESSMENTS

The identification of the potential hazards within a system is one of the most important safeguards against system failure. However, it can also be a very time-consuming and challenging process that is further exacerbated by the complexity of a system, especially in cases where there may be a number of interactions and levels of interdependence.

System failures occur due to a number of factors and conditions, each of which functions in concert, thereby resulting in the perpetuation of errors. For example, a motor vehicle crash may be a consequence of the design of the vehicle, the experience of the driver, the road conditions, and/or the weather conditions at the time. In this type of situation, retrospective analyses, such as those conducted by accident investigation authorities, are very useful mechanisms to identify the factors that may have contributed to the failure.

The origins of system failure that are identified through accident investigation are often used as the basis for subsequent analyses. This provides a focal point for investigations and ensures that the investigative resources are directed towards the identification of factors that might prevent the reoccurrence of the failure. To facilitate this process, an extensive checklist is employed that provides a framework for the investigation and ensures that contributing factors are not overlooked.

The disadvantage associated with this approach in isolation is that it is reactive, and lacks the flexibility to test other issues that might prevent occurrences. A more comprehensive approach involves both the investigation of accidents or incidents where there is the greatest potential for safety improvements and the development of system safety initiatives as a proactive approach to the management of system safety.

The broadening of organisational resources towards both proactive and reactive approaches enables the identification of hazards, including latent conditions, that may not be immediately evident in the course of a specific accident or incident investigation. Even in the context of investigations, incidents and accidents should be considered equivalent, as both enable the identification of significant factors that may explain system failures.

An example of the potential significance of incidents in identifying latent conditions can be drawn from the investigation into a relatively minor system failure involving an Ansett Airlines Boeing 747-312. Ansett Flight 881 had departed Sydney, Australia, en route to Osaka, Japan, on October 19, 1994. Approximately one hour after departure, the crew were required to shut down the number one engine due to an abnormal oil pressure reading (Bureau of Air Safety Investigation, 1996). They subsequently elected to return to Sydney and began a normal approach for a landing on Runway 34. However, as the aircraft was approaching touch-down, air traffic controllers noted that the nose landing gear had not extended. Despite radio calls to the flight crew, they were unable to abort the landing.

Although there were no injuries, and the occurrence resulted in relatively minor damage to the aircraft, the subsequent investigation revealed a number of serious latent conditions that existed within Ansett Airlines at the time. The airline had only recently begun international services, with significant internal pressure to complete the transition as quickly as possible. This rapid development led to the implementation of processes that were flawed. In part, this explained the mismanagement of the situation by the flight crew as they lacked the appropriate training in resource management.

Like accidents, incidents and their investigation generate significant immediate and longer-term costs to organisations. Therefore, from an economic perspective, alternative approaches to system safety are necessary that are capable of identifying latent conditions before they culminate in a safety occurrence.

28.2 HUMAN FACTORS AUDITS

A human factors audit is a proactive approach to the identification of latent conditions, the structure of which is based on the traditional audit system developed for financial management. In effect, it is designed to assess the policies and practices of an organisation against an established benchmark. The overall aim of the audit process is to provide a summary of the organisation that is both detailed and accurately reflects the nature of the organisation (Cacciabue, 2005; Drury, 1998).

In the financial domain, an audit generally involves four main elements:

1. A diagnostic investigation
2. A test for transaction
3. A test of balances
4. The formation of an opinion

Diagnostic investigation provides for a general description of the organisation where special areas of concern are targeted for subsequent analysis. Having established the broad characteristics of the organisation, a test for transaction is undertaken where the core business is assessed by tracing the progress of various transactions between the organisation and the customer (Thatcher, Nayak, & Waterson, 2020). This is equivalent to a sampling of the system, and is designed to highlight any deficiencies in the management of information and/or data as it progresses throughout the organisation.

The test for transaction yields a large amount of data that are subjected to analysis during the test of balances. The outcome of this analysis provides the basis for the development of an audit report that constitutes the final stage of the auditing process. Any deficiencies within the system are noted and remedial action is recommended as necessary.

Although this type of approach to auditing has been developed within the financial domain, the principles remain consistent irrespective of the environment within which it is applied (Drury & Dempsey, 2021). The main considerations that need to be addressed in the design of any auditing system include:

1. The process through which the system might be sampled;
2. The particular features of the system that need to be examined;
3. The process through which the outcomes of the sampling process might be analysed; and
4. The most effective means by which the results of the audit might be communicated.

28.3 STANDARDS

One of the prerequisites for an effective auditing system is an understanding of what constitutes 'good practice'. This provides a benchmark against which the performance of an organisation can be evaluated. In many contexts, published standards are used to establish these expectations of performance.

Standards exist for a wide range of products and systems, and are generally developed on a regional basis by non-statutory authorities (Zuckerman, 1996). In the United States, for example, the responsibility for the implementation of standards is vested with the American National Standards Institute (ANSI). Equivalent organisations exist in Australia (Standards Australia), and in the United Kingdom (British Standards Institute). These are typically non-profit organisations that provide standards for the development of systems ranging from displays to medical equipment.

The process of developing standards normally involves a series of consultations with interested parties to develop a consensus (Chandler, 1997; Zuckerman, 1996). However, as Chandler (1997) notes, reaching a consensus is a very difficult process, particularly when the interested parties may have very different, competing agendas. In many cases, standards organisations will be required to reach a compromise for which there is general agreement. The failure to do so may result in an unwillingness to adhere to standards, thereby defeating the intention of the process.

Although there are a number of standards (such as financial standards) that have been legislated, adherence to most standards is voluntary. Therefore, in the absence of any punitive measures for non-compliance, standards organisations have little choice but to establish other mechanisms that will ensure compliance (Kruithof & Ryall, 1994). One of these strategies is a compromise, although the outcome may not necessarily be consistent with the principles of 'good practice'. An example of a compromise in the development of standards can be drawn from the International Accounting Standards Committee (IASC) where there was an attempt to shift from a standard that permitted a free choice of accounting treatments to one where the choice was limited (IASC, 1989). Part of the reasoning for this shift in policy was the admission that, in the past, a free choice was permitted to ensure a degree of compliance to standards (IASC, 1989).

The difficulties faced by the IASC reflect challenges faced by any organisation tasked with the development of international standards. At a regional level, the development of standards is somewhat easier, since the range of opinions is likely to be relatively consistent. However, at the international level, opinions can differ markedly, depending on a number of factors, including cultural dimensions, the structure of the economy, and the nature of government regulation.

The responsibility for the development of international standards rests with the International Organisation for Standardisation (ISO), a non-profit organisation that comprises representation from various regional bodies, including ANSI and the British Standards Institute (BSI) (Stimson, 1998). This organisation is responsible for the development of standards at an international level, although it is important to recognise that differences may exist between international standards and those at a regional level. For example, Lewis and Griffin (1998) note that there is a discrepancy between the ISO standard for evaluating human exposure to whole-body vibration and the standard advocated by BSI.

Whole-body vibration is most commonly experienced when travelling in vehicles, where the entire vehicle shifts in response to changes in the environment. The standards developed by the ISO and BSI were designed to measure the extent of this vibration, and the likelihood that a particular level of vibration would result in injury. The difference between the two standards occurs primarily in the calculation of the average response, and the frequency weighting that defines the rate at which accelerations occur. According to Lewis and Griffin (1998), calculations using the two standards result in differences in the estimated limits of exposure to whole-body vibration. Consequently, a level of caution needs to be exercised when implementing standards to ensure that they represent 'world-best practice'.

28.4 INTERNATIONAL STANDARDS ORGANISATION (ISO) AND HUMAN FACTORS

ISO 9000 refers to a series of international standards selected from amongst the most appropriate standards available from organisations such as ANSI and BSI (Álvarez-Santos, Miguel-Dávila, Herrera, & Nieto, 2018; Kruithof & Ryall, 1994). It comprises five main documents that are intended to provide a basis for the development of a strong customer focus, and mechanisms for quality control. For example, in the case of document and data control, an organisation meeting the requirements for ISO 9000 should ensure that all of the information required to perform a task successfully is available efficiently and at the appropriate time (Kruithof & Ryall, 1994).

The main limitation associated with the ISO 9000 protocol is that it is generally limited to product control and development, rather than the broader issues associated with quality control, such as occupational health and safety (Lim & Prakash, 2017). Although there is no reason why occupational health and safety requirements should not be included as part of a mechanism for quality control, it is not included as a significant part of this process (Li & Guldenmund, 2018).

Rather than part of the ISO 9000, standards for occupational health and safety systems are included in ISO 45001 (ISO, 2018). This standard is intended to provide a framework for the management of risks and is informed, at a management level, by plans and strategies that are intended to identify and assess risks that might impact occupational health and safety within an organisation (Salguero-Caparrós, Pardo-Ferreira, Martínez-Rojas, & Rubio-Romero, 2020).

The occupational health and safety performance of an organisation is largely determined by the context. Consistent with other international standards, there is a recognition that organisations operate within environments that differ in the regulatory requirements with which they must comply, the external and internal issues with which they are likely to be confronted, and the needs and expectations of stakeholders (Rostykus, Ip, & Dustin, 2016). Together with a comprehensive approach to the management of occupational health and safety that includes consultation with stakeholders, the aim is to establish appropriate performance goals to which the organisation should aspire.

In ISO 45001, the application of human factors principles is most clearly evident in the process of hazard identification. However, it is part of a larger program of planning designed to evaluate both risks and opportunities. The tools that are implemented as part of this program of information acquisition are not specified, to allow organisations to select tools that are contextually appropriate. Nevertheless, the process is underwritten by objectives that, in part, must be measurable and capable of being monitored.

In addressing opportunities together with risks, ISO 45001 is founded on the principle that optimising investments in the identification and management of occupational health and safety risks has tangible outcomes for an organisation. In part, this reflects the difficulty associated with the assessment of human factors and other safety-related interventions in complex operational environments. For example, where the financial practices and procedures of an organisation can be readily inspected and evaluated, the extent to which an organisation implements appropriate human factors principles can be more difficult to establish. The solution is to publish a series of expectations that relate to the design and implementation of the interventions, rather than target outcomes exclusively.

The advantage associated with the implementation of evidential standards is that it overcomes the inclination to rely simply on assurances. Therefore, it complies with one of the main principles associated with the auditing process: the notion of independent verification (Chandler, 1997). Where information is not immediately observable, an addition can be made to audit reports to the effect that the information provided was based on assurances from management or other members of the organisation. This approach enables the acquisition of potentially useful information, while ensuring that the reviewer remains fully aware of any qualifications pertaining to the accuracy of the information.

28.5 STANDARDISATION AND 'BEST PRACTICE'

Although an adherence to standards prescribed by the International Standards Organisation is considered industry 'best-practice', it does not necessarily guarantee that an ISO-accredited organisation will differ substantially from a non-ISO-accredited organisation, particularly in the mechanisms for quality control. For example, Carr, Mak, & Needham (1997) examined a number of organisations in New Zealand, and noted that there was very little difference in the practices for quality management and the reporting systems between those organisations that were ISO-accredited and those that had not been ISO-accredited. However, differences were evident in the process of quality assessment, and the extent to which quality was perceived as more important than cost efficiency. Nevertheless, the results indicate that an adherence to standards is not necessarily a panacea for an organisation to achieve substantial improvements in production and quality control. Rather, standards represent a minimum benchmark to ensure a consistent level of performance across a range of organisations.

Given the purpose of standardisation, the aims associated with the auditing process need to be kept in perspective. The fact that a system meets a minimum requirement does not necessarily mean that errors are likely to be avoided. For example, the minimum legal requirement for hearing protection in the United States is 90 decibels (dB). However, lower levels of noise also have the potential to impact human performance, particularly in communication and attention (Drury, 1998). Therefore, accepting a noise level of 90 dB will not necessarily prevent human performance errors arising from excessive levels of noise.

28.6 DESIGNING AN AUDITING PROTOCOL

An auditing protocol is a mechanism designed to facilitate the acquisition and management of data arising from an auditing process. It is normally developed in the form of a survey or checklist, and is therefore subject to the principles of survey design. The protocol should yield outcomes that are valid and reliable, and should be presented in a format that is both sensitive and usable.

Validity refers to the extent to which an instrument accurately measures the phenomenon under investigation. For example, a question concerning the amount of coffee consumed within an organisation will not necessarily yield information pertaining to the appropriateness of the quality control measures within the production process. Therefore, the design of a human factors audit protocol requires the identification of those factors that are indicative of an adherence to appropriate human factors principles. Some of these factors may include:

- The temperature in the workplace;
- The duration and frequency of work breaks;
- The level and frequency of staff training;
- The amount of documentation available;
- The usability of documentation;
- Access to appropriate equipment;
- Access to safety equipment;
- The level of safety-related training; and
- The level of vibration experienced in the workplace.

Human Factors Assessments 347

In addition to validity, the development of an appropriate auditing protocol is also dependent on the principle of reliability. In the context of human factors audits, reliability refers to the extent to which a protocol produces consistent results for a given organisation, irrespective of the assessor. Therefore, the questions that comprise a protocol must be clear, concise, and must remove as much subjectivity as possible from the evaluative process (Drury & Dempsey, 2021).

One of the most effective mechanisms to reduce the level of subjectivity amongst human factors audits is to employ questions that seek details pertaining to clearly observable events and processes. This ensures that an accurate assessment is recorded, rather than one that requires an interpretation as to whether or not a feature meets a minimum criterion. For example, an audit statement that refers to a minimum requirement might be written as a 'closed question' of the form:

Is the seat height at an appropriate level above the floor?

Alternatively, a statement that *does not* refer to a minimum requirement might be written as an 'open question' of the form:

What is the height of the seat above the floor?

The latter provides both a description of the feature in absolute terms, and provides guidance to indicate the extent to which improvements must be made to ensure compliance with the appropriate standards. Therefore, it is sensitive to changes within the operational environment and embodies a level of usability, since it avoids the difficulties associated with interpreting whether each factor meets a minimum standard.

28.7 SAMPLING PROCEDURES

Due to financial and time constraints, it will rarely be possible to subject an entire organisation to all aspects of an auditing process. Rather, samples will need to be taken at various strategic points with the data arising from these samples used to draw inferences pertaining to the organisation as a whole.

However, despite the cost-effectiveness of sampling techniques, the effectiveness of this process depends on the method(s) of sampling and the nature of the data acquired. For example, Willett and Page (1996) note a number of cases where items have been included or excluded from an auditing process, depending on whether they are perceived as either 'problematic' or overly complex. Clearly, this approach is unacceptable, and Drury (1998) suggests a more appropriate strategy where areas within an organisation are selected at random for consideration. This ensures that all aspects of the organisation have an equivalent probability of being included in the auditing process.

Despite the objectivity afforded by random selection, this type of strategy may not be appropriate for the purposes of auditing, since it does not necessarily enable comparisons between different sets of data. For example, it may be necessary to compare the performance of the catering and maintenance divisions of a company. A random sample of items from within the organisation will not necessarily provide the data necessary for this type of purposeful analysis.

In those cases where comparative data are required, a stratified technique may be employed, where broad areas of interest are identified initially (see Figure 28.1), and items from within these areas are subsequently selected randomly. An alternative

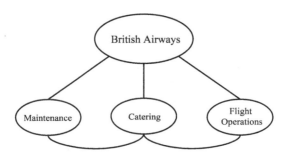

FIGURE 28.1 A diagram illustrating a comparison on the basis of a stratified sampling procedure. In this case, the areas of interest are identified and items with each area are selected for investigation at random.

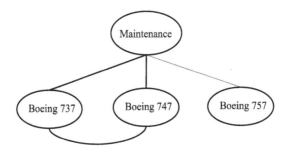

FIGURE 28.2 A diagram illustrating a comparison on the basis of a cluster sampling procedure. In this case, the areas of interest are identified and items with each area may be selected for investigation on either a random or non-random basis.

approach involves the selection of a cluster of items from within a selected area of interest, based on prior knowledge or research. This is referred to as a cluster-sampling technique, and it is a useful strategy that minimises the requirement to examine each element of a system (Drury & Dempsey, 2021). For example, a maintenance department may comprise three distinct areas of operation. However, previous experience may have indicated that sufficient information pertaining to all three areas can be derived from an analysis of two 'exemplar' areas (see Figure 28.2). The consequence is a more efficient analysis that has the potential to yield meaningful data.

The main disadvantage associated with both the stratified and cluster techniques is that they are not entirely random. They are also subject to a level of psychological 'justification' that is unlikely to occur during a purely randomised approach. However, auditing procedures can be both time-consuming and costly. Therefore, it may be necessary to exercise some level of judgement concerning the most appropriate strategy under the circumstances.

TABLE 28.1
An Example of Part of the Workplace Risk Checklist for Upper Extremity

A	B	C	D	E	F
			Time		
Risk Factor Category	Risk Factors	2 to 4 Hours	4+ to 8 Hours	8+ Hours	Score
Repetition (Finger, Wrist, Elbow, Shoulder)	1. Identical or Similar Motions Performed Every Few Seconds	1	3		
	2. Intensive Keying	1	3		
	3. Intermittent Keying	0	1		
Hand Force (Repetitive or Static)	1. Grip More than 10 lb Load	1	3		
	2. Pinch More than 2 lb	1	3		

Source: Adapted from Karwowski and Marras (1997).

28.8 HUMAN FACTORS AND OCCUPATIONAL HEALTH

There are a number of similarities between the disciplines of human factors and occupational health, the most significant of which is the focus on error prevention and the development of systems for hazard mitigation. However, human factors initiatives tend to target the management and prevention of injuries as part of a broader process of developing an optimal relationship between humans and technology. In addition, where the application of occupational health and safety is generally restricted to the workplace, human factors tends to be applied across a broader range of environments.

Despite the differences between human factors and occupational health and safety, many of the investigative strategies employed during an examination of occupational health and safety issues will also have application within a human factors audit. For example, Karwowski and Marras (1997) describe a checklist for the identification of risk factors that also provided implicit guidelines to improve the relationship between the operator and the working environment (see Table 28.1). The risk factors identified included:

1. The same or similar motion, executed every few seconds;
2. A fixed or awkward posture;
3. The use of vibrating or impact tools;
4. The use of forceful hand exertions; and/or
5. Unassisted manual handling.

The relative impact of each risk factor is determined by the length of time performing the task in the absence of a break. An equivalent number is listed for the duration of each risk factor, and these results are summed to provide an overall score. A score of more than five is considered indicative of a problematic task that may lead to injury.

From a human factors perspective, this type of checklist also provides a guide to the designers of systems to ensure that the risk factors for injury are eliminated to the greatest extent possible.

29 Human Reliability Analysis

29.1 RELIABILITY ANALYSIS

One of the most important issues associated with the application of human reliability analysis concerns the type of technique that might be employed. There are a number of approaches, each of which has different assumptions and procedures for the acquisition and integration of task-related information. The process begins with a definition of the task under consideration and an examination of the context within which the task occurs. Subject-matter experts must be identified and their selection needs to be justified on the basis of their experience within the context of the task. Similarly, the techniques for knowledge elicitation must be explained and their selection justified in the subsequent written report. Finally, the information derived from experts must be represented using one of a number of alternative strategies, depending on the nature of the task under consideration. Irrespective of the strategies selected, the aim of this process is to provide an assessment of human reliability that is both as accurate and reliable as possible under the circumstances.

29.2 WHY A HUMAN RELIABILITY ANALYSIS?

There are a number of reasons why a Human Reliability Analysis (HRA) might be conducted within a particular environment. However, the most of significant impetus for the implementation of HRA is generally a requirement to quantify the potential for a system failure. Therefore, HRA might be incorporated as part of a much broader Probabilistic Risk Assessment (PRA) that examines the reliability of both the human and the technological aspects of the system.

Depending on the environment in which they are applied, HRA methodologies have the capability of determining:

- The probability of human error within a system;
- The types of errors that are most likely to occur;
- The rate at which errors are likely to occur;
- The effect of system changes on the probability of error; and
- The types of circumstances under which errors are most likely to occur.

The application of HRA has been most evident in complex socio-technological environments where the outcomes associated with a system failure can be severe. It is a process that has been designed to provide organisational decision-makers with a basis on which to allocate system resources to either prevent errors and/or mitigate their consequences. HRA also provides a mechanism for the management of a system in an otherwise uncertain, dynamic environment.

Despite the potential gains associated with HRA, the strategies are most successful where the operational environment is relatively controlled and where the operators are required to perform clear, sequential tasks. Where an operational environment is dynamic and uncertain, the complexity of the HRA increases to the extent that the resources required to conduct the HRA may outweigh the potential outcomes that may emerge. Consequently, there are some circumstances where HRA may not be the most appropriate approach for the assessment and management of human performance. Alternatives to HRA may include task analysis and work analysis, although a combination of two or more of these approaches is likely to yield results that are the most meaningful and reliable.

29.3 THE ACCURACY OF HRA

In general, the accuracy of the information arising from an HRA will depend on the accuracy of the information included in the model of human performance. Probabilistic data concerning human performance can be determined from three sources: Empirical assessments of performance in controlled environments, assessments derived from previous system failures, and/or assessments made by subject-matter experts. The decision concerning the appropriateness of the data will depend on operational constraints such as the availability of data from previous research, the extent to which previous research outcomes can be transferred to the environment under investigation, the costs associated with conducting new research, and the availability of experts.

Transferring the outcomes of one research environment to another is possible, provided that the behaviour involved is sufficiently generic. For example, one of the aims associated with the development of the Technique for Human Error Rate Prediction (THERP) involved the development of a repertoire of Human Error Probability (HEP) rates associated with generic forms of human behaviour (Swain & Guttman, 1983). The intention was to provide a generic basis from which combinations of HEPs could be arranged to meet the needs of HRAs in a variety of contexts (see Table 29.1).

The extent to which the information derived from generic analyses of human performance can be applied in different situations can be resolved through a reasoned analysis of features of the operating environment. This may require the application of a knowledge elicitation strategy, such as task analysis, to determine both the constraints and the capabilities afforded within the task. For example, the cognitive demands that occur within the cockpit of an aircraft may be quite distinct from those that exist within a control room in a nuclear energy production facility. As a result, a constant value may need to be applied to a base rate of probability to ensure that the HEPs are consistent with the particular characteristics associated with the operational environment.

TABLE 29.1
Examples of Human Error Probabilities Developed for Generic Errors by Swain and Guttman (1983)

Task: Probability of error in selecting the wrong unannunciated display for qualitative or quantitative readings.

Condition	Human Error Probability
Displays are dissimilar to adjacent displays	Negligible
Similar appearing displays that are included as part of well-delineated, function groups	0.0005
Similar appearing displays that are only delineated by labels	0.003

Task: Probability of error in reading and recording quantitative information from an un-annunciated display.

Condition	Human Error Probability
Digital display with fewer than four digits	0.001
Graphs	0.01
Analogue display	0.003

Source: Adapted from Gertman and Blackman (1994, p. 134).

29.4 THE INACCURACY OF HRA

Although there are distinct advantages associated with the quantification of error probabilities, it should be noted that the notion of probability infers a level of uncertainty. For example, Swain and Guttman (1983) suggest that the probability of error should be multiplied by ten where a system display violates a well-established population stereotype and the display is expected to be interpreted under high levels of stress. In this case, the application of a constant value of ten is presumed to account for the relatively complex interaction between population stereotypes, prior experience with a system, and the cognitive demands imposed during the performance of a task.

Adapting to different populations and population stereotypes is an example of the uncertainty that is inevitably associated with HRA. The nature of human performance is such that it is almost impossible to predict, with absolute certainty, the performance of a single individual at a single moment in time. Nevertheless, there are a number of general principles that enable analysts to develop reasonable expectations of the levels of performance that are likely to occur, given certain characteristics of the situation.

Even when prescribed base rates for generic forms of human error are available, HRA often depends on subjective estimates of error probability elicited from subject-matter experts. Given the difficulties associated with subjective judgements of human performance, a reliance on this type of information represents another opportunity for the potential inaccuracy of HRA data. However, there are strategies that might be applied to increase both the accuracy and the reliability of the information derived from SMEs, including the provision of training and the application of a constant value that takes into account previously established errors in judgement that may have occurred (Hanea & Nane, 2019).

29.5 HUMAN RELIABILITY ANALYSIS TECHNIQUES

Although there are a number of different techniques available for HRA, the most common approaches include the Technique for Human Error Rate Prediction (THERP) (Swain & Guttman, 1983), Human Error Rate Assessment and Reduction Technique (HEART) (Williams, 1988), and the Success Likelihood Index Method (SLIM) (Embrey, Humphreys, Rosa, Kirwan, & Rea, 1984). Each of these approaches is based on the principle that base error rates are established for human behaviours and that these values are subsequently incorporated to develop an estimate of the overall reliability of the interaction between the operator and the system. However, THERP is the only approach that embodies both a technique for modelling human performance and a methodology for the quantification of human reliability (Pan & Wu, 2020).

The THERP approach to HRA begins with a task analysis, the aim of which is to identify the key elements of the task and the relationships that exist between the components of the task (Gertman & Blackman, 1994). The task analysis is also designed to yield the relevant performance shaping factors (PSF) that are likely to impact human performance and indicate whether they will need to be incorporated as part of the analytical process (Pan & Wu, 2020). Finally, the task analysis has the capability to establish the characteristics of the interfaces that exist between operators and the system and the nature of the interaction that occurs (Bevilacqua & Ciarapica, 2018).

Having completed a task analysis, the THERP approach involves the construction of an event tree that describes, in sequence, the progression of events and responses. Since event trees incorporate both appropriate and inappropriate responses, a probability estimate can be assigned for each branch of the event tree on the basis of generic HEPs. The influence of PSFs is incorporated into the event tree by modifying the base HEP by a constant that will increase or decrease the probability of an occurrence (Dougherty & Fragola, 1988).

The success of THERP as a mechanism for HRA depends on the deconstruction of a complex behaviour into subcomponents for which a HEP can be calculated. A slightly different approach is advocated as part of the HEART approach to HRA in which the starting point is a consideration of the task as a whole (Williams, 1988). A nominal HEP is estimated for the task, and this value is subsequently modified by Error Producing Conditions (EPC) that have the potential to impact the successful completion of the task.

For each EPC, Williams (1988) has determined a corresponding weighted value that reflects the extent to which the task becomes less reliable. For example, a difference between the operating model of the operator and that envisaged by the designer corresponds to a multiplication factor of eight on the nominal HEP. Therefore, where the HEP associated with a task is nominated at 0.005, the impact on the HEP can be calculated as:

$$0.005 \times 8 = 0.04$$

Human Reliability Analysis

In some cases, a HEART weighted value may be modified where only a proportion of the effect is estimated to be involved. For example, if an analyst was to assume that only 20 per cent of the effect may emerge, then the weighted value would be modified to:

$$((8-1) \times 0.2) + 1 = 1.14$$

(This formula takes into account the nonlinear aspect of the influence of different levels of PSFs.) In this case, the overall HEP could be recalculated as:

$$0.005 \times 1.14 = 0.0057$$

The HEART method of HRA tends to be most useful in design, when there is a need to determine the extent to which the performance of the operator will impact the HEP of the task (Kirwan, Kennedy, Taylor-Adams, & Lambert, 1997). In addition, it allows the analyst to determine which of a series of EPCs is likely to have the greatest impact on the performance of the operator. As a result, system defences can be developed that mitigate the consequences of the EPC.

Where THERP and HEART are both designed to yield estimates of the probability of a system failure, the SLIM technique is designed to yield an estimate of the probability of the success of the system. The process involves the use of SMEs to:

- Review the characteristics of each task within a system;
- Identify the potential PSFs associated with a task;
- Assign weights to the PSFs;
- Provide a rating for each task within the system; and
- Calculate the overall Success Likelihood Index (SLI) (Dougherty, 1993).

The process of task review may make use of a number of resources, including the outcomes of a task analysis, transcripts of interviews with operators, and descriptions of the resources available to operators. The aim of this process is to enable SMEs to identify the various mechanisms of error causation and establish the PSFs that may be associated with these errors. Throughout this process, it is important to ensure that the decision-making process is appropriate and that the outcomes can be justified. Koriat, Lichenstein, & Fischoff (1980) suggest that strategies such as listing the arguments for, and against, a decision can improve the calibration of subsequent ratings.

Once the PSFs have been identified, weights are normally assigned, where the most significant PSF is assigned a weighting of 100. This value becomes the standard against which the weights for less significant PSFs are assigned. The resulting weighted values are normalised by dividing each value by the sum of all of the values assigned to the PSFs (Aven, 1992)

In addition to weighting the significance of PSFs, experts also assign a weighted value to the task itself on a scale from 0 to 100. The SLI is calculated subsequently by summing the product of these values by the associated weighted values for the PSFs.

However, due the number of tasks and PSFs that need to be taken into account as part of this process, Embrey et al. (1984) suggest that the SLI be transformed using a calibration equation that incorporates anchor values.

The use of anchor values is one mechanism to ensure that subjective ratings are relatively consistent and accurate amongst a range of SMEs. In the case of SLIM, known values (possibility from THERP tables) of HEPs are specified for two tasks that are employed as the upper and lower boundaries in the process of calibration. However, it should be noted that the use of anchor values can only occur during the estimation process, and ought to be employed whenever subjective estimates of probability are sought, irrespective of the type of HRA technique involved (Sun, Li, Gong, & Xie, 2012).

Subjective estimates of probability are always likely to be susceptible to some level of inherent error, so that some estimates will generally be more accurate than others for a given individual, or for a group of individuals (Lee, Kim, & Jung, 2013). In general, the accuracy of subjective judgements tends to be associated with uncertainty according to a logarithmic function (see Figure 29.1).

To ensure that the probability of success takes into account the variability that may occur between subjective estimates of probability, the SLI value must be transformed according to:

$$\text{Log of Success Probability} = a\text{SLI} + b$$

where a = log of success probability for the first anchor task (a), and b = log of success probability for the second anchor task (b). The value that emerges from this transformation is presumed to correspond to the probability of a successful outcome of a system, given the types of tasks involved, the PSFs that impact the performance of the tasks, and the estimates of probability generated by SMEs.

The final stage of SLIM involves the generation of the range of uncertainty for the log of success probability. This information is designed to guide the users of this information in calculating the accuracy of the final estimate. In this case, experts are asked to estimate the upper and lower bounds to which the log of success probability could range, given that the value derived initially remains simply an estimate.

29.6 THE PROCESS OF HUMAN RELIABILITY ANALYSIS

While there are a number of strategies for conducting HRA, the process generally involves problem definition, information acquisition, data interpretation and representation, and the calculation of the reliability of human error. The aim is to ensure that the process of information acquisition, data management, and interpretation is structured, so that any interpretations that might emerge as part of the process are as accurate and as reliable as possible. As part of this process, any decisions that are made concerning the application of alternative strategies need to be justified so that the impact on the final outcome can be considered.

Human Reliability Analysis

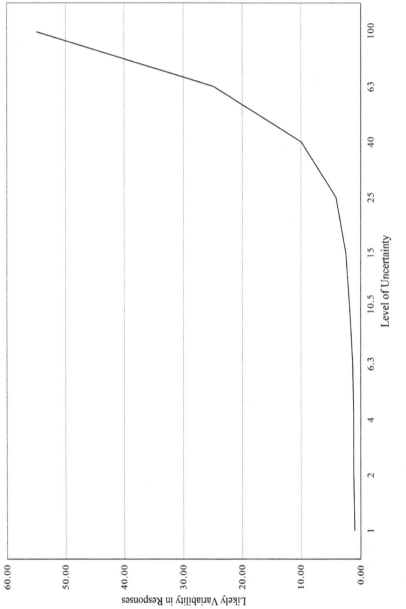

FIGURE 29.1 An illustration of the logarithmic relationship between the level of uncertainty associated with estimates of probability and the likely variability of responses.

29.6.1 DEFINE THE PROBLEM

The first stage associated with most HRA techniques involves defining the problem under investigation. Generally, human reliability problems can range from complex, ill-defined situations that pertain to an entire industry, to relatively structured situations that relate to the behaviour of a small number of individual operators. The characteristics associated with the problem will often provide an indication of the type of strategy that is most suitable under the circumstances.

As a part of the problem definition phase of the HRA, it is important to determine the type of outcome that is expected to emerge from the process. This may require some reference to the nature of the 'customer' for whom the report is intended. For example, a system designer may be interested in identifying the types and severity of errors that are likely to be associated with the use of a particular system as a means of developing appropriate defences. In contrast, the manager of a high-risk organisation may be interested in the probability of human error to ensure that the possibility is incorporated into an overall assessment of the risk of the operation.

29.6.2 DEFINE THE CONTEXT

Once the problem has been defined and the intention of the HRA has been established, it is necessary to define the nature of the environment, independent of the specific task or situation that is the subject of the HRA. This process is consistent with work domain analysis where broad concepts are identified that describe the nature of the environment within which human performance occurs. Typically, a field description of the environment is developed, based on observations and informal discussions with operators (Vicente, 1999).

The purpose of the field description is to provide a context within which to explain the nature of human performance that is to be investigated subsequently. The development of an understanding of the context provides a basis for the development of appropriate knowledge elicitation techniques, assists with the identification and explanation of PSFs, and may elicit information that might otherwise be overlooked through a purely quantitative approach to HRA.

To conduct a field description effectively, it may be useful to consider the characteristics of an operational environment by referring to a model of the cognitive demands that may be involved. For example, Woods (1988) suggests that the cognitive demands associated with system performance can be summarised according to the level of risk involved, the extent to which there are time constraints, the level of uncertainty, and the extent to which the conditions in the operating environment are subject to change. Each of these elements represents a useful indicator of both the type of information that needs to be acquired as part of a field description and the subsequent interpretation of this information.

29.6.3 IDENTIFY SUBJECT-MATTER EXPERTS

By its very nature, expertise is task-specific and an expert in one domain will not necessarily possess the expertise to perform successfully in another domain. Therefore, the

Human Reliability Analysis

identification of an appropriate SME for a specific task can be difficult and time-consuming. In general, a rule of thumb that might be applied is that the expert should possess at least 1,000 hours of direct involvement in the performance of a task, and should be regarded by a range of peers as an expert in the particular field.

Irrespective of the characteristics and experience of the SME, a justification of the selection of the expert should be included in a report arising from an HRA. This description should include a consideration of:

- The level of experience of the SME in terms of direct involvement in the performance of the task;
- The general experience of the SME within the operational domain;
- Any assessments of expertise that may be obtained from peers; and
- Any objective measures of assessment that might be available.

The justification of the selection of an expert is most significant where only one expert is employed for the purposes of knowledge elicitation. In this case, caution should also be exercised in the interpretation of the information derived. Where possible, information from other sources such as empirical research should be used in conjunction with the information derived from subjective estimates.

29.6.4 KNOWLEDGE ELICITATION TECHNIQUES

Knowledge elicitation primarily involves the use of SMEs to assist in the development of an understanding of the nature of the task or situation under investigation. The SME provides the information necessary to deconstruct a complex task into its subcomponents, facilitates the identification of associated PSFs, and, in some cases, provides subjective estimates of the reliability of subcomponents.

There are a number of strategies available for knowledge elicitation including task analysis, cognitive interviews, the critical incident technique, and verbal protocol analysis. Irrespective of the type of strategy that is employed, it remains necessary to justify the selection, and provide an indication of the types of materials that were employed to elicit the information. Consistent with the previous stages of HRA, the justification of the selection of a knowledge elicitation technique as part of an HRA gives the reviewer an opportunity to determine the extent to which the application of a particular type of knowledge elicitation technique may have influenced the types of data acquired.

As part of the knowledge elicitation process, the analyst must have in mind the nature of the data that are expected to be acquired, and the method by which the information will be represented. For organisational psychologists, this requires some level of consideration of the nature of human information processing and cognition. For example, the performance of a task might be represented in the form of a production whereby a condition (IF) and an action (THEN) might be expected to occur. In this case, an expert might be asked to estimate the probability that:

(a) The condition is not detected; and/or
(b) The condition is detected but the action is not executed.

This type of information has implications for the characteristics of the event tree that might be established as part of the process of representation. Therefore, it is important to ensure that the process of knowledge elicitation is guided by a model of cognition that can also provide the basis for the method of representation that occurs subsequently.

29.6.5 Methods of Representation

In HRA, the data arising from a knowledge elicitation process are typically represented in the form of an event tree. However, despite the emphasis of HRA on the quantification of human performance, it might be argued that valuable information may be lost in an effort to reduce otherwise complex situations into purely quantifiable responses. Therefore, a mixed-methods approach to HRA can be advocated where both quantitative and qualitative data are available for analysis.

The qualitative information that might be included in an HRA may simply comprise a series of observations that provide a greater level of explanation for the quantitative information. For example, a qualitative explanation might be provided in a situation where the standard operating procedures could be altered to improve the reliability of human performance. Similarly, case studies might be included to illustrate the particular types of failure that are likely to occur.

30 Human Factors Report Writing

30.1 REPORT WRITING

The process of human factors testing and assessment generally begins with a preliminary evaluation of the extent to which human factors principles or practices are likely to improve human performance. Having determined that a more formal human factors analysis is warranted, an investigation is initiated, based on an appropriate research methodology. The outcomes of the investigation are published in a series of interim reports, and eventually, a final report is produced.

From a management perspective, the preliminary evaluation is a significant milestone in the process of human factors testing. Testing itself is costly, and it needs to be established that the potential outcomes will not detract from existing production, and will outweigh the costs associated with the implementation of the testing process. This requires the development of an informed position, based on the recommendations advocated as part of a human factors test plan.

The test plan is very similar to a research proposal (see Figure 30.2), where a specific problem is identified, an information acquisition strategy is advocated, the costs and benefits are estimated, a timeline for the process is produced, and a series of recommendations is made (Charlton & O'Brien, 2001). The overall aim is to provide an argument that is sufficiently convincing to warrant an investment on the part of an organisation. Therefore, it should include measurable outcomes that are indicative of the Return on Investment (ROI).

30.2 PROBLEM IDENTIFICATION

There are a number of strategies that might be employed to identify human factors-related issues within an operational environment. These include personal experience, accident or incident notifications, the introduction of a new system or design, the outcomes of an audit or benchmarking, and/or a comparative reduction in productivity or perceptions of safety. The objective of this stage of the test is to clearly identify the nature of the problem, introduce the systems that might be investigated, and consider the various issues that are likely to impact either the results or the acquisition of relevant data.

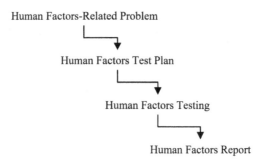

FIGURE 30.1 A summary of the progression from the identification of a human factors-related problem through to the submission of the human factors report.

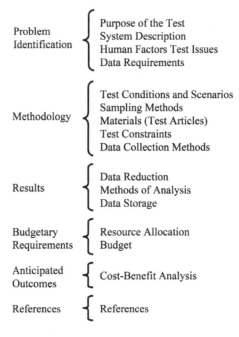

FIGURE 30.2 The main components (left) and subsections (right) that comprise human factors test plans. (Adapted from O'Brien, 1996.)

An important aspect of the problem identification stage involves an analysis of some of the previous initiatives that might have been employed to investigate the same, or similar, issues in the past. In the context of a research proposal, this process is typically referred to as a literature review, and should provide a reasonably broad overview of previous research and development within the domain under investigation. The aim is to provide a logical link, or rationale, between the problem identification stage of the test plan and the methodology stage, where a specific approach is advocated to investigate the problem.

30.3 METHODOLOGY

Given that the test plan may be utilised by personnel other than organisational psychologists, the methodology needs to be described in relatively specific terms. For example, the conditions under which the tests are to be conducted need to be described so that the process can be replicated if required. This includes a description of the intended sample of participants, the methods of recruitment, any testing materials that may be required, and any anticipated limitations to the methodology.

The methodology stage should also include a brief description of the process of data acquisition. For example, if interviews are conducted concerning the usability of a particular product, notes may be taken in preference to the use of a video or audio recording, to safeguard the anonymity of participants. However, this decision may have implications for the subsequent interpretation of the data and needs to be considered as part of the human factors test plan.

It is important to recognise that the human factors test plan is not simply a description of the process for management purposes. It also represents an opportunity to critique and improve the methodology. Therefore, it is as much a discussion document as it is a proposal necessary to acquire the support from management for a particular human factors test.

30.4 RESULTS

Consistent with the development of a research proposal, the results stage of a human factors test plan should incorporate a brief description of the process of data analysis and the strategies for the storage of data. There is also a requirement to describe how the data will be summarised for the purposes of analysis, and how any comparisons will be made. This is also a useful strategy to ensure that the data arising from the test are appropriate for the type of analytical technique that is being proposed.

30.5 BUDGETARY REQUIREMENTS AND ANTICIPATED OUTCOMES

Clearly, the estimated cost associated with a human factors test is one of the most important criteria in determining support for a human factors test. Therefore, it is important to ensure that the estimates are both reasonable and can be justified given the outcomes that are anticipated. This might include a brief consideration of alternative methodologies, the costs associated with which can be compared to the costs associated with the methodology selected.

In some cases, it may be difficult to calculate a precise value for the intended outcomes associated with a human factors test. For example, testing a new design may prevent errors from occurring, where the value of the process is calculated on the basis of the absence of compensation claims against the manufacturer. Where comparative or historical data are unavailable, it can be difficult to argue the precise value of any cost-benefit advantage. Consequently, an argument may need to be developed on the basis of general estimates, rather than specific values. This is a distinction that needs to be made clear as part of the test plan.

Anticipated outcomes may also include intangible assets such as morale and company pride. For example, a number of airlines capitalise financially on the basis of a public perception of safety. Similarly, a human factors-related test might be couched in a form where the outcomes include a public perception that the organisation is a ready participant in safety and/or user-related initiatives, and is intrinsically concerned with the welfare of customers.

30.6 REFERENCES

References constitute an important part of the human factors test plan, since they indicate that prior research has been considered as part of the decision to advocate a particular methodology. In addition, it illustrates to the reader the extent to which the problem has been investigated previously and the nature of the returns that might be anticipated. Any references to standards or design guidelines should also be included as part of this section of the test plan.

30.7 APPROACHES TO HUMAN FACTORS TESTING

Under normal circumstances, the subject of a human factors test will be determined by external factors, such as an incident or accident, or internal factors, such as the introduction of a new system or procedure. This provides the impetus for the development of a test plan and a proactive or reactive investigative process based on a sound and appropriate methodology. The ultimate aim of the investigation is to identify those strategies that might improve human performance within the operational environment.

30.7.1 REACTIVE HUMAN FACTORS TESTING

Where a human factors test is prompted by external factors, the approach is very similar to the process of accident investigation. The first and most important goal is to safeguard the environment within which the accident occurred. For example, if the occurrence was an occupational accident, the area should be sealed, and any relevant information should be acquired as rapidly as possible to avoid contamination from external factors (Ribak & Cline, 1995). This may involve interviews or a physical record of the site in the form of diagrams, photographs, and/or video recordings. The purpose is to ensure that the record is as comprehensive as possible to enable the development of appropriate and accurate conclusions.

Investigations prompted by external factors are often time-constrained and may be subjected to political influence, depending on the seriousness of the occurrence. For example, the political implication of the crash of a Trans World Airlines Boeing 747-100 in 1996 forced the then President of the United States to appoint a committee to examine both the systemic issues and root causes associated with the accident

(McAllister, 1996; Federal Aviation Administration, 1998). A similar initiative was instigated in the aftermath of the explosion aboard the Space Shuttle *Challenger* in 1985 (United States House of Representatives, 1986).

In addition to the political and time constraints that may be imposed on an investigative process, there is also a possibility that litigation may occur amongst the various interested parties. Where litigation occurs, it is likely that a human factors investigator may be called to testify as an expert witness for one or both parties involved in the dispute (Cohen & Larue, 1997). In the case of the loss of the ferry *Estonia* in 1994, litigation proceeded in the absence of an accident investigation report (Kielmas, 1996).

Communicating with interested parties is also an important consideration in the broader process of accident investigation. This is particularly the case where an investigation is complex and may take a considerable amount of time to complete. In the context of aircraft accident investigation, the National Transportation Safety Board (1994) has developed a policy whereby information pertaining to an accident or incident is available electronically and includes preliminary data concerning the occurrence, and the contact details of the investigators. However, where the investigation pertains to a less complex operating environment, an alternative approach might involve the distribution of relevant information in written form through an existing internal distribution network.

The procedure applied in an accident or incident investigation will depend largely on the regulatory environment within which the process occurs. For example, within much of the aviation industry, there is a strong tradition of a non-punitive, independent investigative process of systemic errors (Maurino, Reason, Johnston, & Lee, 1995). Although this approach has considerable merit in the acquisition of relevant information, there is criticism that it may minimise individual responsibility in the occurrence of errors (McCarthy, Healey, Wright, & Harrison, 1997). Consequently, a number of industrial environments, including some airlines, persist with a more punitive approach to the investigation of errors where individual operators are identified and punished (Raviv, Shapira, & Fishbain, 2017).

Once the initial collection of data has been conducted, most investigative approaches to human error involve a sequential classification of these factors against a checklist of possibilities. For example, in the case of an investigation pertaining to the ergonomic features of a task, answers to a number of questions might be sought, including:

- Does the task involve frequent motion greater than six times per minute?
- Does the task require static postures that are expected to be maintained for more than 50 per cent of the task cycle? And/or
- Are awkward postures, such as twisting, required to complete the task?

The responses to these types of questions will highlight those areas where there is an increased risk of error or injury. From an investigative perspective, the identification of risk factors provides the basis for the prioritisation of subsequent data acquisition and analysis.

30.7.2 Proactive Human Factors Testing

In contrast to a reactive approach to human factors testing, a proactive approach is consistent with a typical research process in which a problem is identified, a methodology is adopted to investigate the problem, data are acquired and examined, and conclusions are drawn concerning the nature of the problem. The main advantage associated with this type of approach is the level of control that can be exercised over the environment. As a result, the investigator can determine, with some certainty, the nature of relationships between various factors.

The research process begins with the identification of a general problem, from which a specific question of interest might be derived. For example, a new human–computer interface may have been designed and information may be sought concerning its usability. Given this problem, a specific question might be developed such as: to what extent does the new design increase usability over and above existing designs? This provides the basis for the literature review, a process that would normally have been at least partially completed as part of the development of a human factors test plan.

Having completed a literature review, it might be concluded that, in comparison to existing designs, the style employed in the new design better represents the user's mental model of the system. Therefore, it may be possible to develop a hypothesis, or statement of expectation, concerning the outcomes of the comparison. The main advantage associated with the development of a hypothesis is that it guides the process of data analysis. However, it should be noted that the development of a hypothesis depends on prior research or evidence so that the rationale for the expectation can be justified. Where this level of information is unavailable, retaining the flexibility of a research question is advisable.

Having completed the literature review and developed a research question and/or hypothesis, the methodology is implemented consistent with the human factors test plan. Adherence to the test plan is essential to ensure that the costs associated with the research process remain within budget. However, it should be noted that additional costs may accrue that were unanticipated during the planning stage. In this case, a revision to the test plan needs to be developed and authorised as appropriate.

In addition to the practicalities associated with the development and implementation of the research methodology, there are also a number of ethical issues that need to be considered including the confidentiality of responses, and the development of contingencies in cases where there are adverse consequences for participants. This may involve the provision of counselling support or retraining as necessary.

Following the acquisition of data, the investigator is faced with a decision concerning the most appropriate methods of analysis. This decision will be based on a number of factors, including the nature of the research question or hypothesis on which the assessment was based. A distinction is also necessary concerning the nature of data, since qualitative data, such as verbal or written transcripts, are examined differently to quantitative data in the form of numeric responses.

Given that the outcomes of human factors investigations are likely to be used at a number of different levels within an organisation, it is wise to ensure that any statistical analyses employed are clearly explained. For example, highly complex statistical techniques may not be useful if those stakeholders who are tasked with improving the system fail to interpret the outcomes accurately.

30.8 HUMAN FACTORS REPORTS

Irrespective of the form that it takes, there are a number of principles that will determine the extent to which written information is communicated effectively and efficiently to a reader. Specifically, written forms of communication must be clear, concise, and must maintain and develop an argument that leads to a conclusion. The aim is to ensure that the reader is both convinced by the argument and is provided with the knowledge necessary to implement changes if, and when, required.

30.8.1 Writing Clearly

Writing clearly is not an easy task, particularly when the domain itself is complex. Therefore, there may be a requirement, on behalf of the writer, to simplify the structure or terminology used in the report to ensure that the interactions between various components are clearly understood by the reader. Accident reports are a very useful example, since descriptions of quite complex operating environments are deconstructed into two main sections: factual information and analysis. These sections are further divided into subsections so that the factual information comprises a history of the operation, a description of injuries to personnel, a description of damage, and any other information that may have been available. This process provides a sequential description of the factors involved to highlight a number of key features that may be of particular interest to a reader.

Although the principle of deconstruction is a relatively common method of report writing, it is merely a structural process, and does not necessarily guarantee that the information provided will either be understood or appropriate. Therefore, it is necessary to ensure that the information comprising a report is clear, and relates directly to the issues under discussion. This is particularly the case with report writing, since superfluous information will likely detract from the principal arguments.

30.8.2 Research Reports

Consistent with accident and incident reports, human factors research reports follow a relatively standardised format. They normally include an executive summary or abstract, an introduction, a methodology section, a results section, a conclusions section, a recommendation section, and a references section (see Figure 30.3). The objective of the report is to provide a clear and concise summary of an investigation from the conceptual stage to the conclusion.

The executive summary or abstract is normally listed as the first section of the report. Both are designed to provide a coherent summary of the project, although the abstract is only used where the report is intended to be published in an academic journal. Normally, the executive summary is limited to one page, whereas the abstract is limited to approximately 200 words.

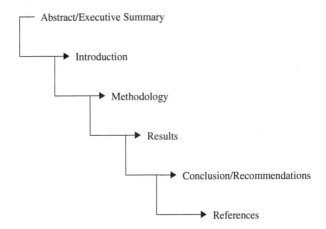

FIGURE 30.3 An illustration of the main components of a human factors research report.

The introduction to the research project follows the executive summary or abstract and should provide information pertaining to the aims of the project, any previous research, and a rationale for the research questions and/or hypotheses. The final stage of the introduction should clearly specify the research questions and/or hypotheses as a prelude to the methodology section.

The methodology section includes subsections where the characteristics of the participants are described, any materials and apparatus are discussed, and the procedure is clearly explained from the perspective of the participants. The main aim associated with this section is to provide sufficient information to ensure that the study can be understood and replicated accurately. In addition, it is used to establish the extent to which the methodology is appropriate for the domain under investigation.

The results section should normally include only those results that pertain directly to the research question or hypothesis. In an industrial environment, this information should be provided in a form that is relatively easy to follow, and is likely to be understood across the potential readership. Figures, such as charts and diagrams, are very useful since differences or relationships can be highlighted.

From a management perspective, the conclusions and recommendations sections are the most important aspects of a research report. The information contained in the report may have significant implications for future management decisions. Consequently, any recommendations that are made must be clear, concise, and must be defensible on the basis of the evidence available.

The final stage of the research report includes a list of references that were referred to within the report. Although the format for referencing differs for different disciplines, human factors reports are generally based on the guidelines advocated by the American Psychological Association.

References

Adams, R., & Thompson, J. (1987). *Aeronautical decision-making for helicopter pilots.* Washington, DC: Federal Aviation Administration (NTIS DOT/FAA/PM-86/45).

Adriaensen, A., Patriarca, R., Smoker, A., & Bergström, J. (2019). A socio-technical analysis of functional properties in a joint cognitive system: a case study in an aircraft cockpit. *Ergonomics, 62,* 1598–1616.

Ahmed, S. I. S. (2019). The impact of emergency operating safety procedures on mitigation the nuclear thermal power plant severe accident. *Annals of Nuclear Energy, 125,* 222–230.

Åhsberg, E., & Fürst, C.J. (2001). Dimensions of fatigue during radiotherapy: an application of the Swedish Occupational Fatigue Inventory (SOFI) on cancer patients. *Acta Oncologica, 40,* 37–43.

Air Accidents Investigation Branch. (1984). *Aircraft Accident Report into the crash of a British Airways Sikorsky S61-N-II off the Isle of Scilly, 16 July, 1983.* London: Author.

Air Accidents Investigation Branch. (1988). *Report on the accident to Boeing 737–236 series I, G-BGJL at Manchester International Airport on 22 August, 1985.* London: Her Majesty's Stationary Office.

Air Accidents Investigation Branch. (1990). *Report on the accident to Boeing 737–400 G-OBME near Kegworth, Leicestershire on 8 January, 1989.* London: Her Majesty's Stationery Office.

Air Accidents Investigation Branch. (1992). *Report on the accident to BAC One-Eleven, G-BJRT over Didcot, Oxfordshire, on 10 June, 1990.* London: Her Majesty's Stationery Office.

Air Accidents Investigation Branch. (2015). *Report on the accident to Airbus A319–131, G-EUOE London Heathrow Airport on 24 May, 2013.* London: Her Majesty's Stationery Office.

Åkerstedt, T. (2000). Consensus statement: Fatigue and accidents in transport operations. *Journal of Sleep Research, 9,* 395–395.

Åkerstedt, T., & Gillberg, M. (1990). Subjective and objective sleepiness in the active individual. *International Journal of Neuroscience, 52,* 29–37.

Allinson, R. E. (1993). *Global disasters: Inquiries into management ethics.* New York: Prentice Hall.

Alruqi, W. M., Hallowell, M. R., & Techera, U. (2018). Safety climate dimensions and their relationship to construction safety performance: a meta-analytic review. *Safety Science, 109,* 165–173.

Alsalem, G., Bowie, P., & Morrison, J. (2018). Assessing safety climate in acute hospital settings: a systematic review of the adequacy of the psychometric properties of survey measurement tools. *BMC Health Services Research, 18,* 353.

Altabbaa, G., Raven, A. D., & Laberge, J. (2019). A simulation-based approach to training in heuristic clinical decision-making. *Diagnosis, 6,* 91–99.

Althubaiti, A. (2016). Information bias in health research: definition, pitfalls, and adjustment methods. *Journal of Multidisciplinary Healthcare, 9,* 211–217.

Álvarez-Santos, J., Miguel-Dávila, J. Á., Herrera, L., & Nieto, M. (2018). Safety management system in TQM environments. *Safety Science, 101,* 135–143.

American National Standards Institute. (1991). *American national standard for environmental and facility safety signs.* New York: Author.

Andersen, L. P., Nørdam, L., Joensson, T., Kines, P., & Nielsen, K. J. (2018). Social identity, safety climate and self-reported accidents among construction workers. *Construction Management and Economics, 36*, 22–31.

Anderson, C. W., Santos, J. R., & Haimes, Y. Y. (2007). A risk-based input–output methodology for measuring the effects of the August 2003 northeast blackout. *Economic Systems Research, 19*, 183–204.

Anderson, J. R. (1983). Procedural learning. In J. R. Anderson (Ed.), *The architecture of cognition* (pp. 215–260). Cambridge, MA: Harvard University Press.

Anderson, J. R. (1987). Skill acquisition: Compilation of weak-method problem-solutions. *Psychological Review, 94*, 192–210.

Anderson, J. R. (1993). *Rules of the mind.* Hillsdale, NJ: Lawrence Erlbaum.

Andre, A., & Wickens, C. D. (1995). When users want what's not best for them. *Ergonomics in Design, 3(4)*, 10–13.

Andre, A. D., Wickens, C. D., Moorman, L., & Boschelli, M. M. (1991). Display formatting techniques for improving situational awareness in the aircraft cockpit. *International Journal of Aviation Psychology, 1*, 205–218.

Anthony, K., Wiencek, C., Bauer, C., Daly, B., & Anthony, M. K. (2010). No interruptions please: impact of a no interruption zone on medication safety in intensive care units. *Critical Care Nurse, 30*, 21–29.

Arashin, K. A. (2010). Using the Synergy Model to guide the practice of rapid response teams. *Dimensions of Critical Care Nursing, 29*, 120–124.

Araújo, D., Hristovski, R., Seifert, L., Carvalho, J., & Davids, K. (2019). Ecological cognition: expert decision-making behaviour in sport. *International Review of Sport and Exercise Psychology, 12*, 1–25.

Armstrong, T. J., Foulke, J. A., Joseph, B. S., & Goldstein, S. A. (1993). Investigation of cumulative trauma disorders in a poultry processing plant. *American Industrial Hygiene Association Journal, 43*, 103–116.

Arora, V., Johnson, J., Lovinger, D., Humphrey, H. J., & Meltzer, D. O. (2005). Communication failures in patient sign-out and suggestions for improvement: A critical incident analysis. *BMJ Quality and Safety, 14*, 401–407.

Aurelio, D. N., & Newman, T. M. (1997, April). Pardon my right turn. *Ergonomics in Design*, 24–27.

Australian Transport Safety Bureau. (2001). *Boeing 747–438, VH-OJH Bangkok, Thailand 23 September 1999* (Investigation Report 199904538). Canberra: Author.

Auton, J. C., Wiggins, M. W., Searle, B. J., Loveday, T., & Rattanasone, N. X. (2013). Prosodic cues used during perceptions of nonunderstandings in radio communication. *Journal of Communication, 63*, 600–616.

Auton, J. C., Wiggins, M. W., Searle, B. J., & Rattanasone, N. X. (2017). Utilization of prosodic and linguistic cues during perceptions of nonunderstandings in radio communication. *Applied Psycholinguistics, 38*, 509–539.

Aven, T. (1992). *Reliability and risk analysis.* London: Elsevier.

Ayers, F. (2021, April). Airline-style checklist flows for private pilots. *Plane and Pilot*, 16–17.

Aziz, A., Ahmed, S., Khan, F., Stack, C., & Lind, A. (2019). Operational risk assessment model for marine vessels. *Reliability Engineering and System Safety, 185*, 348–361.

Bacon, C. T., McCoy, T. P., & Henshaw, D. S. (2020). Failure to rescue and 30-day in-hospital mortality in hospitals with and without crew-resource-management safety training. *Research in Nursing and Health, 43*, 155–167.

Baddeley, A.D. (1998). *Human memory: Theory and practice.* Boston, MA: Allyn & Bacon.

References

Bagnara, S., DiMartino, C., Lisanti, B., Mancini, G., & Rizzo, A. (1991). A human error taxonomy based on cognitive engineering. In G. Apostolakis (Ed.). *Probabilistic safety assessment and management* (Vol. 1, pp. 513–518). New York: Elsevier.

Baker, S. D. (2007). Followership: the theoretical foundation of a contemporary construct. *Journal of Leadership and Organizational Studies, 14*, 50–60.

Ball, L. J., Evans, J. St.B. T., & Dennis, I. (1994). Cognitive processes in engineering design: a longitudinal study. *Ergonomics, 37*, 1753–1786.

Banks, V. A., Plant, K. L., & Stanton, N. A. (2019). Driving aviation forward; contrasting driving automation and aviation automation. *Theoretical Issues in Ergonomics Science, 20*, 250–264.

Banks, V. A., & Stanton, N. A. (2016). Keep the driver in control: automating automobiles of the future. *Applied Ergonomics, 53*, 389–395.

Banks, V. A., & Stanton, N. A. (2019). Analysis of driver roles: modelling the changing role of the driver in automated driving systems using EAST. *Theoretical Issues in Ergonomics Science, 20*, 284–300.

Baram, M. (1993). Industrial technology, chemical accidents, and social control. In B. Wilpert & T. Qvale (Eds.), *Reliability and safety in hazardous work systems* (pp. 223–225). Hove, UK: Lawrence Erlbaum.

Barlay, S. (1990). *The final call*. London: Arrow.

Bass, B. M. (1998). *Transformational leadership: Industry, military, and educational impact.* Mahwah, NJ: Erlbaum

Bass, B. M. (1999). Two decades of research and development in transformational leadership. *European Journal of Work and Organizational Psychology, 8*, 9–32.

Bass, B. M., & Avolio, B. J. (1993). Transformational leadership and organizational culture. *Public Administration Quarterly, 17*, 112–121.

Bates, D. W., Leape, L. L., & Petrycki, S. (1993). Incidence and preventability of adverse drug events in hospitalised patients. *Journal of General Internal Medicine, 8*, 289–294.

Bayram, M., Üngan, M. C., & Ardıç, K. (2017). The relationships between OHS prevention costs, safety performance, employee satisfaction and accident costs. *International Journal of Occupational Safety and Ergonomics, 23*, 285–296.

Begault, D. R. (1993). Head-up auditory displays for traffic collision avoidance system advisories: a preliminary investigation. *Human Factors, 35*, 707–717.

Beh, H. C., & McLaughlin, P. (1997). Effect of long flights on the cognitive performance of flightcrew. *Perceptual and Motor Skills, 84*, 319–322.

Benbunan-Fich, R. (2017). The ethics of online research with unsuspecting users: from A/B testing to C/D experimentation. *Research Ethics, 13*, 200–218.

Benner Jr, L. (2019). Accident investigation data: users' unrecognized challenges. *Safety Science, 118*, 309–315.

Bennett, S. (2004). The 1st July 2002 mid-air collision over Öberlingen, Germany: a holistic analysis. *Risk Management, 6*, 31–49.

Bennett, S. A. (2019). The training and practice of crew resource management: recommendations from an inductive in vivo study of the flight deck. *Ergonomics, 62(2)*, 219–232.

Bent, J. (1997). Training for new technology: practical perspectives. In R. S. Jensen & L. Rakovan (Eds.), *Proceedings of the Ninth International Symposium on Aviation Psychology* (pp. 1184–1189). Columbus, OH: Ohio State University Press.

Beresford, N. A., Fesenko, S., Konoplev, A., Skuterud, L., Smith, J. T., & Voigt, G. (2016). Thirty years after the Chernobyl accident: what lessons have we learnt? *Journal of Environmental Radioactivity, 157*, 77–89.

Berliner, C., Angell, D., & Shearer, J. W. (1964). Behaviors, measures and instruments for performance evaluation in simulated environments. Paper presented at the Symposium and Workshop on the Quantification of Human Performance. Albuquerque, NM.

Bevilacqua, M., & Ciarapica, F. E. (2018). Human factor risk management in the process industry: a case study. *Reliability Engineering and System Safety, 169*, 149–159.

Billings, C. E., & Cheaney, E. S. (1981). Dimensions of the information transfer problem. In C. E. Billings & E. S. Cheaney (Eds.), *Information transfer problems in the aviation system* (NASA Technical Paper 1875, pp. 9–14). Washington, DC: National Aeronautics and Space Administration.

Billings, C. E., & Reynard, W. D. (1984). Human factors in aircraft incidents: results of a 7-year study. *Aviation, Space, and Environmental Medicine, 55*, 960–965.

Birch, N. (1988). *Passenger protection technology in aircraft accident fires*. Aldershot, UK: Gower.

Bisbey, T. M., Reyes, D. L., Traylor, A. M., & Salas, E. (2019). Teams of psychologists helping teams: the evolution of the science of team training. *American Psychologist, 74*, 278–289.

Blake, R. R. & Mouton, J. S. (1978). *The managerial grid*. Houston, TX: Gulf.

Blake, R. R., & Mouton, J. S. (1982). Theory and research for developing a science of leadership. *Journal of Applied Behavioral Science, 18*, 275–291.

Blanchard, A. L., Welbourne, J., Gilmore, D., & Bullock, A. (2009). Followership styles and employee attachment to the organization. *The Psychologist-Manager Journal, 12*, 111–131.

Blijleven, V., Koelemeijer, K., & Jaspers, M. (2019). SEWA: a framework for sociotechnical analysis of electronic health record system workarounds. *International Journal of Medical Informatics, 125*, 71–78.

Bliss, J. P., Harden, J. W., & Dischinger Jr, H. C. (2013). Task shedding and control performance as a function of perceived automation reliability and time pressure. *Proceedings of the Human Factors and Ergonomics Society Annual Meeting, 57(1)*, 635–639.

Boag, C., Neal, A., Loft, S., & Halford, G. S. (2006). An analysis of relational complexity in an air traffic control conflict detection task. *Ergonomics, 49*, 1508–1526.

Boermans, S. M., Kamphuis, W., Delahaij, R., Van den Berg, C., & Euwema, M. C. (2014). Team spirit makes the difference: the interactive effects of team work engagement and organizational constraints during a military operation on psychological outcomes afterwards. *Stress and Health, 30*, 386–396.

Bortolussi, M. R., Hart, S. G., & Shively, R. J. (1989). Measuring moment-to-moment pilot workload using synchronous presentations of secondary tasks in a motion-based trainer. *Aviation, Space and Environmental Medicine, 60*, 124–129.

Botting, R. M., & Johnson, C. W. (1998). A formal and structured approach to the use of task analysis in accident modelling. *International Journal of Human–Computer Studies, 49*, 223–244.

Bowden, Z. E., & Ragsdale, C. T. (2018). The truck driver scheduling problem with fatigue monitoring. *Decision Support Systems, 110*, 20–31.

Boyle, L. N., Tippin, J., Paul, A., & Rizzo, M. (2008). Driver performance in the moments surrounding a microsleep. *Transportation Research Part F: Traffic Psychology and Behaviour, 11*, 126–136.

Bradley, E. A. (1995). Determination of human error patterns: the use of published results of official enquiries into system failures. *Quality and Reliability Engineering International, 11*, 411–427.

Brady, S. (2017). Beyond the limits of imagination: what do the Comet aircraft failures teach us? *The Structural Engineer, 95*, 26–28.

Bragg, J. R., Prince, R. C., Harner, E. J., & Atlas, R. M. (1994). Effectiveness of bioremediation for the Exxon Valdez oil spill. *Nature, 368*, 413–418.
Braithwaite, G. R. (2001). Aviation rescue and firefighting in Australia: is it protecting the customer? *Journal of Air Transport Management, 7*, 111–118.
Braithwaite, J., Wears, R. L., & Hollnagel, E. (2015). Resilient health care: turning patient safety on its head. *International Journal for Quality in Health Care, 27*, 418–420.
Branscombe, N. R. (1988). Conscious and unconscious processing of affective and cognitive information. In K. Fiedler & T. Forges (Eds.), *Affect, cognition and behaviours* (pp. 3–24). Toronto: C. J. Holgrele.
Brehmer, B. (1988). Organisation for decision-making in complex systems. In L. P. Goodstein, H. B. Anderson, & S. E. Olsen (Eds.), *Tasks, errors and mental models* (pp. 116–127). London: Taylor & Francis.
Brennan, P. A., De Martino, M., Ponnusamy, M., White, S., De Martino, R., & Oeppen, R. S. (2020). Avoid, trap, and mitigate: an overview of threat and error management. *British Journal of Oral and Maxillofacial Surgery, 58*, 146–150.
Bridges, W. G., Kirkman, J. Q., & Lorenzo, D. K. (1994, May). Include human errors in process hazard analysis. *Chemical Engineering Progress*, 74–82.
Brindley, P. G., & Reynolds, S. F. (2011). Improving verbal communication in critical care medicine. *Journal of Critical Care, 26*, 155–159.
Broadribb, M. P. (2015). What have we really learned? Twenty five years after Piper Alpha. *Process Safety Progress, 34*, 16–23.
Brooker, P. (2005). Reducing mid-air collision risk in controlled airspace: lessons from hazardous incidents. *Safety Science, 43*, 715–738.
Broom, M. A., Capek, A. L., Carachi, P., Akeroyd, M. A., & Hilditch, G. (2011). Critical phase distractions in anaesthesia and the sterile cockpit concept. *Anaesthesia, 66*, 175–179.
Brown, I. (1997). Corporate resource management. In M. Wiggins, I. Henley, & P. Anderson (Eds.), *Aviation education beyond 2000* (pp. 68–74). Sydney: UWS Macarthur Press.
Brown, I. D. (1995). Accident reporting and analysis. In J. R. Wilson & E. N. Corlett (Eds.), *Evaluation of work* (pp. 969–992). London: Taylor & Francis.
Bruinsma, W. E., Becker, S. J., Guitton, T. G., Kadzielski, J., & Ring, D. (2015). How prevalent are hazardous attitudes among orthopaedic surgeons? *Clinical Orthopaedics and Related Research, 473*, 1582–1589.
Bryman, A. (1986). *Leadership and organisations*. London: Routledge & Kegan Paul.
Buch, G. & Diehl, A. (1984). An investigation of the effectiveness of pilot judgement training. *Human Factors, 26*, 557–564.
Buchholz, S., & Roth, T. (1987). *Creating the high performance team*. New York: John Wiley & Sons.
Buckle, P., Clarkson, P. J., Coleman, R., Ward, J., & Anderson, J. (2006). Patient safety, systems design and ergonomics. *Applied Ergonomics, 37*, 491–500.
Bureau of Air Safety Investigation. (1990). *Human factors in airline maintenance: A study of incident reports*. Canberra: Department of Transport & Communications.
Bureau of Air Safety Investigation. (1994). *Aircraft accident report (9301743)*: Piper PA31-350 Chieftain, at Young, NSW, 11 June, 1993. Canberra: Department of Transport & Communications.
Bureau of Air Safety Investigation. (1996). *Accident investigation report: Boeing 747–312 VH-INH Sydney (Kingsford Smith) Airport, New South Wales, 19 October 1994* (Report No. 9403038). Canberra: Department of Transport.
Bureau Enquêtes Accidents. (1992). *Rapport de la commission d'enquête sur l'accident survenu le 20 janvier 1992 près du Mont Sainte-Odile (Bas Rhin) Á L'airbus A 320 Immatriculé F-Gged Exploité par la Compagnie Air Inter*. Paris, FR: Author.

Burgess-Limerick, R. (2018). Participatory ergonomics: evidence and implementation lessons. *Applied Ergonomics*, *68*, 289–293.

Burian, B. K., Clebone, A., Dismukes, K., & Ruskin, K. J. (2018). More than a tick box: medical checklist development, design, and use. *Anesthesia and Analgesia*, *126*, 223–232.

Burke, R., & Kass, A. (1996). Retrieving stories for case-based teaching. In D. B. Leake (Ed.), *Case-based reasoning* (pp. 93–110). Menlo Park, CA: AAAI Press/MIT Press.

Byrne, Z. S., Hayes, T. L., McPhail, S. M., Hakel, M. D., Cortina, J. M., & McHenry, J. J. (2014). Educating industrial–organizational psychologists for science and practice: where do we go from here? *Industrial and Organizational Psychology*, *7*, 2–14.

Cabon, P., Deharvengt, S., Grau, J. Y., Maille, N., Berechet, I., & Mollard, R. (2012). Research and guidelines for implementing Fatigue Risk Management Systems for the French regional airlines. *Accident Analysis and Prevention*, *45*, 41–44.

Cacciabue, P. C. (2005). Human error risk management methodology for safety audit of a large railway organisation. *Applied Ergonomics*, *36*, 709–718.

Cacciabue, P. C., & Hollnagel, E. (1995). Simulation and cognition. In J. M. Hoc, P. C. Cacciabue, & E. Hollnagel (Eds.), *Expertise and technology* (pp. 55–73). Hillsdale, NJ: Lawrence Erlbaum.

Cacciabue, P. C., & Vivalda, C. (1991). A dynamic methodology for evaluating human error probabilities. In G. Apostolakis (Ed.), *Probabilistic safety assessment and management* (Vol. 1, pp. 507–512). New York: Elsevier.

Cahill, J., & Mc Donald, N. (2006). Human computer interaction methods for electronic flight bag envisionment and design. *Cognition, Technology and Work*, *8*, 113–123.

Cairns, T. D., Hollenback, J., Preziosi, R. C., & Snow, W. A. (1998). A study of Hersey and Blanchard's situational leadership theory. *Leadership and Organization Development Journal*, *19*, 113–116.

Callari, T. C., McDonald, N., Kirwan, B., & Cartmale, K. (2019). Investigating and operationalising the mindful organising construct in an Air Traffic Control organisation. *Safety Science*, *120*, 838–849.

Cameron, I., Mannan, S., Németh, E., Park, S., Pasman, H., Rogers, W., & Seligmann, B. (2017). Process hazard analysis, hazard identification and scenario definition: are the conventional tools sufficient, or should and can we do much better? *Process Safety and Environmental Protection*, *110*, 53–70.

Caimi, G., Chudak, F., Fuchsberger, M., Laumanns, M., & Zenklusen, R. (2011). A new resource-constrained multicommodity flow model for conflict-free train routing and scheduling. *Transportation Science*, *45*, 212–227.

Campbell, K. S., White, C. D., & Johnson, D. E. (2003). Leader–member relations as a function of rapport management. *Journal of Business Communication (1973)*, *40*, 170–194.

Carayon, P. (2006). Human factors of complex sociotechnical systems. *Applied Ergonomics*, *37*, 525–535.

Carayon, P., Hancock, P., Leveson, N., Noy, I., Sznelwar, L., & Van Hootegem, G. (2015). Advancing a sociotechnical systems approach to workplace safety: developing the conceptual framework. *Ergonomics*, *58*, 548–564.

Carayon, P., Wooldridge, A., Hose, B. Z., Salwei, M., & Benneyan, J. (2018). Challenges and opportunities for improving patient safety through human factors and systems engineering. *Health Affairs*, *37(11)*, 1862–1869.

Card, S. K., Moran, T. P., & Newell, A. (1983). *The psychology of human–computer interaction*. Hillsdale, NJ: Lawrence Erlbaum.

Cardosi, K. M., & Murphy, E. D. (1995). *Human factors in the design and evaluation of air traffic control systems* (DOT/FAA/RD-95/3). Washington, DC: Federal Aviation Administration, Office of Aviation Research.

Carr, S., Mak, Y. T., & Needham, J. E. (1997). Differences in strategy, quality management practices and performance reporting systems between ISO accredited and non-ISO accredited companies. *Management Accounting Research, 8*, 383–403.

Carr, T. H., McCauley, C., Sperber, R. D., & Parmelee, C. M. (1982). Words, pictures, and priming: on semantic activation, conscious identification, and the automaticity of information processing. *Journal of Experimental Psychology: Human Perception and Performance, 8*, 757.

Carretta, A., Fiordelisi, F., & Schwizer, P. (2017). *Risk culture in banking*. Cham: Springer International Publishing.

Carretta, T. S., Perry Jr, D. C., & Ree, M. J. (1996). Prediction of situational awareness in F-15 pilots. *International Journal of Aviation Psychology, 6*, 21–41.

Carroll, J. M. (1997). Human–computer interaction: psychology as a science of design. *Annual Review of Psychology, 48*, 61–83.

Carter, R. (1991). 737-400 at Kegworth on 8 January, 1989: Surviving the crash. Paper presented at the 22nd Annual Seminar of the International Society of Air Safety Investigators, Canberra, Australia.

Casey, S. (1998). *Set phasers on stun*. Santa Barbara, CA: Aegean.

Casey, T., Griffin, M. A., Flatau Harrison, H., & Neal, A. (2017). Safety climate and culture: Integrating psychological and systems perspectives. *Journal of Occupational Health Psychology, 22*, 341–353.

Catchpole, K., & Russ, S. (2015). The problem with checklists. *BMJ: Quality and Safety, 24*, 545–549.

Catchpole, K., Sellers, R., Goldman, A., McCulloch, P., & Hignett, S. (2010). Patient handovers within the hospital: translating knowledge from motor racing to healthcare. *BMJ Quality and Safety, 19*, 318–322.

Catt, S. E., Miller, D. S., & Hindi, N. M. (2005). Don't misconstrue communication cues: understanding MISCUES can help reduce widespread and expensive miscommunication. *Strategic Finance*, 51–56.

Champlin, R. E., Kastenberg, W. E., & Gale, R. P. (1988). Radiation accidents and nuclear energy: medical consequences and therapy. *Annals of Internal Medicine, 109*, 730–744.

Chandler, R.A. (1997). Conflict, compromise and conquest in setting auditing standards: the case of the small company qualification. *Critical Perspectives on Accounting, 8*, 411–429.

Chandrasekaran, B. (1990, Winter). Design problem solving: a task analysis. *AI Magazine*, 59–71.

Chang, Y. H., & Liao, M. Y. (2009). The effect of aviation safety education on passenger cabin safety awareness. *Safety Science, 47*, 1337–1345.

Chang, Y. H., & Wang, Y. C. (2010). Significant human risk factors in aircraft maintenance technicians. *Safety Science, 48*, 54–62.

Chapanis, A. (1996). *Human factors in systems engineering*. New York: John Wiley.

Chapanis, A. (1999). *The Chapanis chronicles*. Santa Barbara, CA: Aegean Publishing.

Charles, R. L., & Nixon, J. (2019). Measuring mental workload using physiological measures: a systematic review. *Applied Ergonomics, 74*, 221–232.

Charlton, S. G., & O'Brien, T.G. (2001). *Handbook of human factors testing and evaluation* (2nd ed.). Boca Raton, FL: CRC Press.

Charness, N. (1991). Expertise in chess: the balance between knowledge and search. In K. A. Ericsson & J. Smith (Eds.), *Towards a general theory of expertise* (pp. 39–63). Cambridge: Cambridge University Press.

Chaturvedi, A. K. (2010). Aviation combustion toxicology: an overview. *Journal of Analytical Toxicology, 34*, 1–16.

Chaturvedi, A. K., Smith, D. R., & Canfield, D. V. (2001). Blood carbon monoxide and hydrogen cyanide concentrations in the fatalities of fire and non-fire associated civil aviation accidents, 1991–1998. *Forensic Science International, 121*, 183–188.

Chavaillaz, A., Wastell, D., & Sauer, J. (2016). System reliability, performance and trust in adaptable automation. *Applied Ergonomics, 52*, 333–342.

Chechile, R. A., Eggleston, R. G., Fleischman, R. N., & Sasseville, A. (1989). Modelling the cognitive content of displays. *Human Factors, 31*, 31–43.

Chemers, M. M., & Skrzypek, G. J. (1972). An experimental test of the contingency model of leadership. *Journal of Personality and Social Psychology, 24*, 172–177.

Chen, Y. F., & Tjosvold, D. (2006). Participative leadership by American and Chinese managers in China: the role of relationships. *Journal of Management Studies, 43*, 1727–1752.

Cheng, A., Donoghue, A., Gilfoyle, E., & Eppich, W. (2012). Simulation-based crisis resource management training for pediatric critical care medicine: a review for instructors. *Pediatric Critical Care Medicine, 13*, 197–203.

Chidester, T. R. (1990). Trends and individual differences in response to short-haul flight operations. *Aviation, Space, and Environmental Medicine, 61*, 132–138.

Choi, J. (2006). A motivational theory of charismatic leadership: envisioning, empathy, and empowerment. *Journal of Leadership and Organizational Studies, 13*, 24–43.

Choudhry, R. M., & Fang, D. (2008). Why operatives engage in unsafe work behavior: investigating factors on construction sites. *Safety Science, 46*, 566–584.

Crichton, M. T., & Flin, R. (2004). Identifying and training non-technical skills of nuclear emergency response teams. *Annals of Nuclear Energy, 31*, 1317–1330.

Christ, R. E. (1972). Review and analysis of colour coding research for visual displays. *Human Factors, 17*, 542–570.

Christ, R. E. (1975). Review and analysis of color coding research for visual displays. *Human Factors, 17*, 542–570.

Christie, B., Scane, R., & Collyer, J. (1995). Evaluation of human–computer interaction at the user interface to advanced IT systems. In J. R. Wilson & E. N. Corlett (Eds.), *Evaluation of human work* (pp. 310–356). London: Taylor & Francis.

Chun, J. U., Cho, K., & Sosik, J. J. (2016). A multilevel study of group-focused and individual-focused transformational leadership, social exchange relationships, and performance in teams. *Journal of Organizational Behavior, 37*, 374–396.

Chute, R. D., & Wiener, E. L. (1994). Cockpit and cabin crews: do conflicting mandates put them on a collision course? *Flight Safety Foundation: Cabin Crew Safety, 29*, 1–8.

Chute, R. D., & Wiener, E. L. (1995). Cockpit–cabin communication: I. A tale of two cultures. *International Journal of Aviation Psychology, 5*, 257–276.

Chute, R. D., & Wiener, E. L. (1996). Cockpit–cabin communication: II. Shall we tell the pilots? *International Journal of Aviation Psychology, 6*, 211–232.

Clark, R. E., & Estes, F. (1996). Cognitive task analysis for training. *International Journal of Educational Research, 25*, 403–417.

Clarke, S. (2013). Safety leadership: a meta-analytic review of transformational and transactional leadership styles as antecedents of safety behaviours. *Journal of Occupational and Organizational Psychology, 86*, 22–49.

Clay-Williams, R., & Colligan, L. (2015). Back to basics: checklists in aviation and healthcare. *BMJ: Quality and Safety, 24*, 428–431.

Clay-Williams, R., Hounsgaard, J., & Hollnagel, E. (2015). Where the rubber meets the road: using FRAM to align work-as-imagined with work-as-done when implementing clinical guidelines. *Implementation Science, 10*, 1–8.

Cohen, H. H., & Larue, C. (1997). Advice from an expert witness. *Ergonomics in Design, 5(4)*, 19–24.

Cohen, S. (1982). A monkey on the back, a lump in the throat. *Inside Sports*, April, (4)4, 20

Cohen, S. G., & Bailey, D. E. (1997). What makes teams work: group effectiveness research from the shop floor to the executive suite. *Journal of Management*, *23*, 239–290.

Cohen, T. N., Francis, S. E., Wiegmann, D. A., Shappell, S. A., & Gewertz, B. L. (2018). Using HFACS-Healthcare to identify systemic vulnerabilities during surgery. *American Journal of Medical Quality*, *33*, 614–622.

Cook, R. I., & Woods, D. D. (1996). Adapting to new technology in the operating room. *Human Factors*, *38*, 593–613.

Cooke, N. J. (1990). Modeling human expertise in expert systems. In R. R. Hoffman (Ed.), *The psychology of expertise* (pp. 29–60). New York: Springer-Verlag.

Cookson, S. (2009). Zagreb and Tenerife: Airline accidents involving linguistic factors. *Australian Review of Applied Linguistics*, *32*, 22.1–22.14.

Coplen, M., & Sussman, D. (2000). Fatigue and alertness in the United States railroad industry part II: fatigue research in the office of research and development at the federal railroad administration. *Transportation Research Part F: Traffic Psychology and Behaviour*, *3*, 221–228.

Corlett, E. N., & Clark, T. S. (1995). *The ergonomics of workspaces and machines*. London: Taylor & Francis.

Corrigan, S., Kay, A., Ryan, M., Ward, M. E., & Brazil, B. (2019). Human factors and safety culture: challenges and opportunities for the port environment. *Safety Science*, *119*, 252–265.

Costley, J., Johnson, D., & Lawson, D. (1989). A comparison of cockpit communication B737–B757. In R. S. Jensen (Ed.), *Proceedings of the Fifth Symposium on Aviation Psychology* (pp. 413–418). Columbus, OH: Ohio State University Press.

Coté, C. J., Notterman, D. A., Karl, H. W., Weinberg, J. A., & McCloskey, C. (2000). Adverse sedation events in pediatrics: a critical incident analysis of contributing factors. *Pediatrics*, *105*, 805–814.

Couto, T. B., Barreto, J. K. S., Marcon, F. C., Mafra, A. C. C. N., & Accorsi, T. A. D. (2018). Detecting latent safety threats in an interprofessional training that combines in situ simulation with task training in an emergency department. *Advances in Simulation*, *3(1)*, 23.

Crampton, J., & Adams, R. (1994). The kind of errors you have to be an expert to make. In *Proceedings of the National Coaching Conference* (pp. 32–38). Canberra: Australian Sports Commission.

Crandall, B., & Getchell-Reiter, K. (1993). Critical decision method: a technique for eliciting concrete assessment indicators from the intuition of NICU nurses. *Advances in Nursing Science*, *16*, 42–51.

Crandall, B., Klein, G. A., & Hoffman, R. R. (2006). *Working minds: A practitioner's guide to cognitive task analysis*. Cambridge, MA: MIT Press.

Crecine, J. P. (1986). Defence resource allocation: Garbage can analysis of C3 procurement. In J. G. March & R. Weissinger-Baylon (Eds.), *Ambiguity and command* (pp. 72–119). Marshfield, MA: Pitman.

Crichton, M. T., & Flin, R. (2004). Identifying and training non-technical skills of nuclear emergency response teams. *Annals of Nuclear Energy*, *31*, 1317–1330.

Croskerry, P. (2002). Achieving quality in clinical decision making: cognitive strategies and detection of bias. *Academic Emergency Medicine*, *9*, 1184–1204.

Croskerry, P. (2013). From mindless to mindful practice: cognitive bias and clinical decision making. *New England Journal of Medicine*, *368*, 2445–2448.

Croskerry, P. (2017). A model for clinical decision-making in medicine. *Medical Science Educator*, *27*, 9–13.

Cruz, C. E. & Della Rocco, P. S. (1995). *Sleep patterns in air traffic controllers working rapidly rotating shifts: A field study* (DOT/FAA/AM-95/12). Washington, DC: Federal Aviation Administration, Office of Aviation Medicine.

Cullis, J., Jones, P., & Lewis, A. (2006). Tax framing, instrumentality and individual differences: are there two different cultures? *Journal of Economic Psychology, 27*, 304–320.

Curcuruto, M., & Griffin, M. A. (2018). Prosocial and proactive 'safety citizenship behaviour' (SCB): the mediating role of affective commitment and psychological ownership. *Safety Science, 104*, 29–38.

Cushing, S. (1994). *Fatal words*. Chicago, IL: University of Chicago Press.

Czaja, S. J. (1997). Systems design and evaluation. In G. Salvendy (Ed.), *Handbook of human factors and ergonomics* (pp. 17–40). New York: John Wiley.

Das, B. (1998). Manufacturing workstation design. In W. Karwowski & G. Salvendy (Eds.), *Ergonomics in manufacturing* (pp. 43–64). Dearborn, MI: Society of Manufacturing Engineers.

Davis, D. R. (1948). *Pilot error: Some laboratory experiments*. London: Air Ministry (AP No. 3139A).

Dawson, D., Chapman, J., & Thomas, M. J. (2012). Fatigue-proofing: a new approach to reducing fatigue-related risk using the principles of error management. *Sleep Medicine Reviews, 16*, 167–175.

De, D., & Sahu, P. K. (2018). A survey on current and next generation aircraft collision avoidance system. *International Journal of Systems, Control and Communications, 9*, 306–337.

de Groot, A. D. (1966). Perception and memory versus thought: some old ideas and recent findings. In B. Kleinmuntz (Ed.), *Problem-solving: Research, method and theory* (pp. 19–50). New York: Wiley.

De Keyser, V. (1990). Temporal decision making in complex environments. In D. E. Broadbent, J. Reason, & A Baddeley (Eds.), *Proceedings of the Royal Society Discussion Meeting* (pp. 569–576). Oxford: Clarendon Press.

De Leeuw, K. E., & Mayer, R. E. (2008). A comparison of three measures of cognitive load: Evidence for separable measures of intrinsic, extraneous, and germane load. *Journal of Educational Psychology, 100*, 223–234.

de Ruijter, A., & Guldenmund, F. (2016). The bowtie method: a review. *Safety Science, 88*, 211–218.

De Wulf, H. (1998). Delayed maintenance blamed for Nigerian 707 engine loss. *Flight International* (9–15 December), 17.

Degani, A. & Wiener, E.L. (1990). *Human factors of flight deck checklists: the normal checklist*. National Aeronautics and Space Administration Contractor Report 177549. Moffet Field, CA: Ames Research Center.

Dekker, S., Cilliers, P., & Hofmeyr, J. H. (2011). The complexity of failure: implications of complexity theory for safety investigations. *Safety Science, 49(6)*, 939–945.

Dekker, S. W., & Breakey, H. (2016). 'Just culture': improving safety by achieving substantive, procedural and restorative justice. *Safety Science, 85*, 187–193.

Dempsey, P. G., & Leamon, T. B. (1995, October). Implementing bent-handled tools in the workplace. *Ergonomics in Design*, 15–21.

Department of Transport and Regional Development. (1996). *Marine Incident Investigation Unit (MIIU) 1991–1995*. Canberra: Author.

Deyo, R. A., Mirza, S. K., & Martin, B. I. (2006). Back pain prevalence and visit rates: estimates from US national surveys, 2002. *Spine, 31*, 2724–2727.

References

Diehl, A. E. (1991). Human performance and systems safety considerations in aviation mishaps. *International Journal of Aviation Psychology, 1*, 97–106.

Donaghy, C., Doherty, R., & Irwin, T. (2018). Patient safety: a culture of openness and supporting staff. *Surgery (Oxford), 36*, 509–514.

Dorner, D., & Schaub, H. (1994). Errors in planning and decision-making and the nature of human information processing. *Applied Psychology: An International Review, 43*, 433–453.

Dougherty, E. (1993). Context and human reliability analysis. *Reliability Engineering and System Safety, 41*, 25–47.

Dougherty, E. M., & Fragola, J. R. (1988). *Human reliability analysis.* New York: John Wiley.

Doughty, R. A. (1993). United Airlines prepares for the worst. In J. A. Gottschalk (Ed.), *Crisis response* (pp. 345–364). Detroit, MI: Visible Ink.

Douglas, H. E., Raban, M. Z., Walter, S. R., & Westbrook, J. I. (2017). Improving our understanding of multi-tasking in healthcare: drawing together the cognitive psychology and healthcare literature. *Applied Ergonomics, 59*, 45–55.

Drach-Zahavy, A., & Somech, A. (2001). Understanding team innovation: the role of team processes and structures. *Group Dynamics: Theory, Research, and Practice, 5*, 111–123.

Drury, C. G. (1998). Auditing ergonomics. In W. Karwowski & G. Salvendy (Eds.), *Ergonomics in manufacturing* (pp. 397–412). Dearborn, MI: Society of Manufacturing Engineers.

Drury, C. G., & Brill, M. (1983). New methods of consumer product accident investigation. *Human Factors and Industrial Design in Consumer Products*, 196–229.

Drury, C. G., & Dempsey, P. G. (2021). Human factors and ergonomics audits. In G. Salvendy and W. Karwowski (Eds.), *Handbook of human factors and ergonomics* (5th ed., pp. 853–879). New York: John Wiley.

Dunjó, J., Fthenakis, V., Vílchez, J. A., & Arnaldos, J. (2010). Hazard and operability (HAZOP) analysis. A literature review. *Journal of hazardous materials, 173(1–3)*, 19–32.

Dunlap, J. H., & Mangold, S. J. (1998). *Leadership/followership recurrent training: Student manual.* Washington, DC: Federal Aviation Administration.

Durso, F. T., Bleckley, M. K., & Dattel, A. R. (2006). Does situation awareness add to the validity of cognitive tests? *Human Factors, 48*, 721–733.

Durso, F. T., & Manning, C. A. (2008). Air traffic control. *Reviews of Human Factors and Ergonomics, 4*, 195–244.

Durso, F. T., & Sethumadhavan, A. (2008). Situation awareness: understanding dynamic environments. *Human Factors, 50*, 442–448.

Dyre, L., Tabor, A., Ringsted, C., & Tolsgaard, M. G. (2017). Imperfect practice makes perfect: error management training improves transfer of learning. *Medical Education, 51(2)*, 196–206.

Dyro, F. M. (1989). *The EEG handbook.* Boston, MA: Little, Brown, & Co.

Eddy, D. (1982). Probabilistic reasoning in clinical medicine: Problems and opportunities. In D. Kahneman, P. Slovic, & A. Tversky (Eds.), *Judgement and uncertainty: Heuristics and biases* (pp. 249–267). Cambridge: Cambridge University Press.

Edkins, G. D. (1998). The Indicate safety program: evaluation of a method to proactively improve airline safety performance. *Safety Science, 30*, 275–295.

Edkins, G. D. (2002). A review of the benefits of aviation human factors training. *Human Factors and Aerospace Safety, 2*, 201–216.

Edmonson, A. C. (1996). Learning from mistakes is easier said than done: group and organisational influences on the detection and correction of human error. *Journal of Applied Behavioural Science, 32*, 5–28.

Edwards, E. (1988). Introductory overview. In E. L. Wiener & D. C. Nagel (Eds.), *Human factors in aviation* (pp. 3–26). San Diego, CA: Academic Press.

Edwards, M., & Edwards, E. (1990). *The aircraft cabin*. Aldershot, UK: Gower.

Edwards, T., Homola, J., Mercer, J., & Claudatos, L. (2017). Multifactor interactions and the air traffic controller: the interaction of situation awareness and workload in association with automation. *Cognition, Technology and Work, 19*, 687–698.

Eifler, T., & Howard, T. J. (2018). The importance of robust design methodology: case study of the infamous GM ignition switch recall. Research in Engineering Design, *29*, 39–53.

Einav, Y., Gopher, D., Kara, I., Ben-Yosef, O., Lawn, M., Laufer, N., Liebergall, M., & Donchin, Y. (2010). Preoperative briefing in the operating room: shared cognition, teamwork, and patient safety. *Chest, 137*, 443–449.

Ekman, S. K., & De Backer, M. (2018). Survivability of occupants in commercial passenger aircraft accidents. *Safety Science, 104*, 91–98.

Elie-Dit-Cosaque, C. M., & Straub, D. W. (2011). Opening the black box of system usage: user adaptation to disruptive IT. *European Journal of Information Systems, 20*, 589–607.

Eljiz, K., Greenfield, D., Derrett, A., & Radmore, S. (2019). Health system redesign: changing thoughts, values, and behaviours for the co-production of a safety culture. *International Journal of Health Planning and Management, 34*, 1477–1484.

Ellis, S. R., McGreevy, M. W., & Hitchcock, R. J. (1987). Perspective traffic display format and airline pilot traffic avoidance. *Human Factors, 29*, 371–382.

Ellul, A., & Yerramilli, V. (2013). Stronger risk controls, lower risk: evidence from US bank holding companies. *Journal of Finance, 68(5)*, 1757–1803.

Embrey, D. E., Humphreys, P. C., Rosa, E. A., Kirwan, B., & Rea, K. (1984). *SLIM-MAUD: An approach to assessing human error probabilities using structured expert judgement* (NUREG/CR-3518). Washington, DC: United States Nuclear Regulatory Commission.

Endsley, M. R. (1995a). Toward a theory of situation awareness in dynamic systems. *Human Factors, 37*, 32–64.

Endsley, M. R. (1995b). Measurement of situation awareness in dynamic systems. *Human Factors, 37*, 65–84.

Endsley, M. R. (2017). From here to autonomy: lessons learned from human–automation research. *Human Factors, 59*, 5–27.

Endsley, M., & Kiris, E. O. (1995). The out-of-the-loop performance problem and level of control in automation. *Human Factors, 37*, 381–394.

Endsley, M. R. & Rodgers, M. D. (1997). *Distribution of attention, situation awareness, and workload in a passive air traffic control task: Implications for operational errors and automation* (DOT/FAA/AM-97/13). Washington, DC: Federal Aviation Administration, Office of Aviation Medicine.

Erjavac, A. J., Iammartino, R., & Fossaceca, J. M. (2018). Evaluation of preconditions affecting symptomatic human error in general aviation and air carrier aviation accidents. *Reliability Engineering and System Safety, 178*, 156–163.

Ernsting, J., & King, P. (1988). *Aviation medicine* (2nd ed.). London: Butterworths.

Estival, D., & Molesworth, B. (2011). Radio miscommunication: EL2 pilots in the Australian General Aviation environment. *Linguistics and the Human Sciences, 5*, 351–378.

Etzioni, A. (1965). Dual leadership in complex organizations. *American Sociological Review, 30*, 688–698.

Evans, R. G., Cardiff, K., & Sheps, S. (2006). High reliability versus high autonomy: Dryden, Murphy and Patient Safety. *Healthcare Policy, 1*, 12–20.

Fahlstedt, M., Halldin, P., & Kleiven, S. (2016). The protective effect of a helmet in three bicycle accidents: a finite element study. *Accident Analysis and Prevention, 91*, 135–143.

Falk, C. F., & Blaylock, B. K. (2012). The H factor: a behavioral explanation of leadership failures in the 2007–2009 financial system meltdown. *Journal of Leadership, Accountability and Ethics, 9(2),* 68–82.

Falkland, E. C., & Wiggins, M. W. (2019). Cross-task cue utilisation and situational awareness in simulated air traffic control. *Applied Ergonomics, 74,* 24–30.

Falzer, P. R. (2018). Naturalistic decision making and the practice of health care. *Journal of Cognitive Engineering and Decision Making, 12,* 178–193.

Faraj, S., & Sproull, L. (2000). Coordinating expertise in software development teams. *Management Science, 46,* 1554–1568.

Farao, J., Malila, B., Conrad, N., Mutsvangwa, T., Rangaka, M. X., & Douglas, T. S. (2020). A user-centred design framework for mHealth. *PloS One, 15(8),* e0237910.

Farmer, E. W. (1988). Stress and workload. In J. Ernsting & P. Lang (Eds.), *Aviation medicine* (2nd ed., pp. 435–444). London: Butterworths & Co.

Fastenmeier, W., & Gstalter, H. (2007). Driving task analysis as a tool in traffic safety research and practice. *Safety Science, 45,* 952–979.

Federal Aviation Administration. (1998). *New safety program unveiled: Safer skies: A focused agenda.* Washington, DC: Author.

Feibush, E., Gagvani, N., & Williams, D. (2000). Visualization for situational awareness. *IEEE Computer Graphics and Applications, 20,* 38–45.

Ferguson, R., Crist, E., & Moffatt, K. (2017). A framework for negotiating ethics in sensitive settings: hospice as a case study. *Interacting with Computers, 29,* 10–26.

Feyer, A. M., Williamson, A. M., & Cairns, D. R. (1997). The involvement of human behaviour in occupational accidents: errors in context. *Safety Science, 25,* 55–65.

Feyerabend, P. (1993). *Against method* (3rd ed.). New York: Verso.

Fiedler, F. E. (1967). *A theory of leadership effectiveness.* New York: McGraw-Hill.

Fiedler, F. E. (1971). Validation and extension of the contingency model of leadership effectiveness: a review of empirical findings. *Psychological Bulletin, 76,* 128–148.

Fiedler, F. E. (1978). The contingency model and the dynamics of the leadership process. *Advances in Experimental Social Psychology, 11,* 59–112.

Finkbeiner, K. M., Russell, P. N., & Helton, W. S. (2016). Rest improves performance, nature improves happiness: assessment of break periods on the abbreviated vigilance task. *Consciousness and Cognition, 42,* 277–285.

Finucane, A. M. (2011). The effect of fear and anger on selective attention. *Emotion, 11,* 970–974.

Fischer, S., Frese, M., Mertins, J. C., & Hardt-Gawron, J. V. (2018). The role of error management culture for firm and individual innovativeness. *Applied Psychology, 67(3),* 428–453.

Fischoff, B. (1975). Hindsight =/ foresight: the effect of outcome knowledge on judgement under uncertainty. *Journal of Experimental Psychology: Human Perception and Performance, 1,* 288–299.

Fischoff, B., & Beyth-Marom, R. (1983). Hypothesis evaluation from a Bayesian perspective. *Psychological Review, 90,* 239–260.

Fisk, A. S., Tam, S. K., Brown, L. A., Vyazovskiy, V. V., Bannerman, D. M., & Peirson, S. N. (2018). Light and cognition: roles for circadian rhythms, sleep, and arousal. *Frontiers in Neurology, 9,* 56.

Fitts, P. M. & Jones, R. E. (1947). *Analysis of factors contributing to 460 "pilot error" experiences in operating aircraft controls.* Dayton, OH: Aero Medical Laboratory, Air Material Command, Wright-Patterson Air Force Base.

Fitts, P. M., & Posner, M. I. (1967). *Human performance.* Westport, CT: Greenwood.

Flanagan, J.C. (1954). The critical incident technique. *Psychological Bulletin, 51*, 327–358.

Fleming, G. (1984). *Hitler and the final solution*. Berkeley, CA: University of California Press.

Flin, R., Martin, L., Goeters, K. M., Hormann, H. J., Amalberti, R., Valot, C., & Nijhuis, H. (2003). Development of the NOTECHS (non-technical skills) system for assessing pilots' CRM skills. *Human Factors and Aerospace Safety, 3*, 97–120.

Flin, R., Patey, R., Glavin, R., & Maran, N. (2010). Anaesthetists' non-technical skills. *British Journal of Anaesthesia, 105*, 38–44.

Flin, R., Slaven, G., & Stewart, K. (1996). Emergency decision making in the offshore oil and gas industry. *Human Factors, 38*, 262–277.

Follett, M. P. (1996). The essentials of leadership. In P. Graham (Ed.), *Mary Parker Follett: Prophet of management* (pp. 163–177). Boston, MA: Harvard Business School Publishing.

Foroughi, C. K., Sibley, C., Brown, N. L., Rovira, E., Pak, R., & Coyne, J. T. (2019). Detecting automation failures in a simulated supervisory control environment. *Ergonomics, 62*, 1150–1161.

Forsyth, D. (1983). *An introduction to group dynamics*. Pacific Grove, CA: Brooks/Cole Publishing Co.

Fourcade, A., Blache, J. L., Grenier, C., Bourgain, J. L., & Minvielle, E. (2012). Barriers to staff adoption of a surgical safety checklist. *BMJ: Quality and Safety, 21*, 191–197.

Foushee, H. C. (1982). The role of communications, socio-psychological, and personality factors in the maintenance of crew coordination. *Aviation, Space, and Environmental Medicine, 53*, 1062–1066.

Foushee, H. C., & Helmreich, R. L. (1988). Group interaction and flight crew performance. In E. L. Weiner & D. C. Nagel (Eds.)., *Human factors in aviation* (pp. 189–227). New York: Academic Press.

França, J. E., Hollnagel, E., dos Santos, I. J. L., & Haddad, A. N. (2019). FRAM AHP approach to analyse offshore oil well drilling and construction focused on human factors. *Cognition, Technology and Work, 22*, 1–13.

Franco, E. L., & Harper, D. M. (2005). Vaccination against human papillomavirus infection: a new paradigm in cervical cancer control. *Vaccine, 23(17–18)*, 2388–2394.

Franks, P., Hay, S., & Mavin, T. (2014). Can competency-based training fly? an overview of key issues for ab initio pilot training. *International Journal of Training Research, 12*, 132–147.

Friedman, L., Leedom, D. K., & Howell, W. C. (1991). Training situational awareness through pattern recognition in a battlefield environment. *Military Psychology, 3*, 105–112.

Friedrich, M., Biermann, M., Gontar, P., Biella, M., & Bengler, K. (2018). The influence of task load on situation awareness and control strategy in the ATC tower environment. *Cognition, Technology and Work, 20*, 205–217.

Fry, M. M., & Dacey, C. (2007). Factors contributing to incidents in medicine administration. Part 1. *British Journal of Nursing, 16*, 556–559.

Funk, K., Lyall, B., Wilson, J., Vint, R., Niemczyk, M., Suroteguh, C., & Owen, G. (1999). Flight deck automation issues. *International Journal of Aviation Psychology, 9*, 109–123.

Furuta, K., & Kondo, S. (1993). An approach to assessment of plant man-machine systems by computer simulation of an operator's cognitive behaviour. *International Journal of Man-Machine Studies, 39*, 473–493.

Galy, E., Paxion, J., & Berthelon, C. (2018). Measuring mental workload with the NASA-TLX needs to examine each dimension rather than relying on the global score: an example with driving. *Ergonomics, 61*, 517–527.

Garde, A. H., Begtrup, L., Bjorvatn, B., Bonde, J. P., Hansen, J., Hansen, Å. M., Härmä, M., Jensen, M. A., Kecklund, G., Kolstad, H. A., Larsen, A. D., Lie, J. A., Moreno, C. R. C.,

Nabe-Nielsen, K., & Sallinen, M. (2020). How to schedule night shift work in order to reduce health and safety risks. *Scandinavian Journal of Work, Environment and Health*, *46*, 557–569.

Gawron, V. J., Schflett, S. G., & Miller, J. C. (1988). Measures on in-flight workload. In R. S. Jensen (Ed.), *Aviation psychology* (pp. 240–287). Aldershot, UK: Gower.

Gemoets, L. A., & Mahmood, M. A. (1990). Effect of the quality of user documentation on user satisfaction with information systems. *Information and Management*, *18*, 47–54.

Gerbert, K., & Kemmler, R. (1986). The causes of causes: determinants and background variables of human factor incidents and accidents. *Ergonomics*, *29*, 1439–1453.

Gero, D. (1993). *Aviation disasters*. Somerset, UK: Patrick Stephens.

Gertman, D. I., & Blackman, H. S. (1994). *Human reliability and safety analysis data handbook*. New York: John Wiley.

Gherardini, F., Renzi, C., & Leali, F. (2017). A systematic user-centred framework for engineering product design in small-and medium-sized enterprises (SMEs). *International Journal of Advanced Manufacturing Technology*, *91*, 1723–1746.

Giambra, L., & Quilter, R. (1987). A two-term exponential description of the time course of sustained attention. *Human Factors*, *29*, 635–644.

Gibson, T. M., & Harrison, M. H. (1984). *Into thin air: A history of aviation medicine in the RAF*. London: Robert Hale.

Gibson, W. H., Megaw, E. D., Young, M. S., & Lowe, E. (2006). A taxonomy of human communication errors and application to railway track maintenance. *Cognition, Technology and Work*, *8(1)*, 57–66.

Gladstein, D. L. (1984). Groups in context: a model of task group effectiveness. *Administrative Science Quarterly*, *29*, 499–517.

Glickerman, A. S., Zimmer, S., Montero, R. C, Guerette, P. J., Campbell, W. J., Morgan, B. B., & Salas, E. (1987). *The evolution of teamwork skills: An empirical assessment with implications for training* (Technical Report NTSC 87–016). Arlington, VA: Office of Naval Research.

Goetsch, D.L. (1993). *Industrial safety and health in the age of high technology*. New York: Macmillan.

Goldschmidt, G., & Weil, M. (1998). Contents and structure in design reasoning. *Design Issues*, *14*, 85–100.

Goodwin, P., & Wright, G. (2010). The limits of forecasting methods in anticipating rare events. *Technological Forecasting and Social Change*, *77*, 355–368.

Goossens, L. H. J., & Glansdorp, C. C. (1998). Operational benefits and risk reduction of marine accidents. *Journal of Navigation*, *51*, 368–381.

Gopher, D., & Braune, R. (1984). On the psychophysics of workload: why bother with subjective measures? *Human Factors*, *26*, 519–532.

Gordon, M., Fell, C. W., Box, H., Farrell, M., & Stewart, A. (2017). Learning health 'safety' within non-technical skills interprofessional simulation education: a qualitative study. *Medical Education Online*, *22*, 1272838.

Gordon, R. P. E. (1998). The contribution of human factors to accidents in the offshore oil industry. *Reliability Engineering and System Safety*, *61*, 95–108.

Graeber, R. C. (1988). Aircrew fatigue and circadian rhythmicity. In E. L. Wiener and D. C. Nagel (Eds.), *Human factors in aviation* (pp. 305–346). San Diego, CA: Academic Press.

Grayson, R. L., & Billings, C. E. (1981). Information transfer between air traffic control and aircraft: communication problems in flight operations. In C. E. Billings & E. S. Cheaney (Eds.), *Information transfer problems in the aviation system* (NASA Technical Paper 1875, pp. 47–61). Washington, DC: National Aeronautics and Space Administration

Green, A. E. (1988). Human factors in industrial risk assessment: some early work. In L. P. Goodstein, H. B. Anderson, & S. E. Olsen (Eds.), *Tasks, errors and mental models* (pp. 193–208). London: Taylor & Francis.

Green, B., Parry, D., Oeppen, R. S., Plint, S., Dale, T., & Brennan, P. A. (2017). Situational awareness: what it means for clinicians, its recognition and importance in patient safety. *Oral Diseases, 23*, 721–725.

Green, R. (1990). Human error on the flight deck. *Philosophical Transactions of the Royal Society of London, B327*, 503–512.

Gregorich, S. E., Helmreich, R. L., & Wilhelm, J. A. (1990). The structure of cockpit management attitudes. *Journal of Applied Psychology, 75*, 682–690.

Gretch, M. R., Horberry, T. J., & Koester, T. (2008). *Human factors in the maritime domain*. Boca Raton, FL; CRC Press,

Grier, R. A., Warm, J. S., Dember, W. N., Matthews, G., Galinsky, T. L., Szalma, J. L., & Parasuraman, R. (2003). The vigilance decrement reflects limitations in effortful attention, not mindlessness. *Human Factors, 45*, 349–359.

Griffey, R. T., Schneider, R. M., Todorov, A. A., Yaeger, L., Sharp, B. R., Vrablik, M. C., Aaronson, E. L., Sammer, C., Nelson, A., Manley, H., and Dalton, P. (2019). Critical review, development, and testing of a taxonomy for adverse events and near misses in the emergency department. *Academic Emergency Medicine, 26*, 670–679.

Gronlund, S. D., Dougherty, M. R. P., Ohrt, D. D., Thomson, G. L., Bleckley, M. K., Bain, D. L., Arnell, F., & Manning, C. A. (1997). *The role of memory in air traffic control* (DOT/FAA/AM-97/22). Washington, DC: Federal Aviation Administration.

Gross, B., Rusin, L., Kiesewetter, J., Zottmann, J. M., Fischer, M. R., Prückner, S., & Zech, A. (2019). Crew resource management training in healthcare: a systematic review of intervention design, training conditions and evaluation. *BMJ Open, 9*, e025247.

Grünwald, P. (2018). Safe probability. *Journal of Statistical Planning and Inference, 195*, 47–63.

Guo, K. L. (2008). DECIDE: a decision-making model for more effective decision making by health care managers. *The Health Care Manager, 27*, 118–127.

Gurses, A. P., Seidl, K. L., Vaidya, V., Bochicchio, G., Harris, A. D., Hebden, J., & Xiao, Y. (2008). Systems ambiguity and guideline compliance: a qualitative study of how intensive care units follow evidence-based guidelines to reduce healthcare-associated infections. *BMJ Quality and Safety, 17*, 351–359.

Haas, E. C., & Edworthy, J. (1996). Designing urgency into auditory warnings using pitch, speed and loudness. *Computing and Control Engineering Journal, 7*, 193–198.

Haber, R. N, & Haber, L. (1997). One size fits all? *Ergonomics in Design, 5*, 10–17.

Hackman, J. R. (1983). *A normative model of work team effectiveness* (Tech. Rep. No. 2). New Haven, CT: Yale University Press.

Hackman, R. J. (1993). Teams, leaders, and organizations: new directions for crew-oriented flight training. In E. L Wiener, B. G. Kanki, & E. L. Helmreich (Eds.), *Cockpit resource management* (pp. 47–70). San Diego, CA: Academic Press.

Haerkens, M. H., Kox, M., Noe, P. M., Van Der Hoeven, J. G., & Pickkers, P. (2018). Crew resource management in the trauma room: a prospective 3-year cohort study. *European Journal of Emergency Medicine, 25(4)*, 281–287.

Hafenbrädl, S., Waeger, D., Marewski, J. N., & Gigerenzer, G. (2016). Applied decision making with fast-and-frugal heuristics. *Journal of Applied Research in Memory and Cognition, 5*, 215–231.

Haig, K. M., Sutton, S., & Whittington, J. (2006). SBAR: a shared mental model for improving communication between clinicians. *Joint Commission Journal on Quality and Patient Safety, 32*, 167–175.

Hale, A. R., Heming, B. H. J., Carthey, J., & Kirwan, B. (1997). Modelling of safety management systems. *Safety Science, 26*, 121–140.

Hallowell, M. R., Hinze, J. W., Baud, K. C., & Wehle, A. (2013). Proactive construction safety control: measuring, monitoring, and responding to safety leading indicators. *Journal of Construction Engineering and Management, 139*, 04013010.

Hammond, K. R., Hamm, R. M., Grassia, J., & Pearson, T. (1987). Direct comparison of the efficacy of intuitive and analytical cognition in expert judgement. *IEEE Transactions on Systems, Man, and Cybernetics, 17*, 753–770.

Hancock, P. A., & Matthews, G. (2019). Workload and performance: associations, insensitivities, and dissociations. *Human Factors, 61*, 374–392.

Hancock, P. A. (1987). Mental workload. In P. A. Hancock (Ed.), *Human factors psychology* (pp. 81–121). North-Holland: Elsevier Science Publishers.

Hanea, A. M., & Nane, G. F. (2019). Calibrating experts' probabilistic assessments for improved probabilistic predictions. *Safety Science, 118*, 763–771.

Hannaford, N., Mandel, C., Crock, C., Buckley, K., Magrabi, F., Ong, M., Allen, S., & Schultz, T. (2013). Learning from incident reports in the Australian medical imaging setting: handover and communication errors. *British Journal of Radiology, 86*, 20120336.

Hansen, T. L. (1987). Management's impact on first-line supervisor effectiveness. *SAM Advanced Management Journal, 52*, 41–45.

Hardy, R. C. (1971). Effect of leadership style on the performance of small classroom groups: a test of the contingency model. *Journal of Personality and Social Psychology, 19*, 367–374.

Harle, P. G. (1994). Investigation of human factors: the link to accident prevention. In N. Johnston, N. McDonald, & R. Fuller (Eds.), *Aviation psychology in practice* (pp. 127–148). Aldershot, UK: Avebury.

Harmon-Jones, E., Gable, P., & Price, T. F. (2012). The influence of affective states varying in motivational intensity on cognitive scope. *Frontiers in Integrative Neuroscience, 6*, 73.

Harrald, J. R., Marcus, H. S., & Wallace, W. A. (1990). The Exxon Valdez: An assessment of crisis prevention and management systems. *Interfaces, 20*, 14–30.

Harrald, J. R., Mazzuchi, T. A., Spahn, J., Van Dorp, R., Merrick, J., Shrestha, S., & Grabowski, M. (1998). Using system simulation to model the impact of human error in a maritime system. *Safety Science, 30*, 235–247.

Hart, S. G., & Loomis, L. L. (1980). Evaluation of the potential format and content of a cockpit display of traffic information. *Human Factors, 22*, 591–604.

Haskell, I. D., & Wickens, C. D. (1993). Two and three-dimensional displays for aviation: a theoretical and empirical comparison. *International Journal of Aviation Psychology, 3*, 87–109.

Havinga, J., De Boer, R. J., Rae, A., & Dekker, S. W. (2017). How did crew resource management take-off outside of the cockpit? A systematic review of how crew resource management training is conceptualised and evaluated for non-pilots. *Safety, 3*(4), 26.

Håvold, J. I. (2010). Safety culture and safety management aboard tankers. *Reliability Engineering and System Safety, 95*, 511–519.

Hefner, J. L., Hilligoss, B., Knupp, A., Bournique, J., Sullivan, J., Adkins, E., & Moffatt-Bruce, S. D. (2017). Cultural transformation after implementation of crew resource management: is it really possible? *American Journal of Medical Quality, 32*, 384–390.

Helmreich, R. L. (1984). Cockpit management attitudes. *Human Factors, 26*, 583–589.

Helmreich, R. L. (1987). Exploring flight crew behaviour. *Social Behaviour, 2*, 63–72.

Helmreich, R. L. (1993). Whither CRM? Future directions in crew resource management. In R. S. Jensen & D. Neumeister (Eds.), *Proceedings of the Seventh International Symposium on Aviation Psychology* (pp. 543–548). Columbus, OH: Ohio State University.

Helmreich, R. L. (2000). On error management: lessons from aviation. *British Medical Journal, 320*, 781–785.

Helmreich, R. L., & Davies, J. M. (1997). Anaesthetic simulation and lessons to be learned from aviation. *Canadian Journal of Anaesthesiology, 44*, 907–912.

Helmreich, R. L., & Davies, J. M. (2004). Culture, threat, and error: lessons from aviation. *Canadian Journal of Anesthesia, 51*, R1–R4.

Helmreich, R. L., Merritt, A. C., & Wilhelm, J. A. (1999). The evolution of crew resource management training. *International Journal of Aviation Psychology, 9*, 19–32.

Helmreich, R. L., & Musson, D. M. (2000). Threat and error management model: components and examples. *British Medical Journal, 9*, 1–23.

Helmreich, R. L., & Wilhelm, J. A. (1991). Outcomes of crew resource training. *International Journal of Aviation Psychology, 1*, 287–300.

Helmreich, R. L., Wilhelm, J. A., Gregorich, S. E. & Chidester, T. R. (1990). Preliminary results from the evaluation of cockpit resource management training: performance ratings of flightcrews. *Aviation, Space, and Environmental Medicine, 61*, 576–579.

Helton, W. S., & Russell, P. N. (2015). Rest is best: the role of rest and task interruptions on vigilance. *Cognition, 134*, 165–173.

Helton, W. S., & Russell, P. N. (2017). Rest is still best: the role of the qualitative and quantitative load of interruptions on vigilance. *Human Factors, 59*, 91–100.

Helton, W. S., & Warm, J. S. (2008). Signal salience and the mindlessness theory of vigilance. *Acta Psychologica, 129*, 18–25.

Hendrick, H. W. (1997a). Organisational design and macroeconomics. In G. Salvendy (Ed.), *Handbook of human factors and ergonomics* (pp. 594–636). New York: John Wiley.

Hendrick, H. W. (1997b, April). Good ergonomics is good economics. *Ergonomics in Design*, 1–15.

Hendrick, H. W. (1998). Lowering costs through ergonomics. In W. Karwowski & G. Salvendy (Eds.), *Ergonomics in manufacturing* (pp. 29–42). Dearborn, MI: Society of Manufacturing Engineers.

Hendrick, H. W. (2000). The technology of ergonomics. *Theoretical Issues in Ergonomics Science, 1*, 22–33.

Hendrick, H. W. (2007). Macroergonomics: the analysis and design of work systems. *Reviews of Human Factors and Ergonomics, 3*, 44–78.

Hendry, D. T., & Hodges, N. J. (2019). Pathways to expert performance in soccer. *Journal of Expertise, 2*, 1–13.

Hendy, K. C., Liao, J., Milgram, P. (1997). Combining time and intensity effects in assessing operators information processing loads. *Human Factors, 39*, 30–47.

Hersey, P., & Blanchard, K. H. (1969). Life cycle theory of leadership. *Training and Development Journal, 23*, 26–34

Hignett, K. C. (1996). *Practical safety and reliability assessment.* London: Chapman & Hall.

Hilbert, M. (2012). Toward a synthesis of cognitive biases: how noisy information processing can bias human decision making. *Psychological Bulletin, 138(2)*, 211–237.

Hitchcock, E. M., Warm, J. S., Matthews, G., Dember, W. N., Shear, P. K., Tripp, L. D., Mayleben, D. W., & Parasuraman, R. (2003). Automation cueing modulates cerebral blood flow and vigilance in a simulated air traffic control task. *Theoretical Issues in Ergonomics Science, 4*, 89–112.

Hobson, J. A. (1989). *Sleep.* New York: Scientific American Library.

Hoc, J. M., & Amalberti, R. (2007). Cognitive control dynamics for reaching a satisficing performance in complex dynamic situations. *Journal of Cognitive Engineering and Decision Making, 1*, 22–55.

References

Hoffman, R. R., Crandall, B., & Shadbolt, N. (1998). Use of the critical decision method to elicit expert knowledge: a case study in the methodology of cognitive task analysis. *Human Factors, 40*, 254–276.

Hoffman, R. R., Shadbolt, N. R., Burton, A. M., & Klein, G. (1995). Eliciting knowledge from experts: a methodological analysis. *Organisational Behaviour and Human Decision Processes, 62*, 129–158.

Hogg, D. N., Folles, K., Strand-Volden, F., & Torralba, B. (1995). Development of a situation awareness measure to evaluate advanced alarm systems in nuclear power plant control rooms. *Ergonomics, 38*, 2394–2413.

Hohenhaus, S. M., & Powell, S. M. (2008). Distractions and interruptions: development of a healthcare sterile cockpit. *Newborn and Infant Nursing Reviews, 8*, 108–110.

Hollnagel, E. (1995). The art of efficient man-machine interaction: improving the coupling between man and machine. In J. M. Hoc, P. C. Cacciabue, & E. Hollnagel (Eds.), *Expertise and technology: Cognition and human–computer cooperation* (pp. 229–242). Hillsdale, NJ: Lawrence Erlbaum.

Hollnagel, E. (2012). *FRAM, the functional resonance analysis method: Modelling complex socio-technical systems*. Aldershot, UK: Ashgate.

Hollnagel, E., Cacciabue, P. C., & Hoc, J. M. (1995). Work with technology: some fundamental issues. In J. M. Hoc, P. C. Cacciabue, & E. Hollnagel (Eds.), *Expertise and technology* (pp. 1–15). Hillsdale, NJ: Lawrence Erlbaum.

Hollnagel, E., & Fujita, Y. (2013). The Fukushima disaster: systemic failures as the lack of resilience. *Nuclear Engineering and Technology, 45*, 13–20.

Hollywell, P. D. (1996). Incorporating human development failures in risk assessments to improve estimates of actual risk. *Safety Science, 22*, 177–194.

Horgen, T., Joroff, M. L., Porter, W. L., & Schon, D. A. (1999). *Excellence by design: Transforming workplace and work practice*. Hoboken, NJ: John Wiley & Sons.

Holzman, R. S., Cooper, J. B., Gaba, D. M., Philip, J. H., Small, S. D., & Feinstem, D. (1995). Anesthesia crisis resource management: real-life simulation training in operating room crises. *Journal of Clinical Anesthesia, 7*, 675–687.

Howard, H. H., & Faust, C. L. (1991). The HIT method: a hazard identification technique. In G. Apostolakis (Ed.), *Probabilistic safety assessment and management* (Vol. 1, pp. 401–406). New York: Elsevier.

Hsu, Y. L., & Liu, T. C. (2012). Structuring risk factors related to airline cabin safety. *Journal of Air Transport Management, 20*, 54–56.

Hu, X., Griffin, M. A., & Bertuleit, M. (2016). Modelling antecedents of safety compliance: incorporating theory from the technological acceptance model. *Safety Science, 87*, 292–298.

Huang, X., Iun, J., Liu, A., & Gong, Y. (2010). Does participative leadership enhance work performance by inducing empowerment or trust? The differential effects on managerial and non-managerial subordinates. *Journal of Organizational Behavior, 31*, 122–143.

Huber, S., Gramlich, J., & Grundgeiger, T. (2020). From paper flight strips to digital strip systems: changes and similarities in air traffic control work practices. *Proceedings of the ACM on Human-Computer Interaction, 4*(CSCW1), 1–21.

Hudson, P. (2003). Applying the lessons of high risk industries to health care. *BMJ: Quality and Safety, 12(suppl 1)*, i7–i12.

Hudson, P. (2007). Implementing a safety culture in a major multi-national. *Safety Science, 45*, 697–722.

Hüffmeier, J., Zerres, A., Freund, P. A., Backhaus, K., Trötschel, R., & Hertel, G. (2019). Strong or weak synergy? Revising the assumption of team-related advantages in integrative negotiations. *Journal of Management, 45*, 2721–2750.

Hughes, D. (1996). Solid-state recorders taking over in U.S. *Aviation Week and Space Technology* (4 November), 80–84.

Hugo, J. V., Kovesdi, C. R., & Joe, J. C. (2018). The strategic value of human factors engineering in control room modernization. *Progress in Nuclear Energy, 108*, 381–390.

Human error principles. (1964, May). *Approach*, 23.

Human Factors Society. (1988). *American national standard for human factors engineering of visual display terminal workstations* (ANSI/HFS 100–1988). Santa Monica, CA: Author.

Hunter, D. R. (2005). Measurement of hazardous attitudes among pilots. *International Journal of Aviation Psychology, 15*, 23–43.

Hussain, A., & Oestreicher, J. (2018). Clinical decision-making: heuristics and cognitive biases for the ophthalmologist. *Survey of Ophthalmology, 63*, 119–124.

Huysamen, K., de Looze, M., Bosch, T., Ortiz, J., Toxiri, S., & O'Sullivan, L. W. (2018). Assessment of an active industrial exoskeleton to aid dynamic lifting and lowering manual handling tasks. *Applied Ergonomics, 68*, 125–131.

Hyde, R. C. (1993). Meltdown on Three Mile Island. In J. A. Gottschalk (Ed.), *Crisis response* (pp. 107–122). Detroit, MI: Visible Ink.

International Accounting Standards Committee. (1989). *Comparability of financial statements*. London: Author.

International Civil Aviation Authority. (1993). *Human factors digest No.7: Investigation of human factors accidents and incidents* (Circular 240-AN/144). Montreal: Author.

International Civil Aviation Organisation. (1986). *Manual of aircraft accident investigation*. Montreal: Author.

International Labour Organisation. (2015). *Investigation of occupational accidents and diseases: A practical guide for labour inspectors*. Geneva: Author.

International Standards Organisation. (2016). *Road vehicles-transport information and control systems-detection-response task (DRT) for assessing attentional effects of cognitive load in driving (17488)*. Geneva: Author.

International Standards Organisation. (2018) *Occupational health and safety management systems: Requirements with guidance for use*. Geneva: Author.

Isaac, A., & Guselli, J. (1996). Technology and the air traffic controller: performance panacea or human hindrance? In B. J. Hayward & A. R. Lowe (Eds.), *Applied aviation psychology* (pp. 273–280). Aldershot, UK: Avebury.

Ismail, Z., Doostdar, S., & Harun, Z. (2012). Factors influencing the implementation of a safety management system for construction sites. *Safety Science, 50*, 418–423.

Isreal, J. (1980). Structured interference in dual-task performance: Behavioural and electrophysiological data. Unpublished Ph.D. dissertation. Champaign, IL: University of Illinois.

Itri, J. N., & Patel, S. H. (2018). Heuristics and cognitive error in medical imaging. *American Journal of Roentgenology, 210*, 1097–1105.

Iversen, H. (2004). Risk-taking attitudes and risky driving behaviour. *Transportation Research Part F: Traffic Psychology and Behaviour, 7*, 135–150.

Jamieson, G. A., & Vicente, K. J. (2005). Designing effective human–automation–plant interfaces: a control-theoretic perspective. *Human Factors, 47*, 12–34.

Janis, I. L. (1972). *Victims of groupthink*. Boston, MA: Houghton Mifflin.

Jazayeri, E., & Dadi, G. B. (2017). Construction safety management systems and methods of safety performance measurement: a review. *Journal of Safety Engineering, 6*, 15–28.

Jenkins, I. (1988). Safety and human factors in manned space flight systems. In B. A. Sayers (Ed.), *Human factors and decision-making: Their influence on safety and reliability* (pp. 23–38). London: Elsevier.

Jensen, R. S. (1981). Prediction and quickening in perspective flight displays for curved landing approaches. *Human Factors, 23*, 333–364.

Jensen, R. S. (1997). The boundaries of aviation psychology, human factors, aeronautical decision-making, situation awareness, and crew resource management. *International Journal of Aviation Psychology, 7*, 259–268.

Jeong, K. S., Lee, K. W., & Lim, H. K. (2010). Risk assessment on hazards for decommissioning safety of a nuclear facility. *Annals of Nuclear Energy, 37*, 1751–1762.

Jeziorski, A. (1999). FAA to issue directive on 777-200 tail corrosion. *Flight International* (10–16 February), 8.

Jipp, M., & Ackerman, P. L. (2016). The impact of higher levels of automation on performance and situation awareness: a function of information-processing ability and working-memory capacity. *Journal of Cognitive Engineering and Decision Making, 10*, 138–166.

John, B. E., & Kieras, D. E. (1996). The GOMS family of user interface analysis techniques: comparison and contrast. *ACM Transactions on Computer–Human Interaction (TOCHI), 3*, 320–351.

Johns, G. (1988). *Organisational behaviour*. New York: Harper Collins.

Johns, M. W. (1991). A new method for measuring daytime sleepiness: the Epworth sleepiness scale. *Sleep, 14*, 540–545.

Johnston, D., & Morrison, B. W. (2016). The application of naturalistic decision-making techniques to explore cue use in Rugby League playmakers. *Journal of Cognitive Engineering and Decision Making, 10*, 391–410.

Johnston, N. (1995, Spring). Do blame and punishment have a role in organisational risk management? or Et In Arcadia Eramus. Flight Deck, 33–36.

Johnston, N., & Maurino, D. E. (1996). Applied human factors training workshop report. In B. J. Hayward & A. R. Lowe (Eds.), *Applied aviation psychology: Achievement, change and challenge* (pp. 435–446). Aldershot, UK: Avebury.

Johnston, W. A., & Dark, V. J. (1986). Selective attention. *Annual Review of Psychology, 37*, 43–75.

Joice, P., Hanna, G. B., & Cuschieri, A. (1998). Errors enacted during endoscopic surgery: a human reliability analysis. *Applied Ergonomics, 29*, 409–414.

Jones, C. P. L., Fawker-Corbett, J., Groom, P., Morton, B., Lister, C., & Mercer, S. J. (2018). Human factors in preventing complications in anaesthesia: a systematic review. *Anaesthesia, 73*, 12–24.

Jones, D. G., & Endsley, M. R. (1996). Sources of situation awareness errors in aviation. *Aviation, Space and Environmental Medicine, 67*, 507–512.

Jones, M. B. (1974). Regressing group on individual effectiveness. *Organisational Behavior and Human Performance, 11*, 426–451.

Joyce, M. (1998). Ergonomics training and education for workers and managers. In W. Karwowski & G. Salvendy (Eds.), *Ergonomics in manufacturing* (pp. 107–120). Dearborn, MI: Society of Manufacturing Engineers.

Judge, T. A., & Piccolo, R. F. (2004). Transformational and transactional leadership: a meta-analytic test of their relative validity. *Journal of Applied Psychology, 89*, 755.

Kaba, A., Wishart, I., Fraser, K., Coderre, S., & McLaughlin, K. (2016). Are we at risk of groupthink in our approach to teamwork interventions in health care? *Medical Education, 50*, 400–408.

Kaber, D., & Zahabi, M. (2017). Enhanced hazard analysis and risk assessment for human-in-the-loop systems. *Human Factors, 59*, 861–873.

Kacmar, C. J., & Carey, J. M. (1991). Assessing the usability of icons in user interfaces. *Behaviour and Information Technology, 10*, 443–457.

Kaempf, G. L., Klein, G. A., & Thordsen, M. L. (1991). *Applying recognition-primed decision-making to man-machine interface design.* (Report No NAS2–13359). Moffett Field, CA: NASA Ames Research Centre.

Kaempf, G. L., Klein, G. A., Thordsen, M. L., & Wolf, S. (1996). Decision-making in complex naval command-and-control environments. *Human Factors, 38*, 220–231.

Kahn, B. A., & Baron, J. (1995). An exploratory study of choice rules favoured in high-stakes decisions. *Journal of Consumer Psychology, 4*, 305–328.

Kalyuga, S., & Sweller, J. (2018). *Cognitive load and expertise reversal.* In K. A. Ericsson, R. R. Hoffman, A. Kozbelt, & A. M. Williams (Eds.), *Cambridge handbooks in psychology. The Cambridge handbook of expertise and expert performance* (pp. 793–811). Cambridge: Cambridge University Press.

Kang, H. G., Kim, M. C., Lee, S. J., Lee, H. J., Eom, H. S., Choi, J. G., & Jang, S. C. (2009). An overview of risk quantification issues for digitalized nuclear power plants using a static fault tree. *Nuclear Engineering and Technology, 41*, 849–858.

Kang, J. H., Kim, C. W., & Lee, S. Y. (2016). Nurse-perceived patient adverse events depend on nursing workload. *Osong Public Health and Research Perspectives, 7*, 56–62.

Kanis, H. (2000). Questioning validity in the area of ergonomics/human factors. *Ergonomics, 43*, 1947–1965.

Kanki, B. G., and Foushee, H. C. (1989). Communication as group mediator of aircrew performance. *Aviation, Space, and Environmental Medicine, 60*, 402–410.

Kantowitz, B. H., & Campbell, J. L. (1996). Pilot workload and flightdeck automation. In R. Parasuraman & M. Mouloua (Eds.), *Automation and human performance: Theory and applications* (pp. 117–136). Mahwah, NJ: Lawrence Erlbaum.

Kargl, R. G. (1994). *Aircraft accident investigator's guide.* Casper, WY: Endeavour.

Karwowski, W. (2005). Ergonomics and human factors: the paradigms for science, engineering, design, technology and management of human-compatible systems. *Ergonomics, 48*, 436–463.

Karwowski, W., & Marras, W. S. (1997). Work-related musculoskeletal disorders of the upper extremities. In G. Salvendy (Ed.), *Handbook of human factors and ergonomics* (pp. 1124–1173). New York: John Wiley.

Kato, H., Onda, Y., Saidin, Z. H., Sakashita, W., Hisadome, K., & Loffredo, N. (2019). Six-year monitoring study of radiocesium transfer in forest environments following the Fukushima nuclear power plant accident. *Journal of Environmental Radioactivity, 210*, 105817.

Keane, F. (1996). *Season of blood: A Rwandan journey.* London: Penguin.

Kee, D., Jun, G. T., Waterson, P., & Haslam, R. (2017). A systemic analysis of South Korea Sewol ferry accident: striking a balance between learning and accountability. *Applied Ergonomics, 59*, 504–516.

Keijzers, G., Fatovich, D. M., Egerton-Warburton, D., Cullen, L., Scott, I. A., Glasziou, P., & Croskerry, P. (2018). Deliberate clinical inertia: using meta-cognition to improve decision-making. *Emergency Medicine Australasia, 30*, 585–590.

Kelley, R. E. (2008). Rethinking followership. In R. E. Riggio, I. Chaleff, & J. Lipman-Blumen (Eds.), *The Warren Bennis signature series. The art of followership: How great followers create great leaders and organizations* (pp. 5–15). San Francisco, CA: Jossey-Bass/Wiley.

Kelly, D., & Efthymiou, M. (2019). An analysis of human factors in fifty controlled flight into terrain aviation accidents from 2007 to 2017. *Journal of Safety Research, 69*, 155–165.

Kempf, S. J., Azimzadeh, O., Atkinson, M. J., & Tapio, S. (2013). Long-term effects of ionising radiation on the brain: cause for concern? *Radiation and Environmental Biophysics, 52*, 5–16.

References

Kern, T. (1995). *Darker shades of blue: A case study of failed leadership*. Tucson, AZ: Author.

Kerr, M. P., Knott, D. S., Moss, M. A., Clegg, C. W., & Horton, R. P. (2008). Assessing the value of human factors initiatives. *Applied Ergonomics, 39*, 305–315.

Kessel, C. J., & Wickens, C. D. (1982). The transfer of failure-detection skills between monitoring and controlling dynamic systems. *Human Factors, 24*, 49–60.

Keynes, J. M. (1973). *The general theory and after: Part I Preparation* (Complete Works, Vol. 13). London: Macmillan.

Keyson, D. K., & Parsons, K. C. (1990). Designing the user interface using rapid prototyping. *Applied Ergonomics, 21*, 207–213.

Khakzad, N., Reniers, G., & van Gelder, P. (2017). A multi-criteria decision making approach to security assessment of hazardous facilities. *Journal of Loss Prevention in the Process Industries, 48*, 234–243.

Khan, A. H., Hasan, S., & Sarkar, M. A. R. (2018). Analysis of possible causes of Fukushima disaster. *International Journal of Nuclear and Quantum Engineering, 12*, 53–58.

Kharoufah, H., Murray, J., Baxter, G., & Wild, G. (2018). A review of human factors causations in commercial air transport accidents and incidents: From to 2000–2016. *Progress in Aerospace Sciences, 99*, 1–13.

Khastgir, S., Birrell, S., Dhadyalla, G., & Jennings, P. (2018). Calibrating trust through knowledge: introducing the concept of informed safety for automation in vehicles. *Transportation Research Part C: Emerging Technologies, 96*, 290–303.

Kielmas, M. (1996). Business Insurance (30 September), 39.

Kieras, D., & Polson, P. G. (1985). An approach to the formal analysis of user complexity. *International Journal of Man–Machine Studies, 22*, 365–394.

Kieras, D. E., & Bovair, S. (1984). The role of a mental model in learning to operate a device. *Cognitive Science, 8*, 255–273.

Kim, A. R., Park, J., Kim, Y., Kim, J., & Seong, P. H. (2017). Quantification of performance shaping factors (PSFs)' weightings for human reliability analysis (HRA) of low power and shutdown (LPSD) operations. *Annals of Nuclear Energy, 101*, 375–382.

Kim, N. K., Rahim, N. F. A., Iranmanesh, M., & Foroughi, B. (2019). The role of the safety climate in the successful implementation of safety management systems. *Safety Science, 118*, 48–56.

Kim, T. J., & Seong, P. H. (2019). Influencing factors on situation assessment of human operators in unexpected plant conditions. *Annals of Nuclear Energy, 132*, 526–536.

Kim, Y. (2018). Analyzing accountability relationships in a crisis: lessons from the Fukushima disaster. *American Review of Public Administration, 48*, 743–760.

Kim, Y., Park, J., & Jung, W. (2017). A quantitative measure of fitness for duty and work processes for human reliability analysis. *Reliability Engineering and System Safety, 167*, 595–601.

Kirsch, P., Shi, M., & Sprott, D. (2014). RISKGATE: industry sharing risk controls across Australian coal operations. *Australian Journal of Multi-Disciplinary Engineering, 11*(1), 47–58.

Kirwan, B. (1987). Human reliability analysis of an offshore emergency blowdown system. *Applied Ergonomics, 18*, 23–33.

Kirwan, B. (1994). *A guide to practical human reliability assessment*. London: Taylor & Francis.

Kirwan, B., Kennedy, R., Taylor-Adams, S., & Lambert, B. (1997). The validation of three Human Reliability Quantification techniques – THERP, HEART and JHEDI: Part II— Results of validation exercise. *Applied Ergonomics, 28*, 17–25.

Klein, C., DiazGranados, D., Salas, E., Le, H., Burke, C. S., Lyons, R., & Goodwin, G. F. (2009). Does team building work? *Small Group Research, 40*, 181–222.

Klein, G., & Klinger, D. (1991). Naturalistic decision making. *Crew System Ergonomics Information Analysis Centre Newsletter*, *2*, 1–4.

Klein, G., Kaempf, G.L., Wolf, S., Thordsen, M., & Miller, T. (1997). Applying decision requirements to user-centred design. *International Journal of Human–Computer Studies*, *46*, 1–15.

Klein, G., & Wright, C. (2016). Macrocognition: from theory to toolbox. *Frontiers in Psychology*, *7*, 54.

Klein, G. A., Calderwood, R., & Macgregor, D. (1989). Critical decision method for eliciting knowledge. *IEEE Transactions on Systems, Man, and Cybernetics*, *19*, 462–472.

Kleiner, B. M. (2002). Computer-aided macroergonomics for improved performance and safety. *Human Factors and Ergonomics in Manufacturing and Service Industries*, *12*, 307–319.

Kleiner, B. M. (2006). Macroergonomics: analysis and design of work systems. *Applied Ergonomics*, *37*, 81–89.

Kleinmuntz, B. (1990). Why we still use our heads instead of formulas: toward an integrative approach. *Psychological Bulletin*, *107*, 296–310.

Kletz, T. (1994). Protective system failure. In T. Kletz (Ed.), *Learning from accidents* (2nd ed., pp. 10–19). Oxford: Butterworth-Heinemann.

Kluge, A., Silbert, M., Wiemers, U. S., Frank, B., & Wolf, O. T. (2019). Retention of a standard operating procedure under the influence of social stress and refresher training in a simulated process control task. *Ergonomics*, *62*, 361–375.

Koh, R. Y., Park, T., & Wickens, C. D. (2014). An investigation of differing levels of experience and indices of task management in relation to scrub nurses' performance in the operating theatre: analysis of video-taped caesarean section surgeries. *International Journal of Nursing Studies*, *51*, 1230–1240.

Kontogiannis, T., Leva, M. C., & Balfe, N. (2017). Total safety management: principles, processes and methods. *Safety Science*, *100*, 128–142.

Koriat, A., Lichenstein, S., & Fischoff, B. (1980). Reasons for confidence. *Journal of Experimental Psychology: Human Learning and Memory*, *6*, 107–118.

Kortov, V., & Ustyantsev, Y. (2013). Chernobyl accident: causes, consequences and problems of radiation measurements. *Radiation Measurements*, *55*, 12–16.

Kostoff, M., Burkhardt, C., Winter, A., & Shrader, S. (2016). An interprofessional simulation using the SBAR communication tool. *American Journal of Pharmaceutical Education*, *80*, 157.

Kostopoulou, O., Delaney, B. C., & Munro, C. W. (2008). Diagnostic difficulty and error in primary care: a systematic review. *Family Practice*, *25*, 400–413.

Krieger, J. L. (2014). Family communication about cancer treatment decision making: a description of the DECIDE typology. *Annals of the International Communication Association*, *38*, 279–305.

Kroemer, K. H. E. (1997). Engineering anthropometry. In G. Salvendy (Ed.), *Handbook of human factors and ergonomics* (2nd ed., pp. 219–232). New York: Wiley.

Kruithof, J., & Ryall, J. (1994). *The quality standards handbook*. Melbourne: Business Library.

Kumar, P., Gupta, S., Agarwal, M., & Singh, U. (2016). Categorization and standardization of accidental risk-criticality levels of human error to develop risk and safety management policy. *Safety Science*, *85*, 88–98.

Kummetha, V. C., Kondyli, A., & Devos, H. (2021). Evaluating driver comprehension of the roadway environment to retain accountability of safety during driving automation. *Transportation Research Part F: Traffic Psychology and Behaviour*, *81*, 457–471.

Lanceley, A., Savage, J., Menon, U., & Jacobs, I. (2008). Influences on multidisciplinary team decision-making. *International Journal of Gynecologic Cancer*, *18*, 215–222.

Lane, R., Stanton, N. A., & Harrison, D. (2006). Applying hierarchical task analysis to medication administration errors. *Applied Ergonomics, 37*, 669–679.

Langan-Fox, J., Sankey, M. J., & Canty, J. M. (2009). Human factors measurement for future air traffic control systems. *Human Factors, 51*, 595–637.

Larson, C. E., & LaFasto, L. M. (1989). *Teamwork: What must go right/what can go wrong*. Newbury Park, CA: Sage Publications.

Latino, R. J. (2015). How is the effectiveness of root cause analysis measured in healthcare?. *Journal of Healthcare Risk Management, 35*, 21–30.

Lauber, J. K. (1984). Resource management in the cockpit. *Air Line Pilot, 53*, 20–23.

Lauber, J. K. (1986). Cockpit resource management: background and overview. In H. W. Orlady & H. C. Foushee (Eds.), *Cockpit resource management training: Proceedings of the NASA/MAC workshop* (pp. 5–14). San Francisco, CA: NASA.

Laufer, A. (1987). Construction accident cost and management safety motivation. *Journal of Occupational Accidents, 8*, 295–315.

Laureiro-Martínez, D., & Brusoni, S. (2018). Cognitive flexibility and adaptive decision-making: evidence from a laboratory study of expert decision makers. *Strategic Management Journal, 39*, 1031–1058.

Lavender, S. A., Thomas, J. S., Chang, D., & Andersson, G. B. J. (1995). Effect of lifting belts, foot movement, and lift asymmetry on trunk motions. *Human Factors, 37*, 844–853.

Law, E. L. C., van Schaik, P., & Roto, V. (2014). Attitudes towards user experience (UX) measurement. *International Journal of Human–Computer Studies, 72(6)*, 526–541.

Le Blanc, L. A., & Rucks, C. T. (1996). A multiple discriminant analysis of vessel accidents. *Accident Analysis and Prevention, 28*, 501–510.

Le Bot, P. (2004). Human reliability data, human error and accident models: illustration through the Three Mile Island accident analysis. *Reliability Engineering and System Safety, 83*, 153–167.

Learmont, D. (1993, December). Safety in numbers. *Flight International*, 51–55.

Leas, R. R., & Chi, M. T. H. (1993). Analysing diagnostic expertise of competitive swimming coaches. In J. L. Starkes & F. Allard (Eds.), *Cognitive issues in motor expertise* (pp. 75–94). Amsterdam: Elsevier.

Leber, L. L., Roscoe, S. N., & Southward, G. M. (1986). Mild hypoxia and visual performance with night vision goggles. *Aviation, Space, and Environmental Medicine, 57*, 318–324.

Lee, J., Abe, G., Sato, K., & Itoh, M. (2021). Developing human–machine trust: impacts of prior instruction and automation failure on driver trust in partially automated vehicles. *Transportation Research Part F: Traffic Psychology and Behaviour, 81*, 384–395.

Lee, J., & Chung, H. (2018). A new methodology for accident analysis with human and system interaction based on FRAM: case studies in maritime domain. *Safety Science, 109*, 57–66.

Lee, S. J., Kim, J., & Jung, W. (2013). Quantitative estimation of the human error probability during soft control operations. *Annals of Nuclear Energy, 57*, 318–326.

Lerman, S. E., Eskin, E., Flower, D. J., George, E. C., Gerson, B., Hartenbaum, N., Hursh, S., & Moore-Ede, M. (2012). Fatigue risk management in the workplace. *Journal of Occupational and Environmental Medicine, 54*, 231–258.

Lester, L. F., Diehl, A., & Buch, G. (1985). Private pilot judgement training in flight school settings: a demonstration project. In R. S. Jensen (Ed.), *Proceedings of the Third International Symposium on Aviation Psychology* (pp. 353–366). Columbus, OH: Ohio State University Press.

Leveson, N. (2004). A new accident model for engineering safer systems. *Safety Science, 42*, 237–270.

Leveson, N. (2015). A systems approach to risk management through leading safety indicators. *Reliability Engineering and System Safety, 136*, 17–34.

Leveson, N. G. (2016). *Engineering a safer world: Systems thinking applied to safety.* Cambridge, MA: MIT Press.

Levin, I. P, Schneider, S. L., & Gaeth, G. J. (1998). All frames are not created equal: a typology and critical analysis of framing effects. *Organisational Behaviour and Human Decision Processes, 76*, 149–188.

Lewis, C. H., & Griffin, M. J. (1998). A comparison and assessments obtained using alternative standards for predicting the hazards of whole-body vibration and repeated shocks. *Journal of Sound and Vibration, 215*, 915–926.

Li, Y., & Burns, C. M. (2017). Modeling automation with cognitive work analysis to support human–automation coordination. *Journal of Cognitive Engineering and Decision Making, 11*, 299–322.

Li, Y., & Guldenmund, F. W. (2018). Safety management systems: a broad overview of the literature. *Safety Science, 103*, 94–123.

Li, Y., & Mosleh, A. (2019). Dynamic simulation of knowledge-based reasoning of nuclear power plant operator in accident conditions: modeling and simulation foundations. *Safety Science, 119*, 315–329.

Ligon, K. V., Stoltz, K. B., Rowell, R. K., & Lewis, V. J. (2019). An empirical investigation of the Kelley Followership Questionnaire Revised. *Journal of Leadership Education, 18(3)*, 97–112.

Lim, C. Y., Woods, M., Humphrey, C., & Seow, J. L. (2017). The paradoxes of risk management in the banking sector. *British Accounting Review, 49(1)*, 75–90.

Lim, S., & Prakash, A. (2017). From quality control to labor protection: ISO 9001 and workplace safety, 1993–2012. *Global Policy, 8*, 66–77.

Lindblom, C. E. (1959). The science of 'muddling through'. *Public Administration Review, 19*, 78–88.

Linde, C. (1988). The quantitative study of communicative success: politeness and accidents in aviation discourse. *Language in Society, 17*, 375–399.

Lingard, H., & Holmes, N. (2001). Understandings of occupational health and safety risk control in small business construction firms: barriers to implementing technological controls. *Construction Management and Economics, 19(2)*, 217–226.

Lintern, G., & Garrison, W. V. (1992). Transfer effects of scene content and crosswind in landing instruction. *International Journal of Aviation Psychology, 2*, 225–244.

Lintern, G., Roscoe, S. N., & Sivier, J. E. (1990). Display principles, control dynamics, and environmental factors in pilot training and transfer. *Human Factors, 32*, 299–317.

Lippitt, G. L. (1969). *Work groups in organisations. Organisational renewal.* New York: Meredith Corporation.

Liu, Y., Ayaz, H., & Shewokis, P. A. (2017). Mental workload classification with concurrent electroencephalography and functional near-infrared spectroscopy. *Brain–Computer Interfaces, 4*, 175–185.

Loft, S., Sanderson, P., Neal, A., & Mooij, M. (2007). Modeling and predicting mental workload in en route air traffic control: critical review and broader implications. *Human Factors, 49*, 376–399.

Logan, G. D. (1988). Automaticity, resources and memory: theoretical controversies and practical implications. *Human Factors, 30*, 583–598.

Loh, E. (2018). Medicine and the rise of the robots: a qualitative review of recent advances of artificial intelligence in health. *BMJ Leader*, 1–5.

Loh, M. Y., Idris, M. A., Dormann, C., & Muhamad, H. (2019). Organisational climate and employee health outcomes: a systematic review. *Safety Science, 118*, 442–452.

Long, G. M., & Kearns, D. F. (1996). Visibility of text and icon highway signs under dynamic viewing conditions. *Human Factors, 38,* 690–701.

Louhevara, V., Smolander, J., Aminoff, T., & Ilmarinen, J. (1995). Assessing physical workload. In W. Karwowski & G. Salvendy (Eds.), *Ergonomics in manufacturing* (pp. 121–133). Dearborn, MI: Society of Manufacturing Engineers.

Love, P. E., Sing, M. C., Ika, L. A., & Newton, S. (2019). The cost performance of transportation projects: the fallacy of the Planning Fallacy account. *Transportation Research Part A: Policy and Practice, 122,* 1–20.

Luczak, H., Krings, K., Gryglewski, S., & Stawowy, G. (1998). Ergonomics and TQM. In W. Karwowski & G. Salvendy (Eds.), *Ergonomics in manufacturing* (pp. 505–531). Dearborn, MI: Society of Manufacturing Engineers.

Lukaszewski, J. E., & Gmeiner, J. A. (1993). The Exxon Valdez paradox. In J. A. Gottschalk (Ed.), *Crisis* (pp. 185–214). Detroit, MI: Visible Ink.

Lundberg, J., Arvola, M., Westin, C., Holmlid, S., Nordvall, M., & Josefsson, B. (2018). Cognitive work analysis in the conceptual design of first-of-a-kind systems: designing urban air traffic management. *Behaviour and Information Technology, 37,* 904–925.

Lundell, M. A., & Marcham, C. L. (2018). Leadership's effect on safety culture. *Professional Safety, 63,* 36–43.

Lyman, E. G., & Orlady, H. W. (1980). *Fatigue and associated performance decrements in air transport operations.* (NASA Contract NAS2–10060). Mountain View, CA: Battelle Laboratories.

MacKay, W. E. (1999). Is paper safer? The role of paper flight strips in air traffic control. *ACM Transactions on Computer–Human Interaction (TOCHI), 6,* 311–340.

MacKenzie, L., Ibbotson, J. A., Cao, C. G. L., & Lomax, A. J. (2001). Hierarchical decomposition of laparoscopic surgery: a human factors approach to investigating the operating room environment. *Minimally Invasive Therapy and Allied Technologies, 10,* 121–127.

McAllister, J. F. O. (1996). A veep who leaves prints. *Time* (2 September), 36–38.

McCarthy, J. C., Healey, P. G. T., Wright, P. C., & Harrison, M. D. (1997). Accountability of work-related activity in high-consequence work systems: human error in context. *International Journal of Human–Computer Studies, 47,* 735–766.

McClellan, J. E., & Dorn, H. (1999). *Science and technology in world history: An introduction.* Baltimore, MD: Johns Hopkins University Press.

McClelland, J. L. (1988). Connectionist models and psychological evidence. *Journal of Memory and Language, 27,* 107–123.

McColl, N., Auvinen, A., Kesminiene, A., Espina, C., Erdmann, F., de Vries, E., Greinert, R., Harrison, J., & Schüz, J. (2015). European Code against Cancer 4th edition: ionising and non-ionising radiation and cancer. *Cancer Epidemiology, 39,* S93–S100.

McCowen, P. & McCowen, C. (1989, September). Teaching teamwork. *Management Today,* 107–109.

McKenna, H. T., Reiss, I. K., & Martin, D. S. (2017). The significance of circadian rhythms and dysrhythmias in critical illness. *Journal of the Intensive Care Society, 18,* 121–129.

McLean, G. A., Chittum, C. B., Funkhouser, G. E., Fairlie, G. W., & Folk, E. W. (1992). *Effects of seating configuration and number of Type III exits on emergency aircraft evacuation* (NTIS DOT/FAA/AM-92/27). Washington, DC: US Department of Transportation.

McLean, G. A., & George, M. H. (1995). *Aircraft evacuations through Type-III exits II: Effects of individual subject differences* (NTIS DOT/FAA/AM-95/25). Springfield, VA: Federal Aviation Administration.

McLean, G. A., George, M. H., Chittum, C. B., & Funkhouser, G. E. (1995). *Aircraft evacuations through Type II exits: Effects of seat placement at the exit* (NTIS DOT/FAA/AM-95/22). Springfield, VA: Federal Aviation Administration.

McLeod, R. W., & Bowie, P. (2018). Bowtie analysis as a prospective risk assessment technique in primary healthcare. *Policy and Practice in Health and Safety, 16*, 177–193.

Malec, J. F., Torsher, L. C., Dunn, W. F., Wiegmann, D. A., Arnold, J. J., Brown, D. A., & Phatak, V. (2007). The mayo high performance teamwork scale: reliability and validity for evaluating key crew resource management skills. *Simulation in Healthcare, 2*, 4–10.

Mama, H., & Gethin, H. (1998, November). Indian Air Force training slammed. *Flight International*, 24.

Mamede, S., van Gog, T., van den Berge, K., Rikers, R. M., van Saase, J. L., van Guldener, C., & Schmidt, H. G. (2010). Effect of availability bias and reflective reasoning on diagnostic accuracy among internal medicine residents. *Journal of the American Medical Association, 304*, 1198–1203.

Man, A. P., Lam, C. K., Cheng, B. C., Tang, K. S., & Tang, P. F. (2020). Impact of locally adopted simulation-based crew resource management training on patient safety culture: comparison between operating room personnel and general health care populations pre and post course. *American Journal of Medical Quality, 35*, 79–88.

Mancuso, A., Compare, M., Salo, A., & Zio, E. (2017). Portfolio optimization of safety measures for reducing risks in nuclear systems. *Reliability Engineering and System Safety, 167*, 20–29.

Mandal, A.C. (1997, April). Changing standards for school furniture. *Ergonomics in Design*, 28–31.

Mann, L., & Ball, C. (1994). The relationship between search strategy and risky choice. *Australian Journal of Psychology, 46*, 131–136.

Mansfield, D. P., Michell, P. D., & Finucane, M. (1991). The application of risk assessment techniques to offshore oil and gas platforms. In G. Apostolakis (Ed). *Probabilistic safety assessment and management* (Vol. 1, pp. 239–244). New York: Elsevier.

Marcus, A. A., & Nichols, M. L. (1991). Assessing organisational safety in adapting, learning systems: Empirical studies of nuclear power. In G. Apostolakis (Ed.), *Probabilistic safety assessment and management* (Vol. 1, pp. 165–170). New York: Elsevier.

Marquardt, N. (2019). Situation awareness, human error, and organizational learning in sociotechnical systems. *Human Factors and Ergonomics in Manufacturing and Service Industries, 29*, 327–339.

Marquardt, N., Hoebel, M., & Lud, D. (2021). Safety culture transformation: the impact of training on explicit and implicit safety attitudes. *Human Factors and Ergonomics in Manufacturing and Service Industries, 31*, 191–207.

Marshall, C., Nelson, C., & Gardiner, M. M. (1987). Design guidelines. In M. M. Gardiner & B. Christie (Eds.), *Applying cognitive psychology* (pp. 221–278). New York: John Wiley.

Marshall, R., Cook, S., Mitchell, V., Summerskill, S., Haines, V., Maguire, M., Sims, R., Gyi, D., & Case, K. (2015). Design and evaluation: end users, user datasets and personas. *Applied Ergonomics, 46*, 311–317.

Marti, P. (1998). Structured task analysis in complex domains. *Ergonomics, 41*, 1664–1677.

Martinez-Val, R., & Hedo, J. M. (2000). Analysis of evacuation strategies for design and certification of transport airplanes. *Journal of Aircraft, 37*, 440–447.

Matthews, G., Warm, J. S., Reinerman-Jones, L. E., Langheim, L. K., Washburn, D. A., & Tripp, L. (2010). Task engagement, cerebral blood flow velocity, and diagnostic monitoring for sustained attention. *Journal of Experimental Psychology: Applied, 16*, 187.

Matthews, M. D., Eid, J., Johnsen, B. H., & Boe, O. C. (2011). A comparison of expert ratings and self-assessments of situation awareness during a combat fatigue course. *Military Psychology, 23*, 125–136.

Mauborgne, P., Deniaud, S., Levrat, E., Bonjour, E., Micaëlli, J. P., & Loise, D. (2016). Operational and system hazard analysis in a safe systems requirement engineering process–application to automotive industry. *Safety Science*, *87*, 256–268.

Maurino, D. E., Reason, J., Johnston, N., & Lee, R. B. (1995). *Beyond aviation human factors*. Aldershot, UK: Ashgate.

Mayhorn, C. B., Fisk, A. D., & Whittle, J. D. (2002). Decisions, decisions: analysis of age, cohort, and time of testing on framing of risky decision options. *Human Factors*, *44(4)*, 515–521.

Means, B. (1993). Cognitive task analysis as a basis for instructional design. In M. Rabinowitz (Ed.), *Cognitive science foundations of instruction* (pp. 97–118). Hillsdale, NJ: Lawrence Erlbaum.

Means, B., Salas, E., Crandall, B., & Jacobs, T. O. (1993) Training decision makers for the real world. In G. A. Klein, J. Orasanu, R. Calderwood, & C. E. Zsambok (Eds.), *Decision making in action: Models and methods* (pp. 306–326). Norwood, NJ: Ablex.

Meister, D. (1999). *The history of human factors and ergonomics*. Mahwah, NJ: Lawrence Erlbaum.

Merat, N., Seppelt, B., Louw, T., Engström, J., Lee, J. D., Johansson, E., Green, C. A., Katazaki, S., Monk, C., Itoh, M., & McGehee, D. (2019). The 'out-of-the-loop' concept in automated driving: proposed definition, measures and implications. *Cognition, Technology and Work*, *21*, 87–98.

Mercer, S. J., Whittle, C., Siggers, B., & Frazer, R. S. (2010). Simulation, human factors and defence anaesthesia. *BMJ Military Health*, *156(suppl. 4)*, S365–369.

Merritt, A., & Helmreich, R. L. (1996). Human factors on the flight deck: the influence of national culture. *Journal of Cross-Cultural Psychology*, *27*, 5–24.

Merry, A. F., Shipp, D. H., & Lowinger, J. S. (2011). The contribution of labelling to safe medication administration in anaesthetic practice. *Best Practice and Research Clinical Anaesthesiology*, *25*, 145–159.

Meshkati, N. (1991). A framework for enhancement of human and organisational reliability of complex technological systems. In G. Apostolakis (Ed.), *Probabilistic safety assessment and management* (Vol. 1, pp. 711–716). New York: Elsevier.

Metcalf, J. (1986). Decision-making and the Grenada rescue operation. In J. G. March & R. Weissinger-Baylon (Eds.), *Ambiguity and command* (pp. 277–297). Marshfield, MA: Pitman.

Metzger, U., & Parasuraman, R. (2001). The role of the air traffic controller in future air traffic management: an empirical study of active control versus passive monitoring. *Human Factors*, *43*, 519–528.

Meyer, J. (2004). Conceptual issues in the study of dynamic hazard warnings. *Human Factors*, *46*, 196–204.

Milburn, N. J., & Mertens, H. W. (1997). *Evaluation of a range of target blink amplitudes for attention-getting value in simulated air traffic control display*. Washington, DC: Federal Aviation Administration (NTIS DOT/FAA/AM-97/10).

Militello, L. G., & Hutton, R. J. (1998). Applied cognitive task analysis (ACTA): a practitioner's toolkit for understanding cognitive task demands. *Ergonomics*, *41*, 1618–1641.

Mishra, A., Catchpole, K., & McCulloch, P. (2009). The Oxford NOTECHS System: reliability and validity of a tool for measuring teamwork behaviour in the operating theatre. *BMJ Quality and Safety*, *18*, 104–108.

Modarres, M., Zhou, T., & Massoud, M. (2017). Advances in multi-unit nuclear power plant probabilistic risk assessment. *Reliability Engineering and System Safety*, *157*, 87–100.

Moffatt-Bruce, S. D., Hefner, J. L., Mekhjian, H., McAlearney, J. S., Latimer, T., Ellison, C., & McAlearney, A. S. (2017). What is the return on investment for implementation of a crew resource management program at an academic medical center? *American Journal of Medical Quality, 32(1)*, 5–11.

Mogford, R. H. (1997). Mental models and situation awareness in air traffic control. *International Journal of Aviation Psychology, 7*, 331–341.

Molesworth, B. R., & Burgess, M. (2013). Improving intelligibility at a safety critical point: in flight cabin safety. *Safety Science, 51*, 11–16.

Molesworth, B. R., & Estival, D. (2015). Miscommunication in general aviation: the influence of external factors on communication errors. *Safety Science, 73*, 73–79.

Molloy, R., & Parasurman, R. (1996). Monitoring an automated system for a single failure: vigilance and task complexity effects. *Human Factors, 38*, 311–322.

Monan, W. P. (1986). *Human factors in aviation operations: The hearback problem* (NASA Report 177398). Moffett Field, CA: NASA Ames Research Centre.

Monheit, M. A., & Johnston, J. C. (1994). Spatial attention to arrays of multidimensional objects. *Journal of Experimental Psychology: Human Perception and Performance, 20*, 691.

Moray, N. (1981). The role of attention in the detection of errors and the diagnosis of errors in man-machine systems. In J. Rasmussen & W. Rouse (Eds.), *Human detection and diagnosis of system failures* (pp. 185–198). New York: Plenum Press.

Moray, N. (1982). Subjective mental workload, *Human Factors, 24*, 25–40.

Moreno, V. C., Guglielmi, D., & Cozzani, V. (2018). Identification of critical safety barriers in biogas facilities. *Reliability Engineering and System Safety, 169*, 81–94.

Morgan Jr, B. B., Salas, E., & Glickman, A. S. (1993). An analysis of team evolution and maturation. *Journal of General Psychology, 120*, 277–291.

Morgan, B. B., Glickerman, A. S., Woodward, E. A., Blaiwes, A. S., & Salas, E. (1986). *Measurement of team behaviors in a Navy environment* (Tech. Report TR-86–014). Orlando, FL: Naval Training Center, Human Factors Division.

Morgan, M. G., & Henrion, M. (1990). *Uncertainty*. Cambridge: Cambridge University Press.

Morineau, T., & Flach, J. M. (2019). The heuristic version of cognitive work analysis: a first application to medical emergency situations. *Applied Ergonomics, 79*, 98–106.

Morrow, D. G., Leirer, V. O., Carver, L. M., & Decker Tanke, E. (1998). Older and younger adult memory for health appointment information: implications for automated telephone. *Journal of Experimental Psychology: Applied, 4*, 352–374.

Mortimer, R. G., & von Thaden, T. L. (1999). Examination of part 135 controlled flight into terrain accidents as a basis for enhanced ground proximity warning systems. *Proceedings of the Human Factors and Ergonomics Society Annual Meeting, 43(1)*, 11–15.

Moshanski, V. P. (1992). *Commission of inquiry into the Air Ontario crash at Dryden, Ontario*. Toronto: Minister of Supply and Services.

Mosneron-Dupin, F., Reer, B., Heslinga, G., Strater, O., Gerdes, V., Saliou, G., & Ullwer, W. (1997). Human-centred modelling in human reliability analysis: Some trends based on case studies. *Reliability Engineering and System Safety, 58*, 249–274.

Muir, B. M. (1994). Trust in automation: Part I. Theoretical issues in the study of trust and human intervention in automated systems. *Ergonomics, 37*, 1905–1922.

Muir, H. C., Bottomley, D. M., & Marrison, C. (1996). Effects of motivation and cabin configuration on emergency behaviour and rates of egress. *International Journal of Aviation Psychology, 6*, 57–77.

Nader, R., & Smith, W.J. (1993). *Collision course: The truth about airline safety*. Roseville, NSW: McGraw-Hill.

Nagel, D. C. (1988). Human error in aviation operations. In E. L. Wiener & D. C. Nagel (Eds.), *Human factors in aviation* (pp. 263–303). San Diego, CA: Academic Press.

Naikar, N. (2017). Cognitive work analysis: an influential legacy extending beyond human factors and engineering. *Applied Ergonomics, 59*, 528–540.

Naikar, N., & Elix, B. (2016). Integrated system design: promoting the capacity of sociotechnical systems for adaptation through extensions of cognitive work analysis. *Frontiers in Psychology, 7*, 962.

Nan, X., Xie, B., & Madden, K. (2012). Acceptability of the H1N1 vaccine among older adults: the interplay of message framing and perceived vaccine safety and efficacy. *Health Communication, 27*(6), 559–568.

Nance, J. J. (1986). *Blind trust: How deregulation has jeopardised airline safety and what you can do about it.* New York: Morrow.

Nath, N. D., Akhavian, R., & Behzadan, A. H. (2017). Ergonomic analysis of construction worker's body postures using wearable mobile sensors. *Applied Ergonomics, 62*, 107–117.

National Occupational Health and Safety Commission. (1996). *National code of practice for the control of major hazard facilities* (NOHSC:2016). Canberra: Australian Government Publishing Service.

National Transportation Safety Board. (1973). *Aircraft accident report. Eastern Airlines L-1011, Miami, FL, December 29, 1972.* (NTIS NTSB-AAR-73–14). Washington, DC: National Technical Information Service.

National Transportation Safety Board. (1979a). *Aircraft accident report: American Airlines DC-10, N110AA at Chicago, O'Hare Airport* (NTIS NTSB/AAR-79/17). Springfield, VA: National Technical Information Service.

National Transportation Safety Board. (1979b). *Aircraft accident report – United Airlines, Inc., McDonnell Douglas DC-8–61, N8082U, Portland, Oregon, December 28, 1978* (NTIS NTSB/AAR-79/07). Springfield, VA: National Technical Information Service.

National Transportation Safety Board. (1982). *Aircraft accident report: Air Florida Inc., Boeing 737–222, N62AF, collision with 14th St Bridge near Washington National Airport* (NTIS NTSB/AAR-96/03). Springfield, VA: National Technical Information Service.

National Transportation Safety Board. (1988). *Aircraft accident report: Northwest Airlines Inc. McDonnell Douglas DC-9–82, n312RC, Detroit Michigan Wayne County Airport, Romulus, Michigan, August 16, 1987* (NTIS NTSB/AAR-88/05). Springfield, VA: National Technical Information Service.

National Transportation Safety Board. (1989). *Aircraft accident report: B727–232, N473DA, Dallas-Fort Worth International Airport, Texas* (NTSB/AAR-89/04). Washington, DC: Author.

National Transportation Safety Board. (1990a). *Aircraft accident report: United Airlines Flight 811, Boeing 747–122, N4713U, Honolulu, Hawaii, February 24, 1989* (NTIS NTSB/AAR-90/01). Springfield, VA: National Technical Information Service.

National Transportation Safety Board. (1990b). *Aircraft accident report – United Airlines Flight 232, McDonnell Douglas DC-10–10, Sioux Gateway Airport, Sioux City, Iowa, July 19, 1989* (NTIS NTSB/AAR-90/06). Springfield, VA: National Technical Information Service.

National Transportation Safety Board. (1990c). *Aircraft accident report – AVIANCA, The Airline of Columbia, Boeing 707–321B, HK 2016 Fuel Exhaustion, Cove Neck, New York, January 25, 1990.* Springfield, VA: National Technical Information Service.

National Transportation Safety Board. (1991a). *Aircraft accident report. Northwest Airlines Inc. Flights 1482 and 299 runway incursion and collision Detroit Metropolitan/ Wayne County Airport, Romulus, Michigan, December 3, 1990* (Report No. NTSB-AAR-91–5). Washington, DC: National Technical Information Service.

National Transportation Safety Board. (1991b). *Aircraft accident report – Runway collision of USAIR Flight 1493, Boeing 737 and Skywest Flight 5569 Fairchild Metro, Los Angeles International Airport.* Springfield, VA: National Technical Information Service.

National Transportation Safety Board. (1994). *Proceedings of the aircraft accident investigation symposium: Industry recommendations and Safety Board responses.* Washington, DC: Author (NTIS NTSB/RP-94/01).

National Transportation Safety Board. (1996). *Aircraft accident report: Runway departure during attempted takeoff Tower Air Flight 41, Boeing 7474–136, N306FF, JFK International Airport, December 2, 1995.* Washington, DC: Author (NTIS NTSB/AAR-96/04).

National Transportation Safety Board. (1997a). *Aircraft accident report: Wheels-up landing, Continental Airlines flight 1943, Douglas DC-9 N10556, Houston, Texas, February 19, 1996.* Washington, DC: Author (NTIS NTSB/AAR-97/01).

National Transportation Safety Board. (1997b). *Aircraft accident report: Uncontrolled flight into terrain ABX Air Douglas DC-8–63, N827AX, Narrows Virginia, December 22, 1996.* Washington, DC: Author (NTIS NTSB/AAR-97/05).

National Transportation Safety Board. (1998). *Railroad accident report: Collision and derailment of Union Pacific Railroad freight trains 5981 North and 9186 South in Devine, Texas, June 22, 1997.* Washington, DC: Author (NTIS NTSB/RAR-98/02).

National Transportation Safety Board. (2000). *Marine accident report: Ramming of the Eads Bridge by Barges in Tow of the M/V Anne Holly on the Admiral, St. Louis Harbour, Missouri, April 4, 1998.* Washington, DC: Author (NTIS NTSB/MAR-00/01).

National Transportation Safety Board. (2006). *Railroad accident report: Collision between two BNSF Railway Company freight trains near Gunter, Texas May 19, 2004.* Washington, DC: (NTIS NTSB/RAR-06/02).

National Transportation Safety Board. (2010). *Loss of control on approach Colgan Air, Inc. operating as Continental Connection Flight 3407 Bombardier DHC-8-400, N200WQ Clarence Center, New York February 12, 2009.* Washington, DC: Author (NTIS NTSB/AAR-10/01).

Navarro-Martinez, D., Loomes, G., Isoni, A., Butler, D., & Alaoui, L. (2018). Boundedly rational expected utility theory. *Journal of Risk and Uncertainty, 57*, 199–223.

Naweed, A., Chapman, J., Allan, M., & Trigg, J. (2017). It comes with the job: work organizational, job design, and self-regulatory barriers to improving the health status of train drivers. *Journal of Occupational and Environmental Medicine, 59*, 264–273.

Neigel, A. R., Claypoole, V. L., & Szalma, J. L. (2019). Effects of state motivation in overload and underload vigilance task scenarios. *Acta Psychologica, 197*, 106–114.

Neri, D. F., Shappell, S. A., & DeJohn, C. A. (1992). Simulated sustained flight operations and performance, Part I: The effects of fatigue. *Military Psychology, 4*, 137–155.

Newnam, S., Goode, N., Salmon, P., & Stevenson, M. (2017). Reforming the road freight transportation system using systems thinking: an investigation of coronial inquests in Australia. *Accident Analysis and Prevention, 101*, 28–36.

Nielsen, J. (1994). *Usability engineering.* Burlington, MA: Morgan Kaufmann.

Nielsen, J., & Levy, J. (1994). Measuring usability: preference vs. performance. *Communications of the ACM, 37*, 66–75.

Nijstad, B. A., Stroebe, W., & Lodewijkx, H. F. M. (1999). Persistence of brainstorming groups: how do people know when to stop? *Journal of Experimental Social Psychology, 35*, 165–185.

Nikulin, A., & Nikulina, A. Y. (2017). Assessment of occupational health and safety effectiveness at a mining company. *Ecology, Environment and Conservation, 23*, 351–355.

Nishisaki, A., Keren, R., & Nadkarni, V. (2007). Does simulation improve patient safety? Self-efficacy, competence, operational performance, and patient safety. *Anesthesiology Clinics, 25*, 225–236.

Nitzschner, M. M., Nagler, U. K., & Stein, M. (2019). Identifying accident factors in military aviation: applying HFACS to Accident and Incident Reports of the German Armed Forces. *International Journal of Disaster Response and Emergency Management (IJDREM), 2*(1), 50–63.

Norman, D. A. (1993). *Things that make us smart.* Reading, MA: Addison Wesley.

Norman, D. A. (1988). *The design of everyday things.* New York: Doubleday.

Norman, G. R., Brooks, L. R., & Allen, S. W. (1989). Recall by expert medical practitioners and novices as a record of processing attention. *Journal of Experimental Psychology: Learning, Memory, and Cognition, 15*, 1166–1174.

Nuclear Regulatory Commission. (1975). *Reactor safety study: An assessment of accident risks in United States commercial nuclear power plants* (NUREG-75/014). Washington, DC: Author.

O'Brien, T. (2010). Problems of political transition in Ukraine: leadership failure and democratic consolidation. *Contemporary Politics, 16*, 355–367.

O'Brien, T. G. (1996). Preparing human factors test plans and reports. In T. G. O'Brien & S. G. Charlton (Eds.), *Handbook of human factors testing and evaluation* (pp. 117–134). Mahwah, NJ: Lawrence Erlbaum.

O'Connor, P., Campbell, J., Newon, J., Melton, J., Salas, E., & Wilson, K. A. (2008). Crew resource management training effectiveness: a meta-analysis and some critical needs. *International Journal of Aviation Psychology, 18*, 353–368.

O'Connor, T., Papanikolaou, V., & Keogh, I. (2010). Safe surgery, the human factors approach. *The Surgeon, 8*, 93–95.

O'Dea, A., O'Connor, P., & Keogh, I. (2014). A meta-analysis of the effectiveness of crew resource management training in acute care domains. *Postgraduate Medical Journal, 90*, 699–708.

O'Hare, D., & Roscoe, S. (1990). *Flightdeck performance: The human factor.* Ames, IA: Iowa State University Press.

O'Hare, D., Wiggins, M., Batt, R., & Morrison, D. (1994). Cognitive failure analysis for aircraft accident investigation. *Ergonomics, 37*, 1855–1869.

O'Hare, D., Wiggins, M., Williams, A., & Wong, W. (1998). Cognitive task analysis for decision centred design and training. *Ergonomics, 41*, 1698–1718.

O'Kelley, K. (2019). New employees and safety culture: a social cognitive theory perspective. *Professional Safety, 64*, 37–40.

Oc, B. (2018). Contextual leadership: A systematic review of how contextual factors shape leadership and its outcomes. *Leadership Quarterly, 29*, 218–235.

Ogle, A. D., Rutland, J. B., Fedotova, A., Morrow, C., Barker, R., & Mason-Coyner, L. (2019). Initial job analysis of military embedded behavioral health services: tasks and essential competencies. *Military Psychology, 31*, 267–278.

Oken, B. S., Salinsky, M. C., & Elsas, S. (2006). Vigilance, alertness, or sustained attention: physiological basis and measurement. *Clinical Neurophysiology, 117*, 1885–1901.

Okoli, J. O., Weller, G., & Watt, J. (2016). Information processing and intuitive decision-making on the fireground: towards a model of expert intuition. *Cognition, Technology and Work, 18*, 89–103.

Oliveira, M. D., Lopes, D. F., & Bana e Costa, C. A. (2018). Improving occupational health and safety risk evaluation through decision analysis. *International Transactions in Operational Research, 25*, 375–403.

Orasanu, J. (1990). *Shared mental models and crew decision making* (Tech. Rep. No. 46). Princeton, NJ: Princeton University, Cognitive Science Laboratory.

Orasanu, J. (1995). Training for aviation decision-making: the naturalistic decision-making perspective. *Proceedings of the Human Factors and Ergonomics Society 39th Annual Meeting* (pp. 1258–1262). San Diego, CA: Human Factors and Ergonomics Society.

Onofrio, R., & Trucco, P. (2018). Human reliability analysis (HRA) in surgery: Identification and assessment of Influencing Factors. *Safety Science, 110*, 110–123.

Oriol, M. D. (2006). Crew resource management: applications in healthcare organizations. *JONA: The Journal of Nursing Administration, 36*, 402–406.

Orlandi, L., & Brooks, B. (2018). Measuring mental workload and physiological reactions in marine pilots: building bridges towards redlines of performance. *Applied Ergonomics, 69*, 74–92.

Oron-Gilad, T., Ronen, A., & Shinar, D. (2008). Alertness maintaining tasks (AMTs) while driving. *Accident Analysis and Prevention, 40*, 851–860.

Ortmeier, F., Schellhorn, G., Thums, A., Reif, W., Hering, B., & Trappschuh, H. (2003). Safety analysis of the height control system for the Elbtunnel. *Reliability Engineering and System Safety, 81*, 259–268.

Oser, R., McCallum, G. A., Salas, E., & Morgan, B. B. (1989). *Toward a definition of teamwork: An analysis of critical team behaviors* (Tech. Report 89–004). Orlando, FL: Naval Training Systems Center, Human Factors Division.

Oster, C. V., Strong, J. S., & Zorn, C. K. (1992). *Why airplanes crash*. New York: Oxford University Press.

Palmer, S. P., Jago, S. J., & Baty, D. L. (1980). Perception of horizontal aircraft separation on a cockpit display of traffic information. *Human Factors, 22*, 605–620.

Pambreni, Y., Khatibi, A., Azam, S., & Tham, J. (2019). The influence of total quality management toward organization performance. *Management Science Letters, 9*, 1397–1406.

Pammer, K., Raineri, A., Beanland, V., Bell, J., & Borzycki, M. (2018). Expert drivers are better than non-expert drivers at rejecting unimportant information in static driving scenes. *Transportation Research Part F: Traffic Psychology and Behaviour, 59*, 389–400.

Pan, X., Lin, Y., & He, C. (2017). A review of cognitive models in human reliability analysis. *Quality and Reliability Engineering International, 33*, 1299–1316.

Pan, X., & Wu, Z. (2020). Performance shaping factors in the human error probability modification of human reliability analysis. *International Journal of Occupational Safety and Ergonomics, 26*, 538–550.

Pan, Y., & Hildre, H. P. (2018). Holistic human safety in the design of marine operations safety. *Ocean Engineering, 151*, 378–389.

Papadimitriou, E., Schneider, C., Tello, J. A., Damen, W., Vrouenraets, M. L., & Ten Broeke, A. (2020). Transport safety and human factors in the era of automation: what can transport modes learn from each other? *Accident Analysis and Prevention, 144*, 105656.

Papautsky, E. L., Strouse, R., & Dominguez, C. (2020). Combining cognitive task analysis and participatory design methods to elicit and represent task flows. *Journal of Cognitive Engineering and Decision Making, 14*, 288–301.

Parasuraman, R., Molloy, R., & Singh, I. L. (1993). Performance consequences of automation-induced 'complacency'. *International Journal of Aviation Psychology, 3*, 1–23.

Parasuraman, R., Mouloua, M., & Molloy, R. (1996). Effects of adaptive task allocation on monitoring of automated systems. *Human Factors, 38*, 665–679.

Parasuraman, R., & Riley, V. (1997). Humans and automation: use, misuse, disuse, and abuse. *Human Factors, 39*, 230–253.

Park, J., Kim, Y., & Jung, W. (2018). Calculating nominal human error probabilities from the operation experience of domestic nuclear power plants. *Reliability Engineering and System Safety, 170,* 215–225.

Paté-Cornell, M. E. (1993). Learning from the Piper Alpha accident: a postmortem analysis of technical and organizational factors. *Risk Analysis, 13,* 215–232.

Peerally, M. F., Carr, S., Waring, J., & Dixon-Woods, M. (2017). The problem with root cause analysis. *BMJ Quality and Safety, 26,* 417–422.

Perrow, C. (1984). *Normal accidents: Living with high-risk technologies.* New York: Basic Books.

Peruzzini, M., Grandi, F., & Pellicciari, M. (2017). Benchmarking of tools for user experience analysis in Industry 4.0. *Procedia Manufacturing, 11,* 806–813.

Perry, N. C., Stevens, C. J., Wiggins, M. W., & Howell, C. E. (2007). Cough once for danger: icons versus abstract warnings as informative alerts in civil aviation. *Human Factors, 49,* 1061–1071.

Petitta, L., Probst, T. M., & Barbaranelli, C. (2017). Safety culture, moral disengagement, and accident underreporting. *Journal of Business Ethics, 141,* 489–504.

Petitta, L., Probst, T. M., Barbaranelli, C., & Ghezzi, V. (2017). Disentangling the roles of safety climate and safety culture: multi-level effects on the relationship between supervisor enforcement and safety compliance. *Accident Analysis and Prevention, 99,* 77–89.

Petroski, H. (2012). *To forgive design.* Cambridge, MA: Belknap Press

Phillips, E. D. (1996). Legal problems cloud airline safety programs. *Aviation Week and Space Technology* (4 November), 66–68.

Phillips, R. O., Kecklund, G., Anund, A., & Sallinen, M. (2017). Fatigue in transport: a review of exposure, risks, checks and controls. *Transport Reviews, 37,* 742–766.

Pilcher, J. J., & Morris, D. M. (2020). Sleep and organizational behavior: implications for workplace productivity and safety. *Frontiers in Psychology, 11,* 45.

Pines, J. M. (2006). Profiles in patient safety: confirmation bias in emergency medicine. *Academic Emergency Medicine, 13,* 90–94.

Plant, K. L., & Stanton, N. A. (2016). Distributed cognition in search and rescue: loosely coupled tasks and tightly coupled roles. *Ergonomics, 59,* 1353–1376.

Plous, S. (1993). *The psychology of judgement and decision-making.* New York: McGraw-Hill.

Porta, J., Saco-Ledo, G., & Cabañas, M. D. (2019). The ergonomics of airplane seats: the problem with economy class. *International Journal of Industrial Ergonomics, 69,* 90–95.

Poyet, C., & Leplat, J. (1993). Mixed technologies and management of reliability. In B. Wilpert & T. Qvale (Eds.), *Reliability and safety in hazardous work systems* (pp. 133–156). Hove, UK: Lawrence Erlbaum.

Pradarelli, J. C., George, E., Kavanagh, J., Sonnay, Y., Khoon, T. H., & Havens, J. M. (2021). Training novice raters to assess nontechnical skills of operating room teams. *Journal of Surgical Education, 78,* 386–390.

Praetorius, G., Hollnagel, E., & Dahlman, J. (2015). Modelling vessel traffic service to understand resilience in everyday operations. *Reliability Engineering and System Safety, 141,* 10–21.

Preciado, D., Munneke, J., & Theeuwes, J. (2017). Mixed signals: the effect of conflicting reward-and goal-driven biases on selective attention. *Attention, Perception, and Psychophysics, 79,* 1297–1310.

Predmore, S. C. (1991). Microcoding of communications in accident investigation: crew coordination in United 811 and United 232. In R. S. Jensen (Ed.), *Proceedings of the Sixth International Symposium on Aviation Psychology* (pp. 350–355). Columbus, OH: Ohio State University Press.

Preischl, W., & Hellmich, M. (2013). Human error probabilities from operational experience of German nuclear power plants. *Reliability Engineering and System Safety, 109,* 150–159.

Prince, C., Oser, R., Salas, E., & Woodruff, W. (1993). Increasing hits and reducing misses in CRM/LOS scenarios: guidelines for simulator scenario development. *International Journal of Aviation Psychology, 3*, 69–82.

Prinzo, O. V. (1996). *An analysis of approach control/pilot voice communications*. Washington, DC: US Department of Transportation. (NTIS DOT/FAA/AM-96/26).

Proctor, R. W., & Vu, K. P. L. (2016). Principles for designing interfaces compatible with human information processing. *International Journal of Human–Computer Interaction, 32*, 2–22.

Puisa, R., Lin, L., Bolbot, V., & Vassalos, D. (2018). Unravelling causal factors of maritime incidents and accidents. *Safety Science, 110*, 124–141.

Purba, J. H., Lu, J., Zhang, G., & Ruan, D. (2011). Failure possibilities for nuclear safety assessment by fault tree analysis. *International Journal of Nuclear Knowledge Management, 5*, 162–177.

Puryear, J., Ramirez, G., Botard, C., & Kenady, K. (2018). Fracture toughness and brittle failure: A pressure vessel case study. *Process Safety Progress, 37*, 305–310.

Pylkkönen, M., Sihvola, M., Hyvärinen, H. K., Puttonen, S., Hublin, C., & Sallinen, M. (2015). Sleepiness, sleep, and use of sleepiness countermeasures in shift-working long-haul truck drivers. *Accident Analysis and Prevention, 80*, 201–210.

Rabøl, L. I., Andersen, M. L., Østergaard, D., Bjørn, B., Lilja, B., & Mogensen, T. (2011). Descriptions of verbal communication errors between staff: an analysis of 84 root cause analysis-reports from Danish hospitals. *BMJ Quality and Safety, 20*, 268–274.

Raby, J.S. (1987). Human factors education and the airline pilot. In R. S. Jensen (Ed.), *Proceedings of the Fourth International Symposium on Aviation Psychology* (pp. 384–391). Columbus, OH: Ohio State University Press.

Rail Accident Investigation Branch. (2014). *Rail accident report: Passenger train collision at Norwich, 21 July 2013 (Report 09/2014)*. Derby, UK: Department of Transport.

Ramkumar, A., Stappers, P. J., Niessen, W. J., Adebahr, S., Schimek-Jasch, T., Nestle, U., & Song, Y. (2017). Using GOMS and NASA-TLX to evaluate human–computer interaction process in interactive segmentation. *International Journal of Human–Computer Interaction, 33*, 123–134.

Ramsay, G., Haynes, A. B., Lipsitz, S. R., Solsky, I., Leitch, J., Gawande, A. A., & Kumar, M. (2019). Reducing surgical mortality in Scotland by use of the WHO Surgical Safety Checklist. *Journal of British Surgery, 106*, 1005–1011.

Rasmussen, J. (1982). Human errors: a taxonomy for describing human malfunction in industrial installations. *Journal of Occupational Accidents, 4*, 311–333.

Rasmussen, J. (1983). Skills, rules, and knowledge: signals, signs, and symbols, and other distinctions in human performance models. *IEEE Transactions on Systems, Man, and Cybernetics, SMC-13*, 257–266.

Rasmussen, J. (1986). A framework for cognitive task analysis in systems design. In E. Hollnagel, G. Mancini, & D. D. Woods (Eds.), *Intelligent decision support in process environments* (pp. 175–196). Berlin: Springer-Verlag.

Rasmussen, J. (1990a). Human error and the problem of causality in the analysis of accidents. In D. E. Broadbent, J. Reason, & A. Baddeley (Eds.), *Proceedings of the Royal Society Discussion Meeting* (pp. 449–462). Oxford: Clarendon Press.

Rasmussen, J., Pedersen, O. M., Carnino, A., Griffon, M., Mancini, C., and Gagnolet, P. (1981). *Classification scheme for reporting events involving human malfunctions* (Riso-M-2240, DK-4000). Rosklide, NL: National Laboratories.

Rasmussen, J., Pejtersen, A. M., & Goodstein, L. P. (1994). *Cognitive systems engineering*. New York: Wiley.

Ratcheva, V. (2008). The knowledge advantage of virtual teams–processes supporting knowledge synergy. *Journal of General Management, 33*, 53–67.
Ratzmann, M., Pesch, R., Bouncken, R., & Climent, C. M. (2018). The price of team spirit for sensemaking through task discourse in innovation teams. *Group Decision and Negotiation, 27*, 321–341.
Raviv, G., Shapira, A., & Fishbain, B. (2017). AHP-based analysis of the risk potential of safety incidents: case study of cranes in the construction industry. *Safety Science, 91*, 298–309.
Razieh, S., Somayeh, G., & Fariba, H. (2018). Effects of reflection on clinical decision-making of intensive care unit nurses. *Nurse Education Today, 66*, 10–14.
Razzouk, R., & Shute, V. (2012). What is design thinking and why is it important? *Review of Educational Research, 82*, 330–348.
Reason, J. (1990). *Human error.* Cambridge: Cambridge University Press.
Reason, J. (1993). Managing the management of risk: new approaches to organisational safety. In B. Wilpert & T. Qvale (Eds.), *Reliability and safety in hazardous work systems* (pp. 7–22). Hove, UK: Lawrence Erlbaum.
Reason, J. (1997). *Managing the risks of organisational accidents.* Aldershot, UK: Ashgate.
Reason, J. (2000). Human error: models and management. *British Medical Journal, 320*, 768–770.
Reason, J. (2002). Combating omission errors through task analysis and good reminders. *BMJ Quality and Safety, 11*, 40–44.
Redding, R. E., & Cannon, J.R. (1991). Expertise in air traffic control (ATC): what is it, and how can we train for it. *Proceedings of the Human Factors Society 36th Annual Meeting* (pp. 1326–1330). San Diego, CA: Human Factors Society.
Rees, A., Wiggins, M. W., Helton, W. S., Loveday, T., & O'Hare, D. (2017). The impact of breaks on sustained attention in a simulated, semi-automated train control task. *Applied Cognitive Psychology, 31*, 351–359.
Reeves, S. (1995). On the care and handling of electronic data recorders: or how to keep the bits byting. *ISASI Forum, 28(3)*, 19–24
Ren, J., Jenkinson, I., Wang, J., Xu, D. L., & Yang, J. B. (2008). A methodology to model causal relationships on offshore safety assessment focusing on human and organizational factors. *Journal of Safety Research, 39*, 87–100.
Reynolds, J. R., & Bayan, F. P. (1999). Achieving the virtual cockpit. *ISASI Forum, 32(4)*, 26–29.
Rhoden, S., Ralston, R., & Ineson, E. M. (2008). Cabin crew training to control disruptive airline passenger behavior: a cause for tourism concern? *Tourism Management, 29*, 538–547.
Ribak, J., & Cline, B. (1995). Ground accidents. In J. Ribak, R. B. Rayman, & P. Froom (Eds.), *Occupational health in aviation* (pp. 201–205). San Diego, CA: Academic Press.
Ribeiro, A. C., Sousa, A. L., Duarte, J. P., & Frutuoso e Melo, P. F. (2016). Human reliability analysis of the Tokai-Mura accident through a THERP–CREAM and expert opinion auditing approach. *Safety Science, 87*, 269–279.
Riggle, J. D., Wadman, M. C., McCrory, B., Lowndes, B. R., Heald, E. A., Carstens, P. K., & Hallbeck, M. S. (2014). Task analysis method for procedural training curriculum development. *Perspectives on Medical Education, 3*, 204–218.
Roberts, A. P., & Stanton, N. A. (2018). Macrocognition in submarine command and control: a comparison of three simulated operational scenarios. *Journal of Applied Research in Memory and Cognition, 7*, 92–105.
Roberts, A. P., Stanton, N. A., & Fay, D. (2017). Land ahoy! Understanding submarine command and control during the completion of inshore operations. *Human Factors, 59*, 1263–1288.

Roberts, K. H., Stout, S. K., & Halpern, J. J. (1994). Decision dynamics in two high reliability military organisations. *Management Science, 40*, 614–624.

Rodgers, M. D., & Duke, D. A. (1993). SATORI: Situation assessment through the re-creation of incidents. *Journal of Air Traffic Control, 35*, 10–14.

Rodgers, M. D., Mogford, R. H., & Mogford, L. S. (1998). *The relationship of sector characteristics to operational errors* (DOT/FAA/AM-98/14). Washington, DC: Federal Aviation Administration.

Roets, B., & Christiaens, J. (2019). Shift work, fatigue, and human error: an empirical analysis of railway traffic control. *Journal of Transportation Safety and Security, 11*, 207–224.

Rogers, S. P., Spiker, A., & Cicinelli, J. (1985). Luminance contrast requirements for legibility of symbols on computer generated map displays in aircraft cockpits. In R. S. Jensen & J. Adrion (Eds.), *Proceedings of the Third International Symposium on Aviation Psychology* (pp. 175–182). Columbus, OH: Ohio State University.

Rogers, W. A., Cabrera, E. F., Walker, N., Gilbert, D. K., & Fisk, A. D. (1996). A survey of automatic teller machine usage across the adult lifespan. *Human Factors, 38*, 156–166.

Rogers, W. A., & Fisk, A. D. (1997). ATM design and training issues. *Ergonomics in Design, 5*, 4–9.

Rogers, W. A., Fisk, A. D., Mead, S. E., Walker, N., & Cabrera, E. F. (1996). Training older adults to use automated teller machines. *Human Factors, 38*, 425–433.

Roitsch, P. A., Babcock, G. L., & Edmunds, W. W. (1978). *Human factors report on the Tenerife accident.* Washington, DC: Air Line Pilots Association (ALPA) Engineering and Air Safety Department.

Rollenhagen, C., Alm, H., & Karlsson, K. H. (2017). Experience feedback from in-depth event investigations: how to find and implement efficient remedial actions. *Safety Science, 99*, 71–79.

Romer, H., Haastrup, P., & Petersen, H.J.S. (1995). Accidents during marine transport of dangerous goods: distribution of fatalities. *Journal of Loss Prevention in Process Industries, 8*, 29–34.

Roscoe, S. N. (1982). Landing airplanes, detecting traffic, and the dark focus. *Aviation, Space, and Environmental Medicine, 53*, 970–976.

Roscoe, S. N., Johnson, S. L., & Williges, R. C. (1980). Display motion relationships. In S. N. Roscoe (Ed.), *Aviation psychology* (pp. 68–81). Ames, IA: Iowa State University Press.

Roscoe, S. N., & Williges, R. C. (1975). Motion relationships in aircraft attitude and guidance displays: a flight experiment. *Human Factors, 17*, 374–387.

Rose, J., Bearman, C., & Dorrian, J. (2018). The Low-Event Task Subjective Situation Awareness (LETSSA) technique: development and evaluation of a new subjective measure of situation awareness. *Applied Ergonomics, 68*, 273–282.

Rose, J. A., & Bearman, C. (2012). Making effective use of task analysis to identify human factors issues in new rail technology. *Applied Ergonomics, 43*, 614–624.

Rostykus, W. G., Ip, W., & Dustin, J. A. (2016). Managing ergonomics: applying ISO 45001 as a Model. *Professional Safety, 61*, 34–42.

Roth, E. M., & Mumaw, R. J. (1995). Using cognitive task analysis to define human interface requirements for first-of-a-kind systems. *Proceedings of the Human Factors and Ergonomics Society 39th Annual Meeting* (pp. 520–524). San Diego, CA: Human Factors and Ergonomics Society.

Roth, E. M., & Woods, D. D. (1989). Cognitive task analysis: an approach to knowledge acquisition for intelligent systems design. In G. Guida & C. Tasso (Eds.), *Topics in expert system design* (pp. 233–264). Amsterdam: Elsevier.

Roth, E. M., Woods, D. D., & Pope, H. E. (1992). Cognitive simulations as a tool for cognitive task analysis. *Ergonomics, 35*, 1163–1198.

Rothschild, J. M., Hurley, A. C., Landrigan, C. P., Cronin, J. W., Martell-Waldrop, K., Foskett, C., Burdick, E., Czeiler, C. A., & Bates, D. W. (2006). Recovery from medical errors: the critical care nursing safety net. *Joint Commission Journal on Quality and Patient Safety, 32,* 63–72.

Rouse, W. B., & Boff, K. R. (1997). Assessing cost-benefits of human factors. In G. Salvendy (Ed.), *Handbook of human factors and ergonomics* (pp. 1617–1634). New York: John Wiley.

Rouse, W. B., Cannon-Bowers, J. A., & Salas, E. (1992). The role of mental models in team performance in complex systems. *IEEE Transactions on Systems, Man, and Cybernetics, 22,* 1296–1308.

Rose, J., Bearman, C., Naweed, A., & Dorrian, J. (2019). Proceed with caution: using verbal protocol analysis to measure situation awareness. *Ergonomics, 62,* 115–127.

Rowe, A. L., Cooke, N. J., Hall, E. P., & Halgren, T. L. (1996). Toward an on-line knowledge assessment methodology: building on the relationship between knowing and doing. *Journal of Experimental Psychology: Applied, 2,* 31–47.

Rowe, W. D. (1977). *An anatomy of risk.* New York: Wiley.

Rubio, S., Díaz, E., Martín, J., & Puente, J. M. (2004). Evaluation of subjective mental workload: a comparison of SWAT, NASA-TLX, and workload profile methods. *Applied Psychology, 53,* 61–86.

Ruffell-Smith, H. P. (1979). *A simulator study of the interaction of pilot workload with errors* (NASA TM-78482). Moffett Field, CA: NASA-Ames Research Centre.

Runciman, W. B., Sellen, A., Webb, R. K., Williamson, J. A., Currie, M., Morgan, C., & Russell, W. J. (1993). Errors, incidents and accidents in anaesthetic practice. *Anaesthesia and Intensive Care, 21,* 506–519.

Rundmo, T., & Hale, A. R. (2003). Managers' attitudes towards safety and accident prevention. *Safety Science, 41,* 557–574.

Russ, A. L., Militello, L. G., Glassman, P. A., Arthur, K. J., Zillich, A. J., & Weiner, M. (2019). Adapting cognitive task analysis to investigate clinical decision making and medication safety incidents. *Journal of Patient Safety, 15,* 191–197.

Russo, M. B. (2007). Recommendations for the ethical use of pharmacologic fatigue countermeasures in the US military. *Aviation, Space, and Environmental Medicine, 78,* B119–B127.

Rydenfält, C., Johansson, G., Odenrick, P., Åkerman, K., & Larsson, P. A. (2013). Compliance with the WHO Surgical Safety Checklist: deviations and possible improvements. *International Journal for Quality in Health Care, 25,* 182–187.

Saad, M., Abdel-Aty, M., Lee, J., & Wang, L. (2019). Integrated safety and operational analysis of the access design of managed toll lanes. *Transportation Research Record, 2673,* 127–136.

Sage, A. P. (1995). *Systems Management.* New York: John Wiley.

Sakurahara, T., Mohaghegh, Z., Reihani, S., Kee, E., Brandyberry, M., & Rodgers, S. (2018). An integrated methodology for spatio-temporal incorporation of underlying failure mechanisms into fire probabilistic risk assessment of nuclear power plants. *Reliability Engineering and System Safety, 169,* 242–257.

Salas, E., Burke, C. S., Bowers, C. A., & Wilson, K. A. (2001). Team training in the skies: does crew resource management (CRM) training work? *Human Factors, 43,* 641–674.

Salas, E., Cooke, N. J., & Rosen, M. A. (2008). On teams, teamwork, and team performance: Discoveries and developments. *Human Factors, 50,* 540–547.

Salas, E., Dickinson, T. L., Converse, S. A., & Tannenbaum, S. I. (1992). Toward an understanding of team performance and training. In R. W. Swezey and E. Salas (Eds.), *Teams: Their training and performance* (pp. 3–29). Norwood, NJ: Ablex.

Salas, E., Morgan, B. B., & Glickman, A.S. (1987). The evolution and maturation of operational teams in a training environment. *Proceedings of the Human Factors Society – 31st Annual Meeting – 1987* (pp. 82–86). Santa Monica, CA: Human Factors Society.

Salas, E., Rozell, D., Mullen, B., & Driskell, J. E. (1999). The effect of team building on performance: an integration. *Small Group Research, 30*, 309–329.

Salas, E., Wilson, K. A., Burke, C. S., Wightman, D. C., & Howse, W. R. (2006). A checklist for crew resource management training. *Ergonomics in Design, 14*, 6–15.

Saleh, J. H., & Marais, K. (2006). Reliability: how much is it worth? Beyond its estimation or prediction, the (net) present value of reliability. *Reliability Engineering and System Safety, 91*, 665–673.

Salge, M., & Milling, P. M. (2006). Who is to blame, the operator or the designer? Two stages of human failure in the Chernobyl accident. *System Dynamics Review: The Journal of the System Dynamics Society, 22*, 89–112.

Salguero-Caparrós, F., Pardo-Ferreira, M. C., Martínez-Rojas, M., & Rubio-Romero, J. C. (2020). Management of legal compliance in occupational health and safety. A literature review. *Safety Science, 121*, 111–118.

Sallinen, M., Sihvola, M., Puttonen, S., Ketola, K., Tuori, A., Härmä, M., Kecklund, G., & Åkerstedt, T. (2017). Sleep, alertness and alertness management among commercial airline pilots on short-haul and long-haul flights. *Accident Analysis and Prevention, 98*, 320–329.

Salmon, P., Stanton, N., Walker, G., & Green, D. (2006). Situation awareness measurement: a review of applicability for C4i environments. *Applied Ergonomics, 37*, 225–238.

Salmon, P. M., Hulme, A., Walker, G. H., Waterson, P., Berber, E., & Stanton, N. A. (2020). The big picture on accident causation: a review, synthesis and meta-analysis of AcciMap studies. *Safety Science, 126*, 104650.

Salvendy, G. (1997). *Handbook of human factors* (2nd ed.). New York: John Wiley.

Samel, A., & Wegmann, H.-M. (1988). Circadian rhythm, sleep, fatigue in aircrews operating long-haul routes. In R. S. Jensen (Ed.), *Aviation psychology* (pp. 404–422). Aldershot, UK: Gower.

Sams, T. (1987). Cockpit resource management concepts and training strategies. In R. S. Jensen (Ed.), *Proceedings of the Fourth International Symposium on Aviation Psychology* (pp. 360–371). Columbus, OH: Ohio State University Press.

Sanders, M. S., & McCormick, E. J. (1993). *Human factors in engineering and design.* New York: McGraw-Hill.

Sandilands, B., & Pascoe, R. (1997, March). TAAATs a year away from implementation. *Australian Aviation*, 28–30.

Santos-Reyes, J., & Beard, A. N. (2017). An analysis of the emergency response system of the 1996 Channel tunnel fire. *Tunnelling and Underground Space Technology, 65*, 121–139.

Sarter, N. (2008). Investigating mode errors on automated flight decks: illustrating the problem-driven, cumulative, and interdisciplinary nature of human factors research. *Human Factors, 50*, 506–510.

Sarter, N. B., & Woods, D. D. (1995). How in the world did we ever get into that mode? Mode error and awareness in supervisory control. *Human Factors, 37*, 5–19.

Saunders, F. C., Gale, A. W., & Sherry, A. H. (2015). Conceptualising uncertainty in safety-critical projects: a practitioner perspective. *International Journal of Project Management, 33*, 467–478.

Saus, E. R., Johnsen, B. H., Eid, J., & Thayer, J. F. (2012). Who benefits from simulator training: personality and heart rate variability in relation to situation awareness during navigation training. *Computers in Human Behavior, 28*, 1262–1268.

Savelsbergh, G. J., Van der Kamp, J., Williams, A. M., & Ward, P. (2005). Anticipation and visual search behaviour in expert soccer goalkeepers. *Ergonomics, 48*, 1686–1697.

Sawatzky, S. (2017). Worker fatigue: understanding the risks in the workplace. *Professional Safety, 62*, 45–51.

Scerbo, M. W., Britt, R. C., & Stefanidis, D. (2017). Differences in mental workload between traditional and single-incision laparoscopic procedures measured with a secondary task. *American Journal of Surgery, 213*, 244–248.

Schaafstal, A., Schraagen, J. M., & Van Berl, M. (2000). Cognitive task analysis and innovation of training: the case of structured troubleshooting. *Human Factors, 42*, 75–86.

Schaeffer, N. E. (2012). The role of human factors in the design and development of an insulin pump. *Journal of Diabetes Science and Technology, 6*, 260–264.

Schneider, W. (1985). Training high-performance skills: fallacies and guidelines. *Human Factors, 27*, 285–300.

Schnittker, R., Marshall, S. D., Horberry, T., & Young, K. (2019). Decision-centred design in healthcare: the process of identifying a decision support tool for airway management. *Applied Ergonomics, 77*, 70–82.

Schriver, A. T., Morrow, D. G., Wickens, C. D., & Talleur, D. A. (2008). Expertise differences in attentional strategies related to pilot decision making. *Human Factors, 50*, 864–878.

Schwarz, C., Gaspar, J., & Brown, T. (2019). The effect of reliability on drivers' trust and behavior in conditional automation. *Cognition, Technology and Work, 21*, 41–54.

Seager, L., Smith, D. W., Patel, A., Brunt, H., & Brennan, P. A. (2013). Applying aviation factors to oral and maxillofacial surgery: the human element. *British Journal of Oral and Maxillofacial Surgery, 51*, 8–13.

Seamster, T. L., Boehm-Davis, D. A., Holt, R. W., & Schultz, K. (1998). *Developing advanced crew resource management (ACRM) training: A training manual.* Washington, DC: Federal Aviation Administration.

Seamster, T. L., Redding, R. E., & Kaempf, G. L. (1997). *Applied cognitive task analysis in aviation.* Aldershot, UK: Avebury Technical.

Sedbon, G. (1994). Air Inter grounds pilot for error. *Flight International* (12–18 January), 10.

Segall, N., Bonifacio, A. S., Schroeder, R. A., Barbeito, A., Rogers, D., Thornlow, D. K., ... & Mark, J. B. (2012). Can we make postoperative patient handovers safer? A systematic review of the literature. *Anesthesia and Analgesia, 115*, 102–115.

Seppelt, B. D., & Lee, J. D. (2019). Keeping the driver in the loop: Dynamic feedback to support appropriate use of imperfect vehicle control automation. *International Journal of Human–Computer Studies, 125*, 66–80.

Sexton, J. B., Helmreich, R. L., Neilands, T. B., Rowan, K., Vella, K., Boyden, J., Roberts, P. R., & Thomas, E. J. (2006). The Safety Attitudes Questionnaire: psychometric properties, benchmarking data, and emerging research. *BMC Health Services Research, 6*, 44.

Shadbolt, N., & Burton, M. (1995). Knowledge elicitation: a systemic approach. In J. R. Wilson & E. N. Corlett (Eds.), *Evaluation of human work: A practical ergonomics methodology* (pp. 406–444). London: Taylor & Francis.

Shanteau, J. (1988). Psychological characteristics and strategies of expert decision-makers. *Acta Psychologia, 68*, 203–215.

Sharma, C., Bhavsar, P., Srinivasan, B., & Srinivasan, R. (2016). Eye gaze movement studies of control room operators: a novel approach to improve process safety. *Computers and Chemical Engineering, 85*, 43–57.

Shaud, J. A. (1989). Aircraft coordination training in the U.S. Air Force Training Command. *Aviation, Space and Environmental Medicine, 60*, 601–602.

Shaw, E. P., Rietschel, J. C., Hendershot, B. D., Pruziner, A. L., Miller, M. W., Hatfield, B. D., & Gentili, R. J. (2018). Measurement of attentional reserve and mental effort for cognitive workload assessment under various task demands during dual-task walking. *Biological Psychology, 134*, 39–51.

Shiffrin, R. M., & Schneider, W. (1977). Controlled and automatic human information processing: II. Perceptual learning, automatic attending and general theory. *Psychological Review, 84*, 155–171.

Shikdar, A. A., & Sawaqed, N. M. (2003). Worker productivity, and occupational health and safety issues in selected industries. *Computers and Industrial Engineering, 45*, 563–572.

Shorrock, S. T., & Kirwan, B. (2002). Development and application of a human error identification tool for air traffic control. *Applied Ergonomics, 33*, 319–336.

Silva, T., Cunha, M. P. E., Clegg, S. R., Neves, P., Rego, A., & Rodrigues, R. A. (2014). Smells like team spirit: opening a paradoxical black box. *Human Relations, 67*, 287–310.

Silva, T. R., Hak, J. L., Winckler, M. A. A., & Nicolas, O. (2017). A comparative study of milestones for featuring GUI prototyping tools. *Journal of Software Engineering and Applications, 10*, 564–589.

Simon, A. F., Fagley, N. S., & Halleran, J. G. (2004). Decision framing: moderating effects of individual differences and cognitive processing. *Journal of Behavioral Decision Making, 17*, 77–93.

Simpson, G., & Mason, S. (1995). Economic analysis in ergonomics. In J. R. Wilson & E. N. Corlett (Eds.), *Evaluation of human work: A practical ergonomics methodology* (pp. 1017–1037). London: Taylor & Francis.

Singh, H., Naik, A. D., Rao, R., & Petersen, L. A. (2008). Reducing diagnostic errors through effective communication: harnessing the power of information technology. *Journal of General Internal Medicine, 23*, 489–494.

Singh, I. L., Molloy, R., & Parasuraman, R. (1993). Automation-induced "complacency": Development of the complacency-potential rating scale. *International Journal of Aviation Psychology, 3*, 111–122.

Singleton, W. T. (1984). Application of human error analysis to occupational accident research. *Journal of Occupational Accidents, 6*, 107–115.

Skaugset, L. M., Farrell, S., Carney, M., Wolff, M., Santen, S. A., Perry, M., & Cico, S. J. (2016). Can you multitask? Evidence and limitations of task switching and multitasking in emergency medicine. *Annals of Emergency Medicine, 68*, 189–195.

Skinner, B. F. (1974). *About behaviourism*. London: Cape.

Slovic, P., Griffin, D., & Tversky, A. (1990). Compatibility effects and judgement choice. In R. M. Hogarth (Ed.), *Insights in decision making* (pp. 5–27). Chicago, IL: University of Chicago Press.

Small, A. J., Wiggins, M. W., & Loveday, T. (2014). Cue-based processing capacity, cognitive load and the completion of simulated short-duration vigilance tasks in power transmission control. *Applied Cognitive Psychology, 28*, 481–487.

Smith, A. F., & Plunkett, E. (2019). People, systems and safety: resilience and excellence in healthcare practice. *Anaesthesia, 74*, 508–517.

Smith, C. L., & Borgonovo, E. (2007). Decision making during nuclear power plant incidents: a new approach to the evaluation of precursor events. *Risk Analysis: An International Journal, 27*, 1027–1042.

Smith, G. M. (1993). Self-analysis of LOFT as a strategy for learning CRM in undergraduate flight training. In R. S. Jensen (Ed.), *Proceedings of the Seventh International Symposium on Aviation Psychology* (pp. 533–537). Columbus, OH: Ohio State University.

Smith, G. M., & Hanebuth, C. E. (1995). Differences in CRM evaluation between LOFT facilitators and line check captains. In B. J. Hayward & A. R. Lowe (Eds.), *Applied aviation psychology: Achievement, change and challenge* (pp. 148–158). Aldershot, UK: Ashgate.

References

Smith, J. D., Ellis, S. R., & Lee, E. C. (1984). Perceived threat and avoidance manoeuvres in response to cockpit displays. *Human Factors, 26*, 33–48.

Somech, A. (2003). Relationships of participative leadership with relational demography variables: a multi-level perspective. *Journal of Organizational Behavior, 24*, 1003–1018.

Somech, A. (2005). Directive versus participative leadership: two complementary approaches to managing school effectiveness. *Educational Administration Quarterly, 41*, 777–800.

Sorensen, L. J., & Stanton, N. A. (2016). Keeping it together: The role of transactional situation awareness in team performance. *International Journal of Industrial Ergonomics, 53*, 267–273.

Spiess, B. D. (2013). The use of checklists as a method to reduce human error in cardiac operating rooms. *International Anesthesiology Clinics, 51*(1), 179–194.

Squires, M. A. E., Tourangeau, A., Spence Laschinger, H. K., & Doran, D. (2010). The link between leadership and safety outcomes in hospitals. *Journal of Nursing Management, 18*, 914–925.

Stack, D. W., & Thomas, W. R. (1991). Modelling issues associated with production reactor safety assessment. In G. Apostolakis (Ed.), *Probabilistic safety assessment and management* (Vol. 1, pp. 79–84). New York: Elsevier.

Staff, T., Eken, T., Hansen, T. B., Steen, P. A., & Søvik, S. (2012). A field evaluation of real-life motor vehicle accidents: presence of unrestrained objects and their association with distribution and severity of patient injuries. *Accident Analysis and Prevention, 45*, 529–538.

Stanney, K. M., Maxey, J., & Salvendy, G. (1997). Socially centred design. In G. Salvendy (Ed.), *Handbook of human factors and ergonomics* (pp. 637–656). New York: John Wiley.

Stanton, N., Salmon, P. M., & Rafferty, L. A. (2013). *Human factors methods: a practical guide for engineering and design*. Aldershot, UK: Ashgate.

Stanton, N. A. (2006). Hierarchical task analysis: developments, applications, and extensions. *Applied Ergonomics, 37*, 55–79.

Stanton, N. A. (2014). Representing distributed cognition in complex systems: how a submarine returns to periscope depth. *Ergonomics, 57*, 403–418.

Stanton, N. A., Chambers, P. R., & Piggott, J. (2001). Situational awareness and safety. *Safety Science, 39*, 189–204.

Stanton, N. A., & Harvey, C. (2017). Beyond human error taxonomies in assessment of risk in sociotechnical systems: a new paradigm with the EAST 'broken-links' approach. *Ergonomics, 60*, 221–233.

Stanton, N. A., Salmon, P. M., Walker, G. H., Salas, E., & Hancock, P. A. (2017). State-of-science: situation awareness in individuals, teams and systems. *Ergonomics, 60*, 449–466.

Stanton, N. A., & Young, M. S. (1998). Vehicle automation and driving performance. *Ergonomics, 41*, 1014–1028.

Stanton, N. A., & Young, M. S. (2003). Giving ergonomics away? The application of ergonomics methods by novices. *Applied Ergonomics, 34*, 479–490.

Staunton, J. H. (1996). *Commission of inquiry into the relations between the CAA and Seaview Air*. Canberra: Australian Government Publishing.

Steege, L. M., Pinekenstein, B. J., Rainbow, J. G., & Knudsen, É. A. (2017). Addressing occupational fatigue in nurses: current state of fatigue risk management in hospitals, Part 2. *Journal of Nursing Administration, 47*, 484–490.

Stephenson, S. (2010). The revolution in military affairs: 12 observations on an out-of-fashion idea. *Military Review, 90*, 38.

Sternberg, R. J. (1977). *Intelligence, information process, and analogical reasoning*. Hillsdale, NJ: Lawrence Erlbaum.

Stewart, J. P. (1993). System approaches to risk management focuses resources on most serious hazards. *ICAO Journal, 48(9),* 12–13.

Stewart, S. (1992). *Emergency: Crisis on the flightdeck.* Shrewsbury, UK: Airlife.

Stimson, W. A. (1998). *Beyond ISO 9000: How to sustain quality in a dynamic world.* New York: Amacom.

Stogdill, R. M. (1974). *Handbook of leadership: A survey of the literature.* New York: Free Press.

Stokes, A. F., & Wickens, C. D. (1988). Aviation displays. In E. L. Wiener, & D. C. Nagel (Eds.), *Human factors in aviation* (pp. 387–431). San Diego, CA: Academic Press.

Stolzer, A. J., Friend, M. A., Truong, D., Tuccio, W. A., & Aguiar, M. (2018). Measuring and evaluating safety management system effectiveness using data envelopment analysis. *Safety Science, 104,* 55–69.

Stoop, J., de Kroes, J., & Hale, A. (2017). Safety science, a founding fathers' retrospective. *Safety Science, 94,* 103–115.

Strand, G. O., & Lundteigen, M. A. (2016). Human factors modelling in offshore drilling operations. *Journal of Loss Prevention in the Process Industries, 43,* 654–667.

Strater, O., & Bubb, H. (1999). Assessment of human reliability based on evaluation of plant experience: requirements and implementation. *Reliability Engineering and System Safety, 63,* 199–219.

Strauch, B. (2017). The automation-by-expertise-by-training interaction: why automation-related accidents continue to occur in sociotechnical systems. *Human Factors, 59*(2), 204–228.

Strayer, D. L., Turrill, J., Cooper, J. M., Coleman, J. R., Medeiros-Ward, N., & Biondi, F. (2015). Assessing cognitive distraction in the automobile. *Human Factors, 57,* 1300–1324.

Sturman, D., Wiggins, M. W., Auton, J., Loft, S., Helton, W., Westbrook, J., & Braithwaite, J. (2019). Control room operators' cue utilization predicts cognitive resource consumption during regular operational tasks. *Frontiers in Psychology, 10,* 1967.

Sujan, M. A., Furniss, D., Anderson, J., Braithwaite, J., & Hollnagel, E. (2019). Resilient health care as the basis for teaching patient safety: a Safety-II critique of the World Health Organisation patient safety curriculum. *Safety Science, 118,* 15–21.

Sun, X., Houssin, R., Renaud, J., & Gardoni, M. (2019). A review of methodologies for integrating human factors and ergonomics in engineering design. *International Journal of Production Research, 57*(15–16), 4961–4976.

Sun, Z., Li, Z., Gong, E., & Xie, H. (2012). Estimating human error probability using a modified CREAM. *Reliability Engineering and System Safety, 100,* 28–32.

Sundar, E., Sundar, S., Pawlowski, J., Blum, R., Feinstein, D., & Pratt, S. (2007). Crew resource management and team training. *Anesthesiology Clinics, 25,* 283–300.

Sutherland, A., Canobbio, M., Clarke, J., Randall, M., Skelland, T., & Weston, E. (2020). Incidence and prevalence of intravenous medication errors in the UK: a systematic review. *European Journal of Hospital Pharmacy, 27,* 3–8.

Sutton, G., Liao, J., Jimmieson, N. L., & Restubog, S. L. D. (2011). Measuring multidisciplinary team effectiveness in a ward-based healthcare setting: development of the team functioning assessment tool. *Journal for Healthcare Quality, 33,* 10–24.

Sutton, G., Liao, J., Jimmieson, N. L., & Restubog, S. L. (2013). Measuring ward-based multidisciplinary healthcare team functioning: a validation study of the team functioning assessment tool (TFAT). *Journal for Healthcare Quality, 35,* 36–49.

Svedung, I., & Rasmussen, J. (2002). Graphic representation of accident scenarios: mapping system structure and the causation of accidents. *Safety Science, 40,* 397–417.

Svenson, O. (1979). Process descriptions of decision-making. *Organisational Behaviour and Human Performance, 23,* 86–112.

Swain, A. D., & Guttman, H. E. (1983). *Handbook of human reliability analysis with emphasis on nuclear power plant operations* (NUREG/CR-1278). Washington, DC: United States Nuclear Regulatory Commission.

Swedish Board of Accident Investigation. (1992). *Concerning the crash-worthiness at the accident on a field at Gottröra, 17km Northeast of Arlanda, 27th December, 1991.* Stockholm: Author.

Sweller, J. (2016). Working memory, long-term memory, and instructional design. *Journal of Applied Research in Memory and Cognition, 5*, 360–367.

Sweller, J., & Chandler, P. (1991). Evidence for cognitive load theory. *Cognition and Instruction, 8*, 351–362.

Sweller, J., & Paas, F. (2017). Should self-regulated learning be integrated with cognitive load theory? A commentary. *Learning and Instruction, 51*, 85–89.

Swetonic, M. M. (1993). The death of the asbestos industry. In J. A. Gottschalk (Ed.), *Crisis* (pp. 289–308). Detroit, MI: Visible Ink.

Swuste, P., Groeneweg, J., Van Gulijk, C., Zwaard, W., & Lemkowitz, S. (2018). Safety management systems from Three Mile Island to Piper Alpha, a review in English and Dutch literature for the period 1979 to 1988. *Safety Science, 107*, 224–244.

Swuste, P., van Gulijk, C., Groeneweg, J., Zwaard, W., Lemkowitz, S., & Guldenmund, F. (2020). From Clapham Junction to Macondo, Deepwater Horizon: risk and safety management in high-tech-high-hazard sectors. A review of English and Dutch literature: 1988–2010. *Safety Science, 121*, 249–282.

Takano, K., Sasou, K., & Yoshimura, S. (1997). Structure of operator's mental models in coping with anomalies occurring in nuclear power plants. *International Journal of Human-Computer Studies, 47*, 767–789.

Talluri, B. C., Urai, A. E., Tsetsos, K., Usher, M., & Donner, T. H. (2018). Confirmation bias through selective overweighting of choice-consistent evidence. *Current Biology, 28*, 3128–313.

Tan, W. S., Liu, D., & Bishu, R. (2009). Web evaluation: heuristic evaluation vs. user testing. *International Journal of Industrial Ergonomics, 39*, 621–627.

Tannenbaum, S. I., Beard, R. L., & Salas, E. (1992). Team building and its influence on team developments. In K. Kelley (Ed.) *Issues, theory, and research in psychology* (pp. 117–153). Amsterdam: Elsevier.

Tattersall, A. J., & Hockey, G. R. J. (1995). Level of operator control and changes in heart rate variability during simulated flight maintenance. *Human Factors, 37*, 682–698.

Taylor, B. H., Charlton, S. G., & Canham, L. S. (1996). Operability testing in command and control systems. In T. G. O'Brien & S. G. Charlton (Eds.), *Handbook of human factors testing and evaluation* (pp. 301–311). Mahwah, NJ: Lawrence Erlbaum.

Taylor, J. W., & Munson, K. (1975). *History of aviation.* London: New English Library.

Tear, M. J., Reader, T. W., Shorrock, S., & Kirwan, B. (2020). Safety culture and power: interactions between perceptions of safety culture, organisational hierarchy, and national culture. *Safety Science, 121*, 550–561.

Telfer, R. (1988). Pilot decision-making and judgement. In R. S. Jensen (Ed.), *Aviation psychology* (pp 154–175). Aldershot, Uk: Gower.

Telfer, R. (1993, September). Teaching human factors to the ab-initio student pilot. *ICAO Journal*, 10–13.

Tenhundfeld, N. L., de Visser, E. J., Haring, K. S., Ries, A. J., Finomore, V. S., & Tossell, C. C. (2019). Calibrating trust in automation through familiarity with the autoparking feature of a Tesla Model X. *Journal of Cognitive Engineering and Decision Making, 13*, 279–294.

Teperi, A. M., Puro, V., & Ratilainen, H. (2017). Applying a new human factor tool in the nuclear energy industry. *Safety Science*, *95*, 125–139.

Thatcher, A., Nayak, R., & Waterson, P. (2020). Human factors and ergonomics systems-based tools for understanding and addressing global problems of the twenty-first century. *Ergonomics*, *63*, 367–387.

Theophilus, S. C., Esenowo, V. N., Arewa, A. O., Ifelebuegu, A. O., Nnadi, E. O., & Mbanaso, F. U. (2017). Human factors analysis and classification system for the oil and gas industry (HFACS-OGI). *Reliability Engineering and System Safety*, *167*, 168–176.

Theurel, J., Desbrosses, K., Roux, T., & Savescu, A. (2018). Physiological consequences of using an upper limb exoskeleton during manual handling tasks. *Applied Ergonomics*, *67*, 211–217.

Thomas, M. J. (2004). Predictors of threat and error management: identification of core non-technical skills and implications for training systems design. *International Journal of Aviation Psychology*, *14*, 207–231.

Thompson, D. A. (1995, July). When is a 'warning' not a warning. *Ergonomics in Design*, 25–28.

Thompson, R. C., Agen, R. A., & Broach, D. M. (1998). *Differential training needs and abilities at air traffic control towers: Should all controllers be trained equally?* (DOT/FAA/AM-98/8). Washington, DC: Federal Aviation Administration.

Thomson, D. R., Besner, D., & Smilek, D. (2015). A resource-control account of sustained attention: evidence from mind-wandering and vigilance paradigms. *Perspectives on Psychological Science*, *10*, 82–96.

Thomson, D. R., Besner, D., & Smilek, D. (2016). A critical examination of the evidence for sensitivity loss in modern vigilance tasks. *Psychological Review*, *123*, 70–83.

Thordsen, M. L. (1991). A comparison of two tools for cognitive task analysis: concept mapping and the critical decision method. *Proceedings of the Human Factors Society 35th Annual Meeting* (pp. 283–285). Santa Monica, CA: HFES.

Thorogood, J., & Crichton, M. T. (2014). Threat-and-error management: the connection between process safety and practical action at the worksite. *SPE Drilling and Completion*, *29*, 465–472.

Thorsden, M., Militello, L., & Klein, G. (1993). Determining the decision requirements of complex flight crew tasks. In R. S. Jensen (Ed.), *Proceedings of the 7th Symposium International Symposium on Aviation Psychology* (pp 233–237). Columbus, OH: Ohio State University.

Tindall-Ford, S., Chandler, P., & Sweller, J. (1997). When two sensory modes are better than one. *Journal of Experimental Psychology: Applied*, *3*, 257–287.

Tofel-Grehl, C., & Feldon, D. F. (2013). Cognitive task analysis–based training: a meta-analysis of studies. *Journal of Cognitive Engineering and Decision Making*, *7*, 293–304.

Toff, N. J. (2010). Human factors in anaesthesia: lessons from aviation. *British Journal of Anaesthesia*, *105*, 21–25.

Tognazinni, B. (1992, September). Violence in software. *Sunworld*, 105–107.

Torrance, E. P., Rush, C. H., Kohn, H. B., & Doughty, J. M. (1957). *Factors in fighter-interceptor pilot combat effectiveness* (Technical Report AFPTRC-TR-57-11, ASTIA Document No. AD 146407). San Antonio, TX: Lackland Air Force Base, Air Force Personnel and Training Research Center.

Transport Accident Investigation Commission (1980) *Air New Zealand McDonnell-Douglas DC10-30 ZK-NZP, Ross Island, Antarctica, 28 November 1979, Report 79–139*. Wellington: Author.

Trouche, E., Johansson, P., Hall, L., & Mercier, H. (2016). The selective laziness of reasoning. *Cognitive Science*, *40*, 2122–2136.

Troussier, B., Davoine, P., De Gaudemaris, R., Fauconnier, J., & Phelip, X. (1994). Back pain in school children. A study among 1178 pupils. *Scandinavian Journal of Rehabilitation Medicine*, *26*, 143–146.

Truta, T. S., Boeriu, C. M., Copotoiu, S. M., Petrisor, M., Turucz, E., Vatau, D., & Lazarovici, M. (2018). Improving nontechnical skills of an interprofessional emergency medical team through a one day crisis resource management training. *Medicine*, *97*, 32.

Tsai, H. Y. S., Jiang, M., Alhabash, S., LaRose, R., Rifon, N. J., & Cotten, S. R. (2016). Understanding online safety behaviors: a protection motivation theory perspective. *Computers and Security*, *59*, 138–150.

Tuckerman, B. W. (1965). Developmental sequences in small groups. *Psychological Bulletin*, *63*, 384–399.

Turkelson, C., Aebersold, M., Redman, R., & Tschannen, D. (2017). Improving nursing communication skills in an intensive care unit using simulation and nursing crew resource management strategies. *Journal of Nursing Care Quality*, *32*, 331–339.

Tversky, A., & Fox, C. R. (1995). Weighing risk and uncertainty. *Psychological Review*, *102*, 269–283.

Tversky, A., & Kahneman, D. (1973). Availability: a heuristic for judging frequency and probability. *Cognitive Psychology*, *5*, 207–232.

Tversky, A., & Kahneman, D. (1981). The framing of decisions and the psychology of choice. *Science*, *211*, 453–458.

Tversky, A., & Kahneman, D. (1983). Extensional versus intuitive reasoning: the conjunction fallacy in probability judgement. *Psychological Review*, *90*, 293–315.

Tyson, T. (1989). *Working with groups*. Melbourne: Macmillan Company.

United States Air Force. (1995). *Accident investigation report of the B-52 mishap at Fairchild Air Force Base on 24 June, 1994* (AFR 110–14). Washington, DC: Author.

United States House of Representatives. (1986). *Strategy for safely returning space shuttle to flight status: Hearing before the Subcommittee on Space Science and Applications of the Committee on Science and Technology, House of Representatives, Ninety-ninth Congress, second session*. Washington, DC: Author.

van Dalen, A. S. H. M., Legemaate, J., Schlack, W. S., Legemate, D. A., & Schijven, M. P. (2019). Legal perspectives on black box recording devices in the operating environment. *British Journal of Surgery*, *106*, 14331441.

van Dyck, C., Frese, M., Baer, M., & Sonnentag, S. (2005). Organizational error management culture and its impact on performance: a two-study replication. *Journal of Applied Psychology*, *90*, 1228–1240.

Ventikos, N. P., Stavrou, D. I., & Andritsopoulos, A. (2017). Studying the marine accidents of the Aegean Sea: critical review, analysis and results. *Journal of Marine Engineering and Technology*, *16*, 103–113.

Verwey, W. B., & Veltman, H. A. (1996). Detecting short periods of elevated workload: a comparison of nine workload assessment techniques. *Journal of Experimental Psychology: Applied*, *2*, 270–285.

Verwey, W. B., & Zaidel, D. M. (1999). Preventing drowsiness accidents by an alertness maintenance device. *Accident Analysis and Prevention*, *31*, 199–211.

Vette, G., & McDonald, J. (1983). *Impact Erebus*. Auckland: Aviation Consultants.

Vicente, K. J. (1999). *Cognitive work analysis*. Mahwah, NJ: Lawrence Erlbaum.

Vicente, K. J., Burns, C. M., & Pawlak, W. S. (1997). Muddling through wicked design problems. *Ergonomics in Design*, *5*, 25–30.

Vicente, K. J., Roth, E. M., & Mumaw, R. J. (2001). How do operators monitor a complex, dynamic work domain? The impact of control room technology. *International Journal of Human–Computer Studies*, *54*, 831–856.

Vidmar, P., & Perkovič, M. (2018). Safety assessment of crude oil tankers. *Safety Science, 105*, 178–191.

Vincent, C. J., & Blandford, A. (2014). The challenges of delivering validated personas for medical equipment design. *Applied Ergonomics, 45*, 1097–1105.

Viner, D. (1996). *Accident analysis and risk control*. East Ivanhoe, AUS: Derek Viner.

Virzi, R.A. (1992). Refining the test phase of usability evaluation. How many subjects is enough? *Human Factors, 34*, 457–468.

Viscusi, W. K., & Masterman, C. (2017). Anchoring biases in international estimates of the value of a statistical life. *Journal of Risk and Uncertainty, 54*, 103–128.

Waag, W. L., & Houck, M. R. (1994). Tools for assessing situational awareness in an operational fighter environment. *Aviation, Space, and Environmental Medicine, 65(suppl. 5)*, 13–19.

Wadhera, R. K., Parker, S. H., Burkhart, H. M., Greason, K. L., Neal, J. R., Levenick, K. M., Wiegmann, D.A., & Sundt III, T. M. (2010). Is the 'sterile cockpit' concept applicable to cardiovascular surgery critical intervals or critical events? The impact of protocol-driven communication during cardiopulmonary bypass. *Journal of Thoracic and Cardiovascular Surgery, 139*, 312–319.

Wagenaar, W. A., & Groeneweg, J. (1987). Accidents at sea: multiple causes and impossible consequences. *International Journal of Man-Machine Studies, 27*, 587–598.

Wagennaar, W. A., & Groeneweg, J. (1988). Accidents at sea: multiple causes and impossible consequences. In E. Hollnagel, G. Mancini, & David D. Woods. (Eds.), *Cognitive engineering in complex dynamic worlds* (pp. 133–144). London: Academic Press.

Wagner, D., Birt, J. A., Snyder, M., & Duncanson, J. P. (1996). *Human factors design guide*. Atlantic City, VA: Federal Aviation Administration (NTIS DOT/FAA/CT-96/1).

Wahlström, B. (1991). Influence of organisation and management on human errors. In G. Apostolakis (Ed.), *Probabilistic safety assessment and management* (Vol. 1, pp. 519–524). New York: Elsevier.

Wahlström, B. (2018). Systemic thinking in support of safety management in nuclear power plants. *Safety Science, 109*, 201–218.

Walker, G. H., Gibson, H., Stanton, N. A., Baber, C., Salmon, P., & Green, D. (2006). Event analysis of systemic teamwork (EAST): a novel integration of ergonomics methods to analyse C4i activity. *Ergonomics, 49*, 1345–1369.

Walker, G. H., Stanton, N. A., Kazi, T. A., Salmon, P. M., & Jenkins, D. P. (2009). Does advanced driver training improve situational awareness?. *Applied ergonomics, 40*(4), 678–687.

Walsh, N. E., & Schwartz, K. (1990). The influence of prophylactic orthoses on abdominal strength and low back injury in the workplace. *American Journal of Physical Medicine and Rehabilitation, 69*, 245–250.

Walters, K. N., & Diab, D. L. (2016). Humble leadership: implications for psychological safety and follower engagement. *Journal of Leadership Studies, 10*, 7–18.

Wang, D., Gao, Q., Li, Z., Song, F., & Ma, L. (2017). Developing a taxonomy of coordination behaviours in nuclear power plant control rooms during emergencies. *Ergonomics, 60*, 1634–1652.

Wang, J. (2002). Offshore safety case approach and formal safety assessment of ships. *Journal of Safety Research, 33*, 81–115.

Wang, L., Wu, C., & Sun, R. (2014). An analysis of flight Quick Access Recorder (QAR) data and its applications in preventing landing incidents. *Reliability Engineering and System Safety, 127*, 86–96.

Wang, L., Zhang, J., Dong, C., Sun, H., & Ren, Y. (2019). A method of applying flight data to evaluate landing operation performance. *Ergonomics, 62*, 171–180.

References

Wang, Y. Y., Wan, Q. Q., Lin, F., Zhou, W. J., & Shang, S. M. (2018). Interventions to improve communication between nurses and physicians in the intensive care unit: an integrative literature review. *International Journal of Nursing Sciences*, 5, 81–88.

Warm, J. S., Parasuraman, R., & Matthews, G. (2008). Vigilance requires hard mental work and is stressful. *Human Factors*, 50, 433–441.

Waterson, P., Jenkins, D. P., Salmon, P. M., & Underwood, P. (2017). 'Remixing Rasmussen': the evolution of Accimaps within systemic accident analysis. *Applied Ergonomics*, 59, 483–503.

Wei, J., & Salvendy, G. (2004). The cognitive task analysis methods for job and task design: review and reappraisal. *Behaviour and Information Technology*, 23, 273–299.

Wei, Z. G., Macwan, A. P., & Wieinga, P. A. (1998). A quantitative measure for degree of automation and its relation to system performance and mental load. *Human Factors*, 40, 277–295.

Weigall, F. (2006). *Marine pilot transfers: A preliminary investigation of options* (Marine Safety Research Grant – 200667474). Canberra: Australian Transport Safety Bureau.

Westbrook, J. I., Rob, M. I., Woods, A., & Parry, D. (2011). Errors in the administration of intravenous medications in hospital and the role of correct procedures and nurse experience. *BMJ Quality and Safety*, 20, 1027–1034.

Westli, H. K., Johnsen, B. H., Eid, J., Rasten, I., & Brattebø, G. (2010). Teamwork skills, shared mental models, and performance in simulated trauma teams: an independent group design. *Scandinavian Journal of Trauma, Resuscitation and Emergency Medicine*, 18, 47.

Weyer, J. (2006). Modes of governance of hybrid systems. The mid-air collision at Uberlingen and the impact of smart technology. *Science, Technology and Innovation Studies*, 2, 127–149.

White, B. D., Rowles, J. M., Mumford, C. J., & Firth, J. L. (1990). A clinical survey of head injuries sustained in the M1 Boeing 737 disaster: recommendations to improve aircrash survival. *British Journal of Neurosurgery*, 4, 503–510.

White, M. R., Braund, H., Howes, D., Egan, R., Gegenfurtner, A., van Merrienboer, J. J., & Szulewski, A. (2018). Getting inside the expert's head: an analysis of physician cognitive processes during trauma resuscitations. *Annals of Emergency Medicine*, 72, 289–298.

White, R. M., & Churchill, E. (1971). *The body size of soldiers: U.S. Army Anthropometry – 1966* (Technical Report 72–51-CE). Natick, MA: US Army Natick Laboratories.

Wickens, C. D. (1987). Information processing, decision-making, and cognition. In G. Salvendy (Ed.). *Handbook of human factors* (pp. 72–107). New York: Wiley.

Wickens, C. D. (1991). Processing resources and attention. In D. L. Damos (Ed.), *Multiple-task performance* (pp. 3–34). London: Taylor & Francis.

Wickens, C. D. (2000). Human factors in vector map design: the importance of task-display dependence. *Journal of Navigation*, 53, 54–67.

Wickens, C. D. (2008). Multiple resources and mental workload. *Human Factors*, 50(3), 449–455.

Wickens, C. D., & Flach, J. M. (1988). Information processing. In E. L. Wiener & D. C. Nagel (Eds.), *Human factors in aviation* (pp. 111–155). San Diego, CA: Academic Press.

Wickens, C. D., & Kessel, C. (1980). The processing resource demands of failure detection in dynamic systems. *Journal of Experimental Psychology: Human Perception and Performance*, 6, 564–577.

Wickens, C. D., & Kramer, A. (1985). Engineering psychology. *Annual Review of Psychology*, 36, 307–348.

Wiegmann, D. A., Rich, A., & Zhang, H. (2001). Automated diagnostic aids: the effects of aid reliability on users' trust and reliance. *Theoretical Issues in Ergonomics Science, 2*, 352–367.

Wiegmann, D. A., & Shappell, S. A. (2003). *A human error approach to aviation accident analysis: The human factors analysis and classification system.* Aldershot, UK: Ashgate.

Wiener, E. L. (1988). Cockpit automation. In E. L. Wiener, & D. C. Nagel (Eds.), *Human factors in aviation* (pp. 433–461). San Diego, CA: Academic Press.

Wiener, E. L. (1989). *Human factors of advanced technology (glass cockpit) transport aircraft.* (Report No. 177528). Moffett Field, CA: NASA-Ames Research Centre.

Wiener, E. L., Chidester, T. R., Kanki, B. G., Palmer, E. A., Curry, R. E., & Gregorich, S. E. (1991). *The impact of cockpit automation on crew coordination and communication: I.* (Report No. NCC2–581). Moffett Field, CA: NASA-Ames Research Centre.

Wiener, E. L., Curry, R. E., & Faustina, M. L. (1984). Vigilance and task load: in search of the inverted U. *Human Factors, 26*, 215–222.

Wiggins, M. W., Griffin, B., & Brouwers, S. (2019). The potential role of context-related exposure in explaining differences in water safety cue utilization. *Human Factors, 61*, 825–838.

Wijers, A. A., Okita, T., Mulder, G., Mulder, L. J. M., Lorist, M. M., Poiesz, R., & Scheffers, M. K. (1987). Visual search and spatial attention: ERPs in focussed and divided attention conditions. *Biological Psychology, 25*, 33–60.

Willett, C., & Page, M. (1996). A survey of time budget pressures and irregular auditing practices among newly qualified UK chartered accountants. *British Accounting Review, 28*, 101–120.

Williams, J. C. (1988). A human factors database to influence safety and reliability. In B. A. Sayers (Ed.), *Proceedings of the Safety and Reliability Symposium* (pp. 223–24). London: Elsevier.

Williamson, A., & Friswell, R. (2013). Fatigue in the workplace: causes and countermeasures. *Fatigue: Biomedicine, Health and Behavior, 1*, 81–98.

Wilson, J. R. (1995). A framework and a context for ergonomics methodology. In J. R. Wilson & E. N. Corlett (Eds.), *Evaluation of human work: A practical ergonomics methodology* (pp. 1–40). London: Taylor & Francis.

Wilson, J. R., & Haines, H. H. (1997). Participatory ergonomics. In G. Salvendy (Ed.), *Handbook of human factors and ergonomics* (2nd ed., pp. 490–513). New York: Wiley.

Wilson, J. R., & Rajan, J. A. (1995). Human-machine interfaces for systems control. In J. R. Wilson & E. N. Corlett (Eds.), *Evaluation of human work: A practical ergonomics methodology* (pp. 357–404). London: Taylor & Francis.

Wilson, L. M. (1972). Intensive care delirium: the effect of outside deprivation in a windowless unit. *Archives of Internal Medicine, 130*, 225–226.

Wilson, M. D., Farrell, S., Visser, T. A., & Loft, S. (2018). Remembering to execute deferred tasks in simulated air traffic control: the impact of interruptions. *Journal of Experimental Psychology: Applied, 24*, 360–379.

Wilson, R. L. (1991). Survival factors investigation of Avianca Flight 052. Paper presented at the 22nd Annual Seminar of the International Society of Air Safety Investigators, Canberra, Australia.

Wilson, R. L., & Muir, H. C. (2010). The effect of overwing hatch placement on evacuation from smaller transport aircraft. *Ergonomics, 52*, 931–938.

Wise, J. A. & Wise, B.K. (1984). Humanizing the underground workplace: environmental problems and design solutions. In O. Brown and H. O. Hendricks (Eds.), *First International Symposium on Human Factors in Organizational Design and Management* (pp. 125–135). North Holland: Elsevier Science.

References

Wogalter, M. S., Lim, R. W., & Nyeste, P. G. (2014). On the hazard of quiet vehicles to pedestrians and drivers. *Applied Ergonomics*, *45(5)*, 1306–1312.

Wolfe, T. (1983). *The right stuff*. New York: Farrar, Straus, & Giroux.

Woo, E. H. C., White, P., & Lai, C. W. K. (2016). Ergonomics standards and guidelines for computer workstation design and the impact on users' health: a review. *Ergonomics*, *59*, 464–475.

Wood, R. H., & Sweginnis, R. W. (1995). *Aircraft accident investigation*. Casper, WY: Endeavour Books.

Woodcock, M. (1989). *Team development manual* (2nd ed.). Worcester, UK: Gower.

Woodhouse, M. L., McCoy, R. W., Redondo, D. R., & Shall, L. M. (1995). Effects of back support on intra-abdominal pressure and lumbar kinetics during heavy lifting. *Human Factors*, *37*, 582–590.

Woods, D. D. (1988). Coping with complexity: the psychology of human behaviour in complex systems. In L. P. Goodstein, H. B. Andersen, & S. E. Olsen (Eds.), *Tasks, errors and mental models* (pp. 128–148). London: Taylor & Francis.

Woods, D. D. (1989). Modeling and predicting human error. In J. I. Elkind, S. K. Card, J. Hochberg, & B. M. Huey (Eds.), *Human performance models for computer-aided engineering* (pp. 248–277). Washington, DC: National Academy Press.

Woods, D. D., Johannesen, L. J., Cook, R. I., & Sarter, N. B. (1994). *Behind human error: Cognitive systems, computers, and hindsight* (SOAR 94–01). Dayton, OH: CSERIAC.

Workineh, S. A., & Yamaura, H. (2016). Multi-position ergonomic computer workstation design to increase comfort of computer work. *International Journal of Industrial Ergonomics*, *53*, 1–9.

Wu, B., Yan, X., Wang, Y., & Soares, C. G. (2017). An evidential reasoning-based CREAM to human reliability analysis in maritime accident process. *Risk Analysis*, *37*, 1936–1957.

Wu, Q., Molesworth, B. R., & Estival, D. (2019). An investigation into the factors that affect miscommunication between pilots and air traffic controllers in commercial aviation. *International Journal of Aerospace Psychology*, *29*, 53–63.

Wyler, E., & Bohnenblust, H. (1991). Disaster scaling: a multi-attribute approach based on fuzzy set theory. In G. Apostolakis (Ed.), *Probabilistic safety assessment and management* (Vol. 1, pp. 665–670). New York: Elsevier.

Yan, S., Wei, Y., & Tran, C. C. (2019). Evaluation and prediction mental workload in user interface of maritime operations using eye response. *International Journal of Industrial Ergonomics*, *71*, 117–127.

Yeh, Y., & Wickens, C. D. (1988). Dissociation of performance and subjective measures of workload. *Human Factors*, *30*, 111–120.

Yıldırım, U., Başar, E., & Uğurlu, Ö. (2019). Assessment of collisions and grounding accidents with human factors analysis and classification system (HFACS) and statistical methods. *Safety Science*, *119*, 412–425.

Yıldırım, M., & Mackie, I. (2019). Encouraging users to improve password security and memorability. *International Journal of Information Security*, *18*, 741–759.

Yin, S., Wickens, C. D., Helander, M., & Laberge, J. C. (2015). Predictive displays for a process-control schematic interface. *Human Factors*, *57*, 110–124.

Yin, Z., Liu, Z., & Li, Z. (2021). Identifying and clustering performance shaping factors for nuclear power plant commissioning tasks. *Human Factors and Ergonomics in Manufacturing and Service Industries*, *31*, 42–65.

Young, M. S., & Stanton, N. A. (2002). Malleable attentional resources theory: a new explanation for the effects of mental underload on performance. *Human Factors*, *44*, 365–375.

Yu, N., Ahern, L. A., Connolly-Ahern, C., & Shen, F. (2010). Communicating the risks of fetal alcohol spectrum disorder: effects of message framing and exemplification. *Health Communication*, *25*, 692–699.

Yuris, N., Sturman, D., Auton, J., Giacon, L., & Wiggins, M. W. (2019). Higher cue utilisation in driving supports improved driving performance and more effective visual search behaviours. *Journal of Safety Research*, *71*, 59–66.

Zagonel, A. A., Rohrbaugh, J., Richardson, G. P., & Andersen, D. F. (2004). Using simulation models to address 'what if' questions about welfare reform. *Journal of Policy Analysis and Management*, *23(4)*, 890–901.

Zakay, D., & Wooler, S. (1984). Time pressure, training and decision effectiveness. *Ergonomics*, *27*, 273–284.

Zeier, H. (1994). Workload and psychophysiological stress reactions in air traffic controllers. *Ergonomics*, *37*, 525–539.

Zimmerman, R. (1998). *The story of Apollo 8: The first manned flight to another world.* New York: Four Walls Eight Windows.

Zimolong, B. (1997). Occupational risk management. In G. Salvendy (Ed.), *Handbook of human factors and ergonomics* (2nd ed., pp. 989–1020). New York: Wiley.

Zinkle, S. J., & Was, G. S. (2013). Materials challenges in nuclear energy. *Acta Materialia*, *61*, 735–758.

Zio, E. (2009). Reliability engineering: old problems and new challenges. *Reliability Engineering and System Safety*, *94*, 125–141.

Zuckerman, A. (1996). *International standards desk reference.* New York: Amacom.

Zulkosky, K. D., White, K. A., Price, A. L., & Pretz, J. E. (2016). Effect of simulation role on clinical decision-making accuracy. *Clinical Simulation in Nursing*, *12*, 98–106.

Index

A

A/B testing 247
accident causation 273; energy-based 302, 303; in nuclear systems 301
accident investigation 46, 79, 80, 205, 221–235, 341, 364, 365
Accimap 60
active failure 54–59
active followership 172
adaptability 152, 161, 170, 269
Advanced Crew Resource Management (ACRM) 197–200
Advanced Qualification Program (AQP) 197
Air Traffic Control 15, 17, 19, 104, 120, 143, 175, 176, 187, 193, 224, 235, 282–297
aircraft accidents; Buffalo, New York 141; Calcutta (Kolkata) 222; Chicago 19, 24, 186; Detroit 91, 163, 176; Dryden 184; East Midlands 194, 276; Elba 223; Fairchild Airforce Base 168; Gottröra 278; Honolulu, Hawaii 186; Houston 28; Isles of Scilly 253; Lord Howe Island 21; Los Angeles 227, 279; Madrid 180; Miami 10, 18, 115; Mount Erebus 53, 89; New York 141, 275; Paris 64; Persian Gulf 134; Portland 8; Saulx-les-Chartreau 222; Sioux City, Iowa 19, 187; Strasbourg 113; Sydney 232, 342; Tenerife 7, 9; Uberlingen 285; Washington, DC 177, 234; Zagreb 283
aircraft incidents; Bangkok 56; Oslo 194
accident investigation 46, 53, 79, 80, 205, 221–235, 278, 279, 341, 364
alertness 139, 144, 145
alertness-maintenance 143–145
American National Standards Institute 210, 343
anaesthetists 177, 321, 324
analytical decision-making 123–125
anchoring bias 125, 128
angular acceleration 294
anthropometry 253, 258, 259
archetypes 246
assertiveness 179
attention 6, 97–106, 115, 127, 133, 139, 142, 194, 210, 218, 295, 318, 326, 346; sustained 143–145, 297
attentional resource theory 103
attentional resources 103, 105–108, 144, 183
audit 10, 46, 201, 207, 208, 243, 247, 342–349, 361
Australian Transport Safety Bureau 81, 232
automatic processing 101
automation 192, 265–272, 297, 327
automation-induced failures 265
autopilot 115, 265
availability bias, 125, 127, 128
Aviation Safety Attitude Scale (ASAS) 133
Aviation Safety Reporting System (ASRS) 140, 180, 193, 287

B

baggage handlers 34, 64, 193, 338
base rates 126, 133, 327, 353
Bayes' theorem 130
behavioural analysis 209
behavioural task analysis 76, 336, 337
bias 13, 72, 129, 133, 326, 327; anchoring 128; availability 127; confirmation 127; representative 125, 126
Black Swan scenarios 325
Bowtie method 36, 37
British Standards Institute (BSI) 343, 344
Bureau of Air Safety Investigation (*see also* Australian Transport Safety Bureau) 21, 79, 232, 342

C

Cambridge Cockpit (simulator) 6
causal network model 313–315
causal relationships 31, 85, 164, 215, 230, 315
challenge and response 95
checklists 4, 5, 7, 15, 28, 37, 57, 87, 91–95, 176, 181, 196, 199, 201, 213, 214, 233, 321, 332, 341, 346, 349
chunking 95
circadian rhythms 139–142, 146
cluster-sampling technique 348
Cockpit Display of Traffic Information (CDTI) (*see* Traffic Collision and Avoidance System)
Cockpit Management Attitudes Questionnaire (CMAQ) 192
Cockpit Resource Management (CRM) 191, 193
cockpit voice recorder (CVR) 176, 223, 224, 234, 235
Cognitive Event Tree (COGNET) 215
cognitive load 104, 109, 145
cognitive narrowing 98
Cognitive Task Analysis (CTA) 214–218, 270, 271
Cognitive Work Analysis (CWA) 270–273
colour 291, 295, 296
command and control systems 261, 262, 317, 332

421

Index

communication 8, 10, 15, 19, 22, 23, 28, 29, 42, 43, 45, 48, 54, 57, 61, 118, 119, 134, 149, 151–161, 163, 170–173, 175–189, 191, 194, 196, 198, 199, 201, 208, 221, 225, 228, 232–235, 283, 284, 289, 290, 303, 312, 317, 321, 323, 346, 367
communication errors 179
communication protocols 201
concept demonstration 69, 258
conceptual compatibility 295
confirmation bias 125, 127, 326
continuous process operations 24, 25
controlled flight into terrain (CFIT) 32, 163
controlled processing 101
cooperative automated systems 268
corporate culture 23, 170
cost-benefit analysis 240, 255, 258, 270, 363
cost-effectiveness assessment 13, 14, 68, 242, 243
counter theory 31
countermeasures 11
Crew Resource Management (CRM) 9, 193, 197
Critical Decision Method (CDM) 215, 216, 335, 336
Critical Incident Analysis 334, 335, 338
cross-checking 92, 201, 208
cues 215, 216, 233, 267, 286, 287, 293, 335
culture 22, 23, 29, 31, 194; Just 170, 241; organisational 33, 34, 46, 47, 50, 54, 59, 79, 92, 95, 118, 156, 169–171, 177, 179, 184, 185, 218, 333; safety 32, 169

D

'divided' attention 100, 112
danger 65, 72, 168, 182, 205, 224, 245, 292, 307, 316, 318
DECIDE model 129
decision analysis 135–137
decision-centred design 67
decision-making 13, 16, 32, 42, 43, 51, 59, 67, 72, 123–138, 139, 154, 157, 172, 199, 200, 201, 239, 240, 249, 296, 297, 304, 326, 327, 335, 355
declarative knowledge 101
declarative memory 86, 101, 215
defences 46, 47, 57, 86, 172, 181, 208, 232, 244, 284, 305, 355, 358
degree of automation 57, 269
design 7, 27, 28, 31, 42, 61, 63–74, 76, 87, 99, 207, 222, 243, 245–247, 249, 251, 253–261, 291, 294, 297, 300, 301, 303, 304, 305, 307, 313, 317, 318, 321, 333, 338, 339, 341, 343, 345, 355, 361, 363, 364, 366; *and* checklist 91–95; decision-centred 67; equipment 233, 278; *and* instructional systems 195, 196;

interface 114, 144, 244, 245, 265–266, 290; job 43; minimalist 244; organisational 15, 16; participatory 242; specifications 271; system 43, 48, 134, 196, 212, 239, 242, 288, 289, 291, 339; survey 346; transparency 245; user-centred 66, 242; *and* warnings 210
design cycle 243
design thinking 251
Detection Response Task (DRT) 108
diagnosis 25, 76, 106, 114, 127, 181, 212, 267, 269, 307, 322, 326, 327, 369, 398
dichlorodiphenyltrichloroethane (DDT) 3
display size 293
documentation 233, 244, 245, 287, 294, 323, 333, 346

E

Electronic Centralised Aircraft Monitoring System (ECAM) 95
Electronic Flight Bags (EFB) 294
Emotional Intelligence (EI) 171
enculturation 169
energy accidents; Chernobyl 300, 301, 306, 309; Fukushima 74, 308; Piper Alpha 317
energy-based accident causation model 302; Northeastern Blackout 116
envisioning 170, 171
episodic memory 86
ergonomics 7, 253–262, 339
error detection 25, 269, 297
error management 200, 201, 262, 303, 317
error recovery 266, 269
ethical issues 339, 340
evacuation 20, 99, 276, 278, 280, 281, 308
Event Analysis of Systemic Teamwork (EAST) 61, 62
event tree 215, 217, 272, 354, 360
expertise 42, 117, 145, 155, 183, 243, 271, 288, 332, 335, 358, 359
external threats 201

F

failure analysis 20
fatigue 139–146, 205, 284, 316, 318
fault tree analysis 307, 318, 338
fire hazards 279
flight attendants 9, 20, 22, 157, 184, 185, 193, 277, 278, 281
Flight Data Recorder (FDR) 208, 223, 234, 235
Flight Management Attitudes Questionnaire (FMAQ) 192
Flight Operations Quality Assurance (FOQA) 208
flight progress strip 286, 287
focussed attention 98–100

Index

followership 155, 171, 172
framing effect 131, 132
functional flow analysis, 334, 336
Functional Resonance Analysis Method (FRAM) 62, 324

G

Gaussian (normal) distribution 23, 254–256
Goals, Operators, Methods and Selection (GOMS) model 249
Ground Proximity Warning System (GPWS) 32
groups 15, 19, 20, 149–161, 164–167, 169, 182–184, 192, 208, 246, 247, 316, 331, 353, 383
group-think 72

H

hazard analysis 47, 205–209, 307, 339
Hazard and Operability Analysis (HAZOP) 46
hazard identification 47, 305, 308, 345
Hazard Identification Technique (HIT) 305, 306
hazard patterns 207
hazardous attitudes 132, 133
Head-Up Displays (HUD) 294
healthcare 93, 177–179, 240, 321–327
heart rate variability 106
heuristics 97, 125, 326
Hierarchical Task Analysis (HTA) 323
high reliability organisation (HRO) 19, 20, 50, 75, 77, 134, 299
Human and Organisation Error (HOE) Classification Scheme 42, 43
human error 7, 10, 11, 23, 29, 33, 41–52, 54, 60, 64, 73, 75, 77, 78, 91, 97, 140, 143, 212, 214, 230, 231, 269, 303, 304, 312, 316, 318, 322, 323, 338, 351, 352, 353, 354, 356, 358, 365, 371; modelling 48; probability 214, 352, 353
Human Factors Analysis and Classification System (HFACS) 230
human factors; audits 342–349; reports 361–367; testing 14, 331–339, 361–366
human information processing 85–91, 254, 336, 359
Human Reliability Analysis (HRA) 75–81, 351–360
human-computer interaction (HCI) 247–249

I

incidents 7, 10–13, 18, 30, 31, 36, 41, 45, 51, 57, 65, 71, 91, 97, 140, 146, 170, 176, 180, 181, 186, 187, 205, 206, 218, 221, 229–232, 237, 287, 288, 291, 309, 312, 316, 332, 341, 342, 372

INDICATE 208
Individual Plant Examinations (IPE) 75
Inertial Navigation System (INS) 89
injuries 4, 19, 20, 34, 36, 45, 61, 64, 113, 115, 207, 209, 210, 225, 227, 261, 275–279, 342, 367
Input-Process-Output (IPO) model 151, 156
Instructional Systems Design (ISD) 195, 212
Integrated Model of Team Performance and Training 159, 160
internal threats 201
International Civil Aviation Organisation (ICAO) 221, 223, 224, 228, 229, 230, 231, 233
International Labour Organisation (ILO) 224
International Standards Organisation 344–346
intuitive decision-making 125
ionising radiation 299, 300, 308
ISO 45001 345
ISO 9000 344, 345

J

'just in time' philosophy 25
just culture 170, 241

K

Kelley Followership Questionnaire 172
knowledge elicitation 216, 270, 271, 336, 351, 352, 358, 359, 360
knowledge-based behaviour 42, 217, 268, 269, 304
Knowledge, Skills, and Attitudes (KSA) 289, 290

L

laissez-faire 167–169
lapse 43, 97, 100, 103, 111, 215, 217
latent condition 12, 54–59, 172, 230, 232, 245, 273, 301, 312, 313, 318, 341, 342
Leader–Member Relations scale 165–167
leadership 33, 48, 50, 118, 152, 154–159, 163–173, 199, 321, 323, 332
life-cost 239, 240
Line Oriented Flight Training (LOFT) 9, 89, 286, 289, 290
Line Oriented Simulation (LOS) 158, 186, 227, 279
Liveware-Liveware (L-L) interface 9
load-shedding 54, 100
local factors 18, 29, 54–56
long-term memory 86–88, 98, 114
loosely coupled systems 19, 27, 28
Low-Event Task Subjective Situation Awareness (LETSSA) 120
luminance 90, 104, 227, 291, 294, 295

M

manual handling 260–261, 338, 349
margin of safety 12
marine accidents: France 311; Mississippi River 312; North Sea 317; Portland, Oregon 313; Prince William Sound 314; Zeebrugge 55
medication; administration 177, 241; errors 323, 324
mental effort 106, 107
mental model 114–118, 121, 124, 179, 216, 267, 284, 293, 306, 326, 335, 366
mental representation 76, 114, 182, 286, 307
mental workload 105, 144, 269, 270
mindlessness theory 103
miscommunication 8, 156, 176, 178–181, 184, 188, 189
misconceptions of chance 125, 128
Mission Awareness Rating Scale (MARS) 120
mistakes 44, 187, 215, 217, 241
mitigating language 176, 178, 184
mode error 25, 113, 265
modelling 47, 48, 50, 51, 60–62, 70, 77, 170, 171, 249, 289, 337, 354
morale 43, 45, 52, 57, 161, 256, 332, 333, 364
motivation 24, 57, 97, 127, 132, 139, 151, 164, 167, 172, 175, 327, 332
Multiple Resources Model 88, 102

N

National Aeronautics and Space Administration (NASA) 58, 59, 91, 107, 193
naturalistic decision-making 124
Non-Technical (NOTECHS) skills 199, 200
nuclear accidents; Chernobyl 300, 301, 306, 309; Fukushima 74, 308; Idaho Falls 300
nuclear power 4, 74, 78, 80, 218, 300, 301, 303, 304, 306–308, 309, 310, 338
National Transportation Safety Board (NTSB) 8, 10, 20, 28, 32, 71, 91, 115, 176, 177, 179, 180, 186, 187, 221, 231, 275, 279, 312, 365

O

oil platforms 317, 318
occupational health 47, 57, 280, 308, 316, 332, 344, 345, 349
operational analysis 333
operational concept 69, 256, 258, 333
operational cues 335
operational errors (OE) 54, 179, 180, 181, 287, 288, 290
organisation failure 58, 79
organisational climate 168, 170
organisational culture 29, 31, 33, 46, 47, 50, 54, 59, 79, 92, 95, 118, 168–171, 177, 179, 184, 185, 218, 333
organisational design 15, 16
organisational factors 27, 29, 30, 42, 43, 48–50, 68, 79, 80, 219, 240, 260, 323, 324, 332
organisational norms 23
organisational policies 50
organisational psychology 4, 6, 86, 191
organisational structure 14–16, 18, 80, 134, 159, 165, 221, 310, 333
'out of the loop' syndrome 113, 266

P

participative leadership 152, 172, 173
participatory ergonomics 259, 339
Passenger Protective Breathing Equipment (PPBE) 280
percentile 254–256
perception 4, 42, 43, 78, 249, 294
Performance Shaping Factors (PSF) 78, 79, 214, 218, 354–356, 358, 359
persona 246
personal protective equipment (PPE) 50, 228
perspective display 292, 293, 295
phraseology 156, 180, 181, 185, 188, 284
physiological evaluation 339
planning 14, 21, 72, 128, 176, 224, 262, 267, 269, 312, 318, 323, 333, 345, 366
Position Power 165–167
Precursor, Action, Result, Interpretation (PARI) 216
predictive displays 296
prescriptive decision-making 129
prevalence bias 327
primary sources 225
proactive approach 14, 15, 34, 35, 47, 208, 334, 341, 342, 366
proactive management 14
proactive strategy 37, 46, 51
Probabilistic Risk Assessment (PRA) 74, 75, 351
Probability–Impact Diagram 35, 36
procedural control 285, 286
procedural errors 230
procedural memory 86, 215
procedural models 77, 78
process control 143, 144, 282–284, 287, 288, 289
process tracing 272, 336
productivity 12, 15, 45, 154, 166, 238–240, 245, 332, 361
prosodic cues 188, 189
prospect theory 135, 136
psychomotor control 24, 25

psychophysiological measures 106, 107
psychosocial factor 15, 139
pulse surveys 52, 146

Q

Quick Access Recorders (QAR) 208

R

radiation 205, 224, 299–301, 308, 309, 322, 375
rail accidents; Channel Tunnel 98; Devine, Texas 65; Gunter, Texas 179, 284; Norwich 229
rail control 98, 143, 179, 282, 284
reactive approach 3, 35, 46, 366
read back 63, 66, 189
readiness 167
receptiveness 170, 171, 184
reductionist approach 85, 87, 88
redundancy 19, 46, 55, 60, 62, 66, 73, 93, 186, 213, 284
region of influence 28, 29
regression to the mean 133
reliability 18, 19, 20, 29, 41, 50, 70, 73–81, 114, 120, 134–136, 192, 198, 214, 217, 218, 228, 247, 249, 266, 267, 269, 273, 285, 290, 299, 303, 304, 327, 331, 332, 337, 347, 351–360; analysis 73–80, 135, 218, 351–354; *and* automated systems 269, 270; engineering 73, 74
remedial strategies 11, 14, 18, 41, 48, 50, 116, 144, 146, 170, 179, 198, 199, 253, 262, 278, 306, 313, 323, 326, 332, 338, 343
representative bias 125–127
resident pathogens 57, 58
residual attention 101, 105
residual resources 98, 100, 106
resilience 201, 206, 325, 326
resource management 9, 156–158, 191–201, 321, 322, 342; bridge 317; crew 9, 193, 197
retrospective analyses 222, 331, 341
retrospective approach 46
risk 10, 17, 20, 21, 27–37, 46, 48–51, 59, 61, 65, 74, 78–80, 90, 95, 116, 123, 124, 139, 143, 145, 146, 185, 188, 198, 201, 209, 243, 277, 300, 302, 303, 305, 309, 313, 314, 316, 317, 325, 349, 350, 351, 358, 365; analysis 20, 303, 305; assessment 74, 75, 79, 131, 132, 201, 309, 326, 351; controls 33, 34, 37, 51, 300; estimation 30
risk factors 349, 350, 365
risk management 36, 37, 145
RISKGATE 34
root cause analysis (RCA) 178, 231, 318, 325, 364
rule-based behaviour 42, 44, 76, 77, 79, 215, 217, 267, 305

S

Safety Attitudes Questionnaire 192
Safety II 325
safety audits 10, 201
Safety Citizenship Behaviour (SCB) 59
Safety Management Systems (SMS) 50, 51, 145
SBAR 179
secondary sources 225, 226, 267
secondary tasks 106–108
selective attention 97–99
selection 4, 6, 15, 42, 43, 126; *and* experts 351, 359
semantic memory 86, 188
SHELL(O) model 5, 233
short-term memory 86, 88–91, 98, 104
simulation 48, 49, 69, 70, 80, 117, 120, 158, 206, 226, 227, 259, 288, 322, 325, 326
Situation Awareness Control Room Inventory (SACRI) 120, 121
situation control 164, 165
Situation Present Assessment Method (SPAM) 120
situational awareness 68, 76, 78, 111–121, 149, 151, 156, 157, 175, 196, 199, 200, 287, 287, 312, 321, 335
Situational Awareness Global Assessment Test (SAGAT) 120, 121
Situational Awareness Rating Technique (SART) 120, 121
skill acquisition 42, 197
skill-based behaviour 42, 76, 77, 79, 157, 215, 217, 267, 304, 305
Skill-Rule-Knowledge (SRK) model 41, 42, 77, 215, 273, 304, 305
slip 36, 43, 97, 215, 217
Social Network Analysis 61
standard Operating Procedures (SOP) 53, 156, 197
standardisation 22, 59, 90, 126, 156, 185, 200, 230, 244, 338, 367
standards 50, 66, 109, 168, 171, 197, 198, 208, 210, 222, 229, 247, 259, 286, 296, 324, 338, 343–347, 364
stochastic variability 23, 35, 238
stratification 347, 348
stress 6, 107, 157, 158, 166, 205, 261, 278, 284, 303, 304, 335, 339, 353
stress management 157, 335
structured interview 335
Subject-Matter Experts (SME) 30, 216, 314, 318, 337, 359
subjective data 51, 332
Subjective Expected Utility (SEU) 13, 124

Subjective Workload Assessment Test (SWAT) 107
Success Likelihood Index Methodology (SLIM) 78, 354–356
Surface Movement Coordination (SMC) 232, 233, 286
sustained attention 98, 102, 143, 144, 297
symbolic architecture 215, 216, 292, 295
synergy 149, 182
System 1 123
System 2 123, 307, 326
system failure 12, 15, 17, 24, 28, 29, 35, 46, 47, 51, 54, 56, 59–61, 66, 67, 73, 74, 75, 77, 103, 112, 133, 134, 238, 299, 300, 305, 307, 309, 313, 317–319, 341, 351, 352, 355
system safety 17, 18, 34, 35, 46, 47, 50, 53–62, 64, 80, 140, 172, 193, 195, 300, 309, 313, 325, 341, 342
Systematic Air Traffic Operations Research Initiative (SATORI) 287, 288
Systematic Human Error Reduction and Prediction Approach (SHERPA) 323
systemic approach 18, 54, 232, 300
systems analysis 206
systems approach 10, 11
systems engineering 65–68, 70, 72, 73
systems thinking 10
Systems-Theoretic Accident Modelling and Processes (STAMP) 60, 61

T

task analysis 67, 76, 157, 206, 211–219, 270, 271–273, 323, 336–338, 352, 354, 355, 359
task load 104
Task Load Index (TLX) 107
Task Mental Load (TML) 269
task structure 165–167
task-centred design 66, 70
task-shedding 100
taxonomies of human performance 15, 75, 322
team 9, 52, 58, 72, 92, 117, 120, 121, 149–161, 166, 179, 182–185, 188, 191, 193, 198, 228, 231, 321–323
Team Evolution and Maturation (TEAM) model 151–153
Team Functioning Assessment Tool (TFAT) 323
team-building 150, 151

Technique for Human Error Rate Prediction (THERP) 212–214, 352, 354–356
Test for Existence 228
Test for Influence 229
Test for Validity 229
thalidomide 3
Threat and Error Management (TEM) 200, 201
tightly coupled systems 19, 23, 25, 27, 28, 79, 81
time reliability correlation (TRC) 78
total quality management (TQM) 50, 261, 262
Traffic Collision and Advoidance System (TCAS) 284, 285, 291
train graphs 287
training 4, 6, 9, 11, 12, 21, 28, 33, 42–44, 46, 47, 50, 54, 57, 60, 69–71, 93, 95, 99, 114, 117, 118, 128, 130, 132, 134, 137, 151, 152, 157–160, 170, 185, 191–193, 195, 197–199, 201, 207, 208, 218, 242, 245, 246, 259–261, 265–267, 271, 288, 289, 291, 304, 306, 317, 318, 321, 322, 326, 327, 336, 339, 342, 346, 354
Transactional Leadership Theory 167, 168
Transformational Leadership Theory 167, 168, 171

U

uncertainty 18, 20, 24, 27–37, 124, 135–138, 159, 185, 186, 189, 265, 321, 323, 353, 356–358
unsafe acts 43, 79, 230, 232, 318
usability 4, 237–254, 321, 346, 347, 363, 366
user-centred design 64, 66, 67, 242
user experience (UX) 237–245, 249

V

validity 9, 30, 31, 36, 80, 88, 109, 120, 132, 166, 167, 192, 198, 206, 227, 229, 308, 316, 332, 346, 347
Vessel Traffic Service (VTS) 316
vigilance 102, 114, 120, 144, 266; decrement 103

W

'what-if' scenarios 15, 117, 128
wireframe 250, 251
workload 7, 8, 48, 54, 57, 66, 88, 91, 92, 97–109, 115, 144, 183, 186, 199, 208, 214, 233, 268, 269, 270, 278, 288–290, 295, 296, 323, 333